Advances in Experimental Medicine and Biology

Volume 951

More information about this series at http://www.springer.com/series/5584

Feridoun Karimi-Busheri
Michael Weinfeld
Editors

Biobanking and Cryopreservation of Stem Cells

Springer

Editors
Feridoun Karimi-Busheri
Department of Oncology
University of Alberta
Edmonton, AB, Canada

Michael Weinfeld
Department of Oncology
Cross Cancer Institute and University of Alberta
Edmonton, AB, Canada

ISSN 0065-2598 ISSN 2214-8019 (electronic)
Advances in Experimental Medicine and Biology
ISBN 978-3-319-83298-2 ISBN 978-3-319-45457-3 (eBook)
DOI 10.1007/978-3-319-45457-3

Softcover reprint of the hardcover 1st edition 2016

Printed on acid-free paper

This Springer imprint is published by Springer Nature
The registered company is Springer International Publishing AG
The registered company address is: Gewerbestrasse 11, 6330 Cham, Switzerland

Preface

Biobanking and stem cell technology are relatively new fields in science and are increasingly bound to one another through a myriad of biotechnology, business, basic science, and medicinal networks. Both fields are growing every year, and the global expansion estimate for both is approaching an astonishing investment of close to $200 billion. Inevitably, such a large expansion creates its own challenges that require comprehensive and orderly studies, optimal methods and devices, and standardization of bioprocessing.

Rules governing stem cell growth, stability, preservation, and storage are different from normal cells. Any permanent alterations in the genetic makeup of stored cells could change cell fate and drastically impair discovery research and clinical outcome if therapy is an objective. Current widely used protocols for cryopreservation to some extent are regarded as an "old school" discipline that needs to be revised and revisited for an emerging new field such as stem cell technology. This has put enormous pressure and demands on both the biobanking industry and stem cell technology to adopt new strategies and methodologies.

This book has been prepared with a focus on these challenges and to provide a comprehensive overview of what directions industry and academia need to take to overcome the problems of cryopreservation. To accomplish our objectives a diverse group of researchers and clinicians agreed to contribute their expertise to this book. This volume contains 17 chapters which provide an in-depth study, analysis, and reviews by experts from many research centers and laboratories with different scientific disciplines around the world.

The first three chapters take a more general theoretical approach toward understanding the principles of banking and integrity of the banked specimens at the molecular level. In the fourth chapter, the authors provide resources available for precision medicine in glioblastoma. The next two chapters, Chaps. 5 and 6, in contrast, discuss in detail the protocols and methodology used for human pluripotent stem cells and alternative cryopreservation procedures in the preparation of cord blood cells. Chapters 7, 8, 9, 10, 11, 12, and 13 discuss some of the most promising topics in stem cell research such as mesenchymal stem cells, adipose tissue-derived stem cells, pluripotent stem cells, and oocyte cryopreservation for fertility preservation. Regenerative medicine, which has developed extremely rapidly over the last few years, has generated much excitement among scientists and clinicians, as well as the public. The final four chapters deal with current approaches using stem cells in tissue engineering and regenerative medicine in a variety of tissues or organs such as cardiac regenerative medicine, ocular epithelial limbal stem cells, hair-follicle-associated pluripotency, and banking of dental stem cells.

We express our sincere gratitude to all those who agreed to contribute and assist us in preparing this publication.

Edmonton, Canada

Feridoun Karimi-Busheri
Michael Weinfeld

Contents

Contributors

Yasuyuki Amoh Department of Dermatology, Kitasato University, Sagamihara, Japan

Beng Ti Christopher Ang Singapore Institute for Clinical Sciences, Agency for Science, Technology and Research (A*STAR), Singapore, Singapore

Duke-National University of Singapore Medical School, Singapore, Singapore

Department of Neurosurgery, National Neuroscience Institute, Singapore, Singapore

Department of Physiology, Yong Loo Lin School of Medicine, National University of Singapore, Singapore, Singapore

Vanessa Barbaro Fondazione Banca degli Occhi del Veneto, Venice, Italy

Silvana Bardelli Swiss Institute for Regenerative Medicine, Cardiocentro Ticino Foundation, Lugano, Switzerland

John G. Baust Institute of Biomedical Technology, State University of New York at Binghamton, Binghamton, NY, USA

Department of Biological Sciences, Binghamton University, Binghamton, NY, USA

John M. Baust CPSI Biotech, Owego, NY, USA

Institute of Biomedical Technology, State University of New York at Binghamton, Binghamton, NY, USA

Marina Bertolin Fondazione Banca degli Occhi del Veneto, Venice, Italy

Karen Bieback Institute of Transfusion Medicine and Immunology, Medical Faculty Mannheim, Heidelberg University, Heidelberg, Germany

German Red Cross Blood Service Baden-Württemberg – Hessen, Mannheim, Germany

Cesar V. Borlongan Department of Neurosurgery and Brain Repair, Morsani College of Medicine, University of South Florida, Tampa, FL, USA

Claudia Breda Fondazione Banca degli Occhi del Veneto, Venice, Italy

Annelies Bronckaers Laboratory of Morphology, Biomedical Research Institute (BIOMED), Hasselt University, Diepenbeek, Belgium

Wenluo Cao AntiCancer, Inc, San Diego, CA, USA

Department of Surgery, University of California, San Diego, CA, USA

Department of Anatomy, Second Military Medical University, Shanghai, China

Rusan Catar Department of Nephrology and Intensive Care Medicine, Charité Universitätsmedizin, Berlin, Germany

Jane Ru Choi Department of Biomedical Engineering, Faculty of Engineering, University of Malaya, Kuala Lumpur, Malaysia

William Corwin Department of Immunology and Carole and Ray Neag Comprehensive Cancer Center, University of Connecticut School of Medicine, Farmington, CT, USA

Ram V. Devireddy Department of Mechanical Engineering, Louisiana State University, Baton Rouge, LA, USA

Yörg Dillen Laboratory of Morphology, Biomedical Research Institute (BIOMED), Hasselt University, Diepenbeek, Belgium

Ronald B. Driesen Laboratory of Morphology, Biomedical Research Institute (BIOMED), Hasselt University, Diepenbeek, Belgium

Barbara Ferrari Fondazione Banca degli Occhi del Veneto, Venice, Italy

Stefano Ferrari Fondazione Banca degli Occhi del Veneto, Venice, Italy

Lydia Masako Ferreira Federal University of Sao Paulo, Sao Paulo, SP, Brazil

Sven Geißler Berlin-Brandenburg Center/School for Regenerative Therapies (BCRT/BSRT), Charité Universitätsmedizin, Berlin, Germany

Julius Wolff Institute (JWI), Berlin, Germany

Pascal Gervois Laboratory of Morphology, Biomedical Research Institute (BIOMED), Hasselt University, Diepenbeek, Belgium

Jeffrey M. Gimble Center for Stem Cell Research & Regenerative Medicine, Tulane University, New Orleans, LA, USA

La Cell LLC, New Orleans BioInnovation Center, New Orleans, LA, USA

David T. Harris University of Arizona Health Sciences Biorepository, Tucson, AZ, USA

University of Arizona, Tucson, AZ, USA

Division of Translational Medicine, University of Arizona, Tucson, AZ, USA

GMP Laboratory at the University of Arizona, Tucson, AZ, USA

Celebration Stem Cell Centre, Gilbert, AZ, USA

Na He China National GeneBank-Shenzhen, BGI-Shenzhen, Shenzhen, China

Petra Hilkens Laboratory of Morphology, Biomedical Research Institute (BIOMED), Hasselt University, Diepenbeek, Belgium

Robert M. Hoffman AntiCancer, Inc, San Diego, CA, USA

Department of Surgery, University of California, San Diego, CA, USA

Martin J. Hoogduijn Nephrology and Transplantation, Department of Internal Medicine, Erasmus Medical Center, Rotterdam, The Netherlands

Lech Ignatowicz Division of Dermatology and Venereology, Department of Clinical Sciences, Lund University, Lund, Sweden

Sussannah Kaelber Department of Neurosurgery and Brain Repair, Morsani College of Medicine, University of South Florida, Tampa, FL, USA

Satoshi Kajiura Department of Dermatology, Kitasato University, Sagamihara, Japan

Feridoun Karimi-Busheri Department of Oncology, University of Alberta, Edmonton, AB, Canada

Angela M. Krackhardt III. Medical Department, Hematology and Oncology, Klinikum rechts der Isar, Technische Universität München, München, Germany

Ivo Lambrichts Laboratory of Morphology, Biomedical Research Institute (BIOMED), Hasselt University, Diepenbeek, Belgium

Tina Liang Department of Obstetrics and Gynecology, Faculty of Medicine and Dentistry, University of Alberta, Edmonton, AB, Canada

Trenton Lippert Department of Neurosurgery and Brain Repair, Morsani College of Medicine, University of South Florida, Tampa, FL, USA

Fang Liu Department of Anatomy, Second Military Medical University, Shanghai, China

Yaqiong Liu China National GeneBank-Shenzhen, BGI-Shenzhen, Shenzhen, China

Zala Lužnik Eye Hospital, University Medical Centre, Ljubljana, Slovenia

Takamichi Miyazaki Institute for Integrated Cell-Material Sciences, Kyoto University, Kyoto, Japan

Marco Moccetti Cardiology Department, Cardiocentro Ticino Foundation, Lugano, Switzerland

Guido Moll Berlin-Brandenburg Center/School for Regenerative Therapies (BCRT/BSRT), Charité Universitätsmedizin, Berlin, Germany

Therapeutic Immunology (TIM), Department of Laboratory Medicine, Karolinska Institutet, Stockholm, Sweden

Tarek Motan Division of Reproductive Endocrinology & Infertility, Department of Obstetrics and Gynecology, Faculty of Medicine and Dentistry, University of Alberta, Edmonton, AB, Canada

Hung Nguyen Department of Neurosurgery and Brain Repair, Morsani College of Medicine, University of South Florida, Tampa, FL, USA

Robert Peter III. Medical Department, Hematology and Oncology, Klinikum rechts der Isar, Technische Universität München, München, Germany

Marcela K. Preininger Division of Pediatric Cardiology, Department of Pediatrics, Emory University School of Medicine and Children's Healthcare of Atlanta, Atlanta, GA, USA

Wallace H. Coulter Department of Biomedical Engineering, Georgia Institute of Technology and Emory University, Atlanta, GA, USA

Aghdass Rasouli-Nia Department of Oncology, Cross Cancer Institute and University of Alberta, Edmonton, AB, Canada

Jessica Ratajczak Laboratory of Morphology, Biomedical Research Institute (BIOMED), Hasselt University, Diepenbeek, Belgium

Olle Ringdén Therapeutic Immunology (TIM), Department of Laboratory Medicine, Karolinska Institutet, Stockholm, Sweden

Division of Therapeutic Immunology (TIM, F79), Karolinska University Hospital, Stockholm, Sweden

Maria Carolina Oliveira Rodrigues Division of Clinical Immunology, Ribeirão Preto Medical School, University of São Paulo, Ribeirão Preto, Brazil

Paul R. Sanberg Department of Neurosurgery and Brain Repair, Morsani College of Medicine, University of South Florida, Tampa, FL, USA

Edwin Sandanaraj Department of Research, National Neuroscience Institute, Singapore, Singapore

School of Biological Sciences, Nanyang Technological University, Singapore, Singapore

Singapore Institute for Clinical Sciences, Agency for Science, Technology and Research (A*STAR), Singapore, Singapore

Petra Schollmayer Eye Hospital, University Medical Centre, Ljubljana, Slovenia

Shahensha Shaik Department of Mechanical Engineering, Louisiana State University, Baton Rouge, LA, USA

Monalisa Singh Division of Pediatric Cardiology, Department of Pediatrics, Emory University School of Medicine and Children's Healthcare of Atlanta, Atlanta, GA, USA

Kristi K. Snyder CPSI Biotech, Owego, NY, USA

Institute of Biomedical Technology, State University of New York at Binghamton, Binghamton, NY, USA

Silvia Spoerl III. Medical Department, Hematology and Oncology, Klinikum rechts der Isar, Technische Universität München, München, Germany

Dirk Strunk Experimental and Clinical Cell Therapy Institute, Paracelsus Medical University, Salzburg, Austria

Hirofumi Suemori Institute for Frontier Medical Sciences, Kyoto University, Kyoto, Japan

Changbin Sun China National GeneBank-Shenzhen, BGI-Shenzhen, Shenzhen, China

Si Yan Melanie Tan Department of Research, National Neuroscience Institute, Singapore, Singapore

School of Biological Sciences, Nanyang Technological University, Singapore, Singapore

Carol Tang Department of Research, National Neuroscience Institute, Singapore, Singapore

Humphrey Oei Institute of Cancer Research, National Cancer Centre, Singapore, Singapore

Duke-National University of Singapore Medical School, Singapore, Singapore

Robert Van Buskirk CPSI Biotech, Owego, NY, USA

Institute of Biomedical Technology, State University of New York at Binghamton, Binghamton, NY, USA

Department of Biological Sciences, Binghamton University, Binghamton, NY, USA

Tim Vangansewinkel Laboratory of Morphology, Biomedical Research Institute (BIOMED), Hasselt University, Diepenbeek, Belgium

Wan Kamarul Zaman Wan Safwani Department of Biomedical Engineering, Faculty of Engineering, University of Malaya, Kuala Lumpur, Malaysia

Michael Weinfeld Department of Oncology, Cross Cancer Institute and University of Alberta, Edmonton, AB, Canada

Esther Wolfs Laboratory of Morphology, Biomedical Research Institute (BIOMED), Hasselt University, Diepenbeek, Belgium

Xiying Wu La Cell LLC, New Orleans BioInnovation Center, New Orleans, LA, USA

Chunhui Xu Division of Pediatric Cardiology, Department of Pediatrics, Emory University School of Medicine and Children's Healthcare of Atlanta, Atlanta, GA, USA

Wallace H. Coulter Department of Biomedical Engineering, Georgia Institute of Technology and Emory University, Atlanta, GA, USA

Kar Wey Yong Department of Biomedical Engineering, Faculty of Engineering, University of Malaya, Kuala Lumpur, Malaysia

Jianhui Yue China National GeneBank-Shenzhen, BGI-Shenzhen, Shenzhen, China

Fabiana Zanata Federal University of Sao Paulo, Sao Paulo, SP, Brazil

Center for Stem Cell Research & Regenerative Medicine, Tulane University, New Orleans, LA, USA

Xi Zhang China National GeneBank-Shenzhen, BGI-Shenzhen, Shenzhen, China

Yong Zhang Huahan Gene Research Institute (HHGene), Shenzhen, China

1 Key Issues Related to Cryopreservation and Storage of Stem Cells and Cancer Stem Cells: Protecting Biological Integrity

Feridoun Karimi-Busheri, Aghdass Rasouli-Nia, and Michael Weinfeld

Abstract

Cryopreservation and biobanking of stem cells are becoming increasingly important as stem cell technology and application attract the interest of industry, academic research, healthcare and patient organisations. Stem cell are already being used in the treatment of some diseases and it is anticipated that stem cell therapy will play a central role in future medicine. Similarly, the discovery of both hematopoietic and solid tumor stem cells and their clinical relevance have profoundly altered paradigms for cancer research as the cancer stem cells are considered promising new targets against cancer. Consequently, long-term cryopreservation and banking of normal and malignant stem cells is crucial and will inevitably become a routine procedure that requires highly regulated and safe methods of specimen storage. There is, however, an increasing amount of evidence showing contradictory results on the impact of cryopreservation and thawing of stem cells, including extensive physical and biological stresses, apoptosis and necrosis, mitochondrial injuries, changes to basal respiration and ATP production, cellular structural damage, telomere shortening and cellular senescence, and DNA damage and oxidative stress. Notably, cell surface proteins that play a major role in stem cell fate and are used as the biomarkers of stem cells are more vulnerable to cold stress than other proteins. There are also data supporting the alteration in some biological features and genetic integrity at the molecular level of the post-thawed stem cells. This article reviews the current and future challenges of cryopreservation of stem cells and stresses the need for further rigorous

F. Karimi-Busheri (✉)
Department of Oncology, University of Alberta, Edmonton, AB, Canada
e-mail: fkarimi@ualberta.ca

A. Rasouli-Nia • M. Weinfeld (✉)
Department of Oncology, Cross Cancer Institute and University of Alberta, Edmonton, AB, Canada
e-mail: Michael.Weinfeld@albertahealthservices.ca

F. Karimi-Busheri, M. Weinfeld (eds.), *Biobanking and Cryopreservation of Stem Cells*, Advances in Experimental Medicine and Biology 951, DOI 10.1007/978-3-319-45457-3_1

research on the methodologies for freezing and utilizing cancer stem cells following long-term storage.

Keywords
Cryopreservation • Cancer stem cells • Cell surface proteins • Membrane proteins • Proliferation • Embryonic stem cells • Mesenchymal stem cells • Vitrification • Mammospheres • Lungospheres

Abbreviations

AML	Acute myeloid leukemic
CD	Cluster of differentiation
CSCs	Cancer stem cells
ESCs	Embryonic stem cells
hESC-IPs	Human embryonic stem cell derived islet progenitors
HLA-DR	Human leukocyte antigen – antigen D related
MSCs	Mesenchymal stem cells
PBSC	Peripheral blood stem cell

1.1 Introduction

The term "stem cell" was first coined by the eminent German biologist Ernst Haeckel in 1868 [1]. A century later, a series of discoveries led to the first bone marrow transplant in 1968, followed by the discovery of stem cells in human cord blood, embryonic and adult stem cells, and cancer stem cells that dramatically revolutionized science and clinical sciences. Today, stem cells are at the forefront of modern science and medicine. Despite many challenges and heated debates among public, religious, ethical, government and scientific leaders the overwhelming environment is positive and promising.

Stem cells are found and can be obtained from a variety of cellular sources in the very early stages of development. However they all share the properties of being flexible, versatile and undifferentiated. Human embryonic stem cells (hESCs) obtained from early blastocysts retain the flexibility to become any one of the human body cell types. They present the most promising resource for regenerative medicine and provide a new prospect in a rapidly growing cell therapy industry [2].

Adult stem cells (ASCs), also known as somatic stem cells, are undifferentiated cells with the capacity for self-renewal and differentiation. Their function is primarily to maintain and repair the tissue they are located in. Many different type of adult stem cells have been identified in organs and tissues including hematopoietic, mesenchymal (MSCs), neural, epithelial, skin stem cells and induced pluripotent stem cells (iPSCs) [2]. Potentially, ASCs can play a major role in regenerative medicine and transplantation-based therapy in future.

MSCs can be derived from bone marrow and a variety of other tissues like adipose tissue, placenta, skin, and umbilical cord blood, and dental pulp. MSCs are multipotent cells having self-renewal and multilineage differentiation capabilities into various specialized cell types such as adipocytes, cartilage, bone, tendons, muscle, and skin [3].

Induced pluripotent stem cells generated in the laboratory almost a decade ago [4] could be considered a landmark in the field of stem cells that has opened the door to create stem cells from any other cell type in the body. iPSCs are similar to embryonic stem cells and could be used as a good model to study pluripotency, cell fate, and disease modeling [5].

In late 1990s, Bonnet and Dick reported the isolation of a subpopulation of acute myeloid leukemic (AML) cells capable of self-renewal and initiating human AML in NOD/SCID mice [6].

These cells were exclusively $CD34^+/CD38^-$. This was followed by a large number of studies that led to the isolation of solid tumor stem cells from brain, breast, pancreas, colorectal, and prostate cancer [7–10]. These subpopulations share similar characteristics, notably the ability for self-renewal and differentiation, resistance to conventional chemo- and radiation therapy, and tumorigenicity [11, 12]. They also possess up-regulated signaling proteins such as hedgehog, epidermal growth factor receptor, NOTCH, and Bmi-1 necessary for metastasis [13, 14], and they are capable of producing cytokines, chemokines, and angiogenic factors [15, 16]. Despite many properties reminiscent of normal stem cells, including self-renewal, they exhibit important differences including dysregulation of self-renewal that leads to uncontrolled proliferation [17].

Over time, our understanding of cancer stem cells and their role in tumor formation and metastasis has evolved. It is now recognized that tumors are heterogeneous entities consisting of various type of cells and distinct niches that can support and instruct tumour regrowth. Tumors therefore possess multiple genomes that belong to distinct subclones, including cancer stem cells [18–20]. The rare subpopulation of cells with stemness properties can drive successive rounds of tumor development and growth. They are also equipped with inherent machinery that protects them from conventional chemotherapy and radiation, and may govern patient outcome [19, 21, 22].

The clinical relevance of cancer stem cells is gaining momentum in oncology. Considering these cells constitute a small rare subpopulation that has extensively expanded our vision of development, progression, and metastasis of malignancies adds to the importance of banking and preservation in this emerging field [23–25].

Regardless of their origin, stem cells have brought a wealth of opportunities and promise to science and medicine. Nevertheless, a number of significant challenges, such as cryopreservation and storage, remain to be overcome in order to succeed in bringing stem cell therapy to patients. Cryopreservation is considered an intrinsic factor that can contribute to the stability of stem cells. One critical step in cell therapy bioprocesses is to optimize cellular cryopreservation techniques that minimize alterations in biological function, spontaneous differentiation, and irreversible damages to the cells after freezing and thawing. Current protocols of hematopoietic and mesenchymal stem cell cryopreservation and banking are more or less clinically effective and safe, while there is still considerable controversy and inconsistency about the protocols and methodology used in the storage and safety of hESCs and iPSCS [26].

1.2 Pros and Cons of Stem Cell Banking

Recovery and consequent proliferation of the cells is the most important goal after thawing frozen specimens [23, 27]. During the last few decades a large body of evidence has accumulated indicating that the majority of adult and embryonic stem cells (ESCs) survive and maintain proliferation after cryopreservation and thawing. Neuronal progenitor cells, for example, have been shown to preserve survival and multipotency activity during exposure to different cryoprotectants and length of storage [28]. No long-term storage effect on the phenotype, proliferation, or osteogenic differentiation of human adipose-derived stem cells has been observed in a standard 5–10 % dimethyl sulfoxide (DMSO) protocol [29–31]. Similar results have also been reported on the impact of cryopreservation and the consequent thawing and survival of adult human brain sub-ventricular zone precursor cells [32], human embryonic stem cell-derived cardiomyocytes [33], and fetal liver stem/progenitor cells [34]. It is generally accepted that cryopreserved adult stem cells can proliferate and be expanded and manipulated for clinical purposes. However, there is also increasing evidence that cryopreservation and prolonged storage of stem cells induce extensive physical and biological stresses due to both intracellular and extracellular ice formed during cooling. The stress can lead to disassembly of the cell cytoskeleton and other cellular structural damage, mitochondrial injuries causing faulty basal respiration and ATP production, oxidative stress and DNA damage, apoptosis and necrosis, and telomere shortening and cellular senescence [27, 35–43].

Different cell types appear to respond differently to cold stress. Embryonic stem cells, for example, are more fragile and vulnerable to differentiation in cold stress, which can result in a decline in their recovery and survival when thawed [44, 45]. Adult stem cells, on the other hand, seem more tolerant under long term cryopreservation and even suitable for use in clinical grade [46]. Studies of cryopreservation on amniotic fluid derived stem cells and human limbal epithelial stem cells have found alterations in the molecular profiles of the stem cells [47, 48]. The finding of the latter study is particularly significant since the cryopreserved and expanded human limbal epithelial stem cells express no cell surface receptor human leukocyte antigen - antigen D related (HLA-DR) protein. This indicates that the expanded cells are nonimmunogenic in nature and raises interesting questions about the survival of these cells in an allogenic environment and their potential use in allogeneic cell therapy [48].

Similarly, cancer stem cell long-term preservation can result in the accumulation of karyotypic, genotypic, and phenotypic changes [49]. Our knowledge of the effect of cold stress on cancer stem cells is limited and, due to technical issues and lack of standard preservation methods and techniques, banking of these cells could result in alterations in many biological functions of the cell, including lipid composition of cellular membranes and biomarkers specific to these cells, as well as suppressing the rate of protein synthesis and cell proliferation [23, 50, 51]. Developing conditions specific for cancer stem cell storage will be important to avoid or reduce these drawbacks.

1.3 Vitrification

The application of cryopreservation on human tissue goes back to the mid-1950s when modern sperm preservation started [52]. This was followed decades later by rapid expansion of the methods for cryopreservation of other tissues such as human embryos, hESCs, ovarian tissue, and hematopoetic cells [53, 54]. In the majority of cases the technology for freezing and recovery of cells or tissues has been shown to be safe [55, 56], but, nevertheless, cell injury arising from conventional slow-freezing cryopreservation associated with ice formation and other stresses is inevitable [57].

A new ultra-fast freezing method called vitrification was introduced in 1985 to overcome ice crystallization during traditional cryopreservation [58]. Vitrification seems to have less effect on cell physiology and is less traumatic to the meiotic spindle than slow freezing methods [59]. These advantages make vitrification the standard method of cryopreservation for gametes as well as complex tissues.

There are an increasing number of reports on the application of vitrification in mesenchymal stem cells (MSCs), including amnion-derived MSCs, as slow freezing and consequent ice formation is so damaging that the cells are not able to survive [60]. An examination of vitrification of human umbilical cord Wharton's jelly showed that the expression of surface antigens is retained during vitrification in comparison to non-vitrified samples [58]. A report on the effect of varied methods of vitrification on human embryonic stem cell derived islet progenitors (hESC-IPs) demonstrated that the cells exhibit robust tolerance to exposure to varied vitrification protocols and maintain expression of some of the critical islet progenitor markers [61].

Vitrification, though it retains cell viability, has its own drawbacks that may impose therapeutic limitation for patient treatment or high-throughput drug screening procedures [62]. As an example, human embryonic stem cells are preserved as colonies and not single monolayer cells that could interfere with their inherent heterogeneity and variability in size [62].

To date, there are very few studies on the application of vitrification procedure for cryopreservation of cancer stem cells. Chong et al. [49] used vitrification for brain tumor-initiating cells derived from patients with glioblastoma multiforme. Vitrified neurospheres maintained their biological phenotypes and genetic profiles including self-renewal, multi-potency, and tumorigenic ability in NOD/SCID mice, without any evidence of genetic modification of the cells. This opens the door for further investigation into the utilization of vitrification of cancer stem cells.

1.4 Cell Surface Markers Under Cold Stress

Cancer stem cell isolation from patient specimens requires fresh tumor tissue and immediate processing [63]. The next step is obtaining specific and reliable molecular signatures by which these cells are defined and isolated. Morphology, immunophenotyping, tumorigenicity, and biomarkers are the most common methods used to identify cancer stem cells [23]. Biomarkers, however, are the hallmark signatures in identifying cancer stem cells. Despite the lack of any truly specific molecules, currently used biomarkers of cancer stem cells have been beneficial for research in the stem cell community and oncology. Cluster of differentiation (CD) proteins expressed on the surface of cells, are often the dominant molecules that serve as markers of normal and cancer stem cells (Table 1.1) and as such have been key players in stem cell research [93].

CDs were initially proposed to be used for the classification of the monoclonal antibodies against various cell surface molecules on leukocytes at various stages of differentiation. Later, however, the CD nomenclature was used in a much broader context to cover endothelial, stromal, and many other cell types as "human cell differentiation molecules", rather than only antibodies [94]. Currently, over 440 unique clusters and subclusters have been identified [95] and two thirds of drug targets are directed towards these proteins [96].

Cell surface proteins are more exposed to the external environment, are less abundant than other cellular proteins and more hydrophobic, which makes them less soluble than cytosolic and intraorganellar proteins [97]. They also often undergo extensive post-translational modifications, such as glycosylation, addition of lipid moieties, and phosphorylation that modulate their function [96]. These proteins seem also to be more vulnerable to long term preservation, and cold stress has been shown to disrupt the nuclear membrane [50, 98]. Recently, Gordiyenko and colleagues have reported that the retention of cell surface properties is dependent on the method of cryopreservation. It was shown that in comparison to other cryoreagents, PEG-1500 maintains the surface properties of frozen erythrocytes better than other agents, such as DMSO, after thawing [99]. A report on the cryopreservation of human dental pulp stem cells also indicated that length of time of preservation could affect membrane cell surface marker expression possibly because freezing may alter the 3-D shape of the proteins [100].

1.5 Cryopreservation Impact on Molecular Integrity of Cancer Stem Cells

A critical step in cryopreservation is not only survival and subsequent cell proliferation but maintenance of cell integrity and potential biomarker signature, particularly if clinical application is desired [27, 101]. In this regard, there are only a limited number of publications that up until now have dealt directly with the cryopreservation of cancer stem cells [49, 50].

A significant down regulation in gene expression has been observed in global gene expression analysis of fresh versus post-thawed lung and breast cancer stem cells [50]. Comparison of changes in gene expression between lung and breast cancer stem cells indicated that no gene was shared between the two sets [50]. There was, however, similarity between the two populations of differentially expressed genes in terms of their biological functions, distribution and canonical pathways, including genes involved in genetic disorders, cancer, cell death, cellular growth and proliferation, and tissue development. Flow cytometric analysis of some of the major cell surface biomarkers of lung cancer stem cells (lungo-

Table 1.1 Potentially useful cell surface biomarkers of cancer stem cells

Genes	Description	Function	Malignancy	References
CD24	Cell adhesion molecule encodes a sialoglycoprotein that is expressed in mature granulocytes and in many B cells	Modulates B-cell activation responses	Breast, head and neck, pancreatic	[64]
CD38	Marker of cell activation; novel multifunctional ectoenzyme widely expressed in cells and tissues	Cell adhesion, signal transduction and calcium signaling	Lung, leukemia	[50, 65]
CD326(EpCAM)	Homophillic Ca2+-independent cell adhesion molecule; carcinoma-associated antigen	Epithelial cell adhesion molecule	Breast, liver, pancreatic, colorectal	[66, 67]
CD133(PROM1)	Five-transmembrane domain glycoprotein	Potential role in the organization of plasma membrane topology	Brain, lung, thyroid, colorectal	[67–69]
CD29(ITGB1)	Integrin Beta-1 protein, receptors for collagen	Involved in cell adhesion and in a variety of processes including embryogenesis, hemostasis, tissue repair, immune response and metastasis	Vascular	[70]
CD44	Cell-surface glycoprotein	Cell–cell interactions, cell adhesion and migration.	Breast, head and neck, pancreatic, lung, gastric, thyroid, esophageal, colorectal	[68, 69, 71]
CD166(ALCAM)	Transmembrane type-1 glycoprotein, activated leucocyte cell adhesion molecule	T-cell activation and proliferation, angiogenesis, hematopoiesis and axon fasciculation	Colorectal, hematopoietic	[69, 72]
CD9(THY1)	Glycosylphosphatidylinositol-anchored membrane glycoprotein	Involved in signal transduction	brain, liver, lung, leukemia	[73, 74]
CD20(MS4A1)	Activated-glycosylated phosphoprotein expressed on the surface of all B-cells	Acts as a calcium channel in the cell membrane	Melanoma	[75]
Cd13(ANPEP)	Member of the super family of zinc-binding metalloproteinases	Cellular processes such as mitosis, invasion, cell adhesion, angiogenesis, radiation resistance, and antiapoptosis	Liver	[76]
CD66c(CEACAM6)	Carcinoembryonic antigen-related cell adhesion molecule 6; cell surface glycoprotein	Cell adhesion and tumor markers in serum immunoassay determinations of carcinoma	Colorectal cancer	[77]
CD117	Receptor tyrosine kinase protein encoded by the KIT gene	Binds to stem cell factor	Ovarian cancer	[78]
CD34	Cell surface glycoprotein	Cell-cell adhesion factor	Leukemia	[65]

Genes	Description	Function	Malignancy	References
CD33(SIGLEC-3)	A transmembrane receptor	Differentiation antigen	Leukemia	[79]
CD123(interleukin-3 receptor)	Leukemia-associated antigen	A soluble cytokine important in the immune system	Leukemia	[80]
CD13(ANPEP)	Membrane peptidases	Involved in the final digestion of peptides in the small intestine and other cell types	Leukemia, hepatocellular carcinoma	[81, 82]
CD96	Type I membrane protein belongs to the immunoglobulin superfamily	May play a role in the adhesive interactions of activated T and NK cells	Leukemia	[83–86]
CD47	Transmembrane protein	Ligand for signal regulatory protein alpha expressed on phagocytic cells	Leukemia	[87]
CD32(FCGR2A)	Encodes one member of a family of immunoglobulin Fc receptor genes	Involved in the process of phagocytosis and clearing of immune complexes	Leukemia	[88, 89]
CD25(IL2RA)	Receptor for interleukin 2	Specific immunomodulatory B-cell subset ready to proliferate upon IL-2 stimulation	Leukemia	[89, 90]
CD90(THY1)	Cell surface glycoprotein	Involve in cell adhesion and cell communication	Leukemia, liver	[87, 91]
CD26(DPPIV)	Intrinsic membrane glycoprotein and a serine exopeptidase	Involved in the costimulatory signal essential for T-cell receptor	Leukemia	[92]
CD45(PTPRC)	Protein tyrosine phosphatase, receptor type C	Signaling molecules regulating cell growth, differentiation, mitotic cycle, and oncogenic transformation	Liver	[91]

Table 1.2 The expression of cell surface biomarkers in fresh and 12 months freeze-thaw H460 and A549 non-small cell lung cancer derived stem cells

Markers	Cell line	Fresh (%)	Frozen (%)
MUC1	H460	91.60	77.45
(CD227)	A549	92.72	21.54
EpCam	H460	21.66	19.12
(CD326)	A549	73.16	36.43
Fas	H460	99.66	97.16
(CD95)	A549	98.70	98.41

spheres) revealed a pattern of down regulation (Table 1.2) following freezing that indicates susceptibility of the stem cell membrane to cryopreservation [50].

1.6 Self Renewal and Cryopreservation

Data on the effect of cryopreservation and self-renewal ability of stem cells are still scarce. Self-renewal is a unique property of stem cells by which stem cells divide and make clones of themselves, thus preserving the stem cell pool throughout life. Dysregulation of stem cells could potentially transform normal adult stem cells into malignant cells capable of driving tumorigenesis through clonal expansion of stem/progenitor cells with frequent genetic or epigenetic modifications [23].

Shima et al. [102] evaluated the viability and functions of peripheral blood stem cells (PBSC) after more than 10 years of cryopreservation and did not observe any effect on the self-renewal potential of $CD34^{+}$ cells. Similar results have also been reported on the maintenance of self-renewal ability in breast cancer stem cells [23]. In this study cells were thawed at different times and the self-renewal ability of the cells was compared with fresh breast cancer stem cells (mammospheres). The difference in self-renewal of cells was not statistically significant regardless of the length of time of cryopreservation. Since self-renewal is a promising target in anti-tumor drug targeting the impact of cryopreservation, storage and the thawing of the cells could be significant.

1.7 Senescence and Cryopreservation

Normal cells do not proliferate indefinitely and in culture they gradually lose their ability to divide and progress through the cell cycle but remain viable for a while. Most of the senescent cells go through drastic changes including irreversible growth arrest, resistance to apoptosis, and significant changes in expression of genes such as p21 and p16, two known cell-cycle inhibitors [103]. All these are vital mechanisms that constrain the malignant progression of many tumors [104] and at the same time are potentially important for the response to therapy [105].

To show that cryopreservation could induce senescence Honda et al. [39] demonstrated that freezing and subsequently thawing retinal pigment epithelial cells severely impaired proliferation when compared with fresh non-cryopreserved cells. The authors also observed a notable excess of single-strand DNA breaks and an acceleration of telomere shortening in cryopreserved cells immediately after thawing that confirmed the telomerase-induced senescence [39]. Pollock et al. [106] also observed growth arrest in the post thaw samples of mesenchymal stromal cells with high senescence an indication that senescence appears to correlate with poor post-thaw function. In contrast, Choudhery and Harris [107] did not detect a significant effect on senescent features of mesenchymal stem cells following cryopreservation.

The only report, to our knowledge, of investigating any possible association between cryopreservation and senescence induction in cancer stem cells is the observation of a significant increase in the rate of senescence-mediated pathways measured by senescence-associated β-galactosidase assay in cryopreserved mammospheres [23, 108].

1.8 Conclusion

An overwhelming majority of reports confirm that cryopreservation of stem cells and cancer stem cells is reliable and safe. The frozen cells after long-term storage are also capable of preserving their basic repair mechanisms and grow and proliferate normally. There are, however, several studies indicating alterations in some biological features and genetic integrity at the molecular level in post-thawed cell populations. Both global gene expression and molecular analysis of stem cells and cancer stem cells, in particular of cell surface markers, provide evidence of significant down regulation following long term storage of the cells. Different cell types differ significantly in response to cold stress and freeze-thaw processes, which sheds doubt on whether a standard freezing protocol would be optimal for all types of stem cells. In the case of cancer stem cells this is instructive since if they are more sensitive to cryopreservation then the discrepancies in the results obtained by different investigators examining biomarkers might, in fact, be in part attributable to the freezing process. Accordingly, it is highly advisable to employ rigorous methodology for freezing and utilizing cancer stem cells following long-term storage.

References

1. Ramalho-Santos M, Willenbring H (2007) On the origin of the term "stem cell". Cell Stem Cell 1(1):35–38
2. National Institute of Health. Stem Cell Information. http://stemcells.nih.gov/info/basics/pages/basics4.aspx. Accessed 14 Apr 2016
3. Keating A (2012) Mesenchymal stromal cells: new directions. Cell Stem Cell 10(6):709–716
4. Takahashi K, Yamanaka S (2006) Induction of pluripotent stem cells from mouse embryonic and adult fibroblast cultures by defined factors. Cell 126(4):663–676
5. Karagiannis P, Eto K (2016) Ten years of induced pluripotency: from basic mechanisms to therapeutic applications. Development 143(12):2039–2043
6. Bonnet D, Dick JE (1997) Human acute myeloid leukemia is organized as a hierarchy that originates from a primitive hematopoietic cell. Nat Med 3(7):730–737
7. Ignatova TN, Kukekov VG, Laywell ED, Suslov ON, Vrionis FD, Steindler DA (2002) Human cortical glial tumors contain neural stem-like cells expressing astroglial and neuronal markers in vitro. Glia 39(3):193–206
8. Al-Hajj M, Wicha MS, Benito-Hernandez A, Morrison SJ, Clarke MF (2003) Prospective identification of tumorigenic breast cancer cells. Proc Natl Acad Sci U S A 100(11):3983–3988
9. Lonardo E, Hermann PC, Heeschen C (2010) Pancreatic cancer stem cells – update and future perspectives. Mol Oncol 4(5):431–442
10. Lang SH, Frame FM, Collins AT (2009) Prostate cancer stem cells. J Pathol 217(2):299–306
11. Milas L, Hittelman WN (2009) Cancer stem cells and tumor response to therapy: current problems and future prospects. Semin Radiat Oncol 19(2):96–105
12. Eyler CE, Rich JN (2008) Survival of the fittest: cancer stem cells in therapeutic resistance and angiogenesis. J Clin Oncol 26(17):2839–2845
13. Mimeault M, Batra SK (2010) New advances on critical implications of tumor- and metastasis-initiating cells in cancer progression, treatment resistance and disease recurrence. Histol Histopathol 25(8):1057–1073
14. Simeone DM (2008) Pancreatic cancer stem cells: implications for the treatment of pancreatic cancer. Clin Cancer Res 14(18):5646–5648
15. Levina V, Marrangoni AM, DeMarco R, Gorelik E, Lokshin AE (2008) Drug-selected human lung cancer stem cells: cytokine network, tumorigenic and metastatic properties. PLoS One 3(8):e3077
16. Hermann PC, Huber SL, Herrler T, Aicher A, Ellwart JW, Guba M, Bruns CJ, Heeschen C (2007) Distinct populations of cancer stem cells determine tumor growth and metastatic activity in human pancreatic cancer. Cell Stem Cell 1(3):313–323
17. O'Brien CA, Kreso A, Jamieson CH (2010) Cancer stem cells and self-renewal. Clin Cancer Res 16(12):3113–3120
18. Junttila MR, de Sauvage FJ (2013) Influence of tumour micro-environment heterogeneity on therapeutic response. Nature 501(7467):346–354
19. Kreso A, Dick JE (2014) Evolution of the cancer stem cell model. Cell Stem Cell 14(3):275–291
20. McGranahan N, Swanton C (2015) Biological and therapeutic impact of intratumor heterogeneity in cancer evolution. Cancer Cell 27(1):15–26
21. Gargini R, Cerliani JP, Escoll M, Antón IM, Wandosell F (2015) Cancer stem cell-like phenotype and survival are coordinately regulated by Akt/FoxO/Bim pathway. Stem Cells 33(3):646–660
22. Eppert K, Takenaka K, Lechman ER, Waldron L, Nilsson B, van Galen P, Metzeler KH, Poeppl A, Ling V, Beyene J et al (2011) Stem cell gene expression programs influence clinical outcome in human leukemia. Nat Med 17(9):1086–1093
23. Karimi-Busheri F, Zadorozhny V, Shawler DL, Fakhrai H (2010) The stability of breast cancer progenitor cells during cryopreservation: maintenance of proliferation, self-renewal, and senescence characteristics. Cryobiology 60(3):308–314
24. Karimi-Busheri F, Rasouli-Nia A (2015) Integration, networking, and global biobanking in the age of new biology. Adv Exp Med Biol 864:1–9
25. Monteiro J, Fodde R (2010) Cancer stemness and metastasis: therapeutic consequences and perspectives. Eur J Cancer 46(7):1198–1203

26. Lee JE, Lee DR (2011) Human embryonic stem cells: derivation, maintenance and cryopreservation. Int J Stem Cells 4(1):9–17
27. Bissoyi A, Nayak B, Pramanik K, Sarangi SK (2014) Targeting cryopreservation-induced cell death: a review. Biopreserv Biobank 12(1):23–34
28. Xue T, Qiao L, Qiu J, Lu L, Zha D, Wei L (2008) Proliferation, multipotency and neuronal differentiation of cryopreserved neural progenitor cells derived from the olfactory neuroepithelium of the adult rat. Cell Biol Int 32(8):950–958
29. Gonda K, Shigeura T, Sato T, Matsumoto D, Suga H, Inoue K, Aoi N, Kato H, Sato K, Murase S, Koshima I, Yoshimura K (2008) Preserved proliferative capacity and multipotency of human adipose-derived stem cells after long-term cryopreservation. Plast Reconstr Surg 121(2):401–410
30. Grimshaw MJ, Cooper L, Papazisis K, Coleman JA, Bohnenkamp HR, Chiapero-Stanke L, Taylor-Papadimitriou J, Burchell JM (2008) Mammosphere culture of metastatic breast cancer cells enriches for tumorigenic breast cancer cells. Breast Cancer Res 10(3):R52
31. Yong KW, Pingguan-Murphy B, Xu F, Abas WA, Choi JR, Omar SZ, Azmi MA, Chua KH, Wan Safwani WK (2015) Phenotypic and functional characterization of long-term cryopreserved human adipose-derived stem cells. Sci Rep 5:9596
32. Ayuso-Sacido A, Roy NS, Schwartz TH, Greenfield JP, Boockvar JA (2008) Longterm expansion of adult human brain subventricular zone precursors. Neurosurgery 62(1):223–229
33. Norstrom A, Akesson K, Hardarson T, Hamberger L, Bjorquist P, Sartipy P (2006) Molecular and pharmacological properties of human embryonic stem cell-derived cardiomyocytes. Exp Biol Med (Maywood) 231(11):1753–1762
34. Oertel M, Menthena A, Chen YQ, Shafritz DA (2006) Properties of cryopreserved fetal liver stem/progenitor cells that exhibit long-term repopulation of the normal rat liver. Stem Cells 24(10):2244–2251
35. Baust JG, Gao D, Baust JM (2009) Cryopreservation: an emerging paradigm change. Organogenesis 5(3):90–96
36. Baust JM, Corwin WL, VanBuskirk R, Baust JG (2015) Biobanking: the future of cell preservation strategies. Adv Exp Med Biol 864:37–53
37. Bischof JC, Wolkers WF, Tsvetkova NM, Oliver AE, Crowe JH (2002) Lipid and protein changes due to freezing in Dunning AT-1 cells. Cryobiology 45(1):22–32
38. Hosu BG, Mullen SF, Critser JK, Forgacs G (2008) Reversible disassembly of the actin cytoskeleton improves the survival rate and developmental competence of cryopreserved mouse oocytes. PLoS One 3(7):e2787
39. Honda S, Weigel A, Hjelmeland LM, Handa JT (2001) Induction of telomere shortening and replicative senescence by cryopreservation. Biochem Biophys Res Commun 282(2):493–498
40. Tatone C, Di Emidio G, Vento M, Ciriminna R, Artini PG (2010) Cryopreservation and oxidative stress in reproductive cells. Gynecol Endocrinol 26(8):563–567
41. Keane KN, Calton EK, Cruzat VF, Soares MJ, Newsholme P (2015) The impact of cryopreservation on human peripheral blood leucocyte bioenergetics. Clin Sci (Lond) 128(10):723–733
42. Bray SE, Paulin FE, Fong SC, Baker L, Carey FA, Levison DA, Steele RJ, Kernohan NM (2010) Gene expression in colorectal neoplasia: modifications induced by tissue ischaemic time and tissue handling protocol. Histopathology 56(2):240–250
43. Del Bo' C, Fracassetti D, Lanti C, Porrini M, Riso P (2015) Comparison of DNA damage by the comet assay in fresh versus cryopreserved peripheral blood mononuclear cells obtained following dietary intervention. Mutagenesis 30(1):29–35
44. Berz D, McCormack EM, Winer ES, Colvin GA, Quesenberry PJ (2007) Cryopreservation of hematopoietic stem cells. Am J Hematol 82(6):463–472
45. Ware CB, Nelson AM, Blau CA (2005) Controlled-rate freezing of human ES cells. Biotechniques 38(6):879–883
46. Thirumala S, Goebel WS, Woods EJ (2009) Clinical grade adult stem cell banking. Organogenesis 5(3):143–154
47. Seo JM, Sohn MY, Suh JS, Atala A, Yoo JJ, Shon YH (2011) Cryopreservation of amniotic fluid-derived stem cells using natural cryoprotectants and low concentrations of dimethylsulfoxide. Cryobiology 62(3):167–173
48. Vasania VS, Prasad P, Gill RK, Mehta A, Viswanathan C, Sarang S, Majumdar AS (2011) Molecular and cellular characterization of expanded and cryopreserved human limbal epithelial stem cells reveal unique immunological properties. Exp Eye Res 92(1):47–56
49. Chong YK, Toh TB, Zaiden N, Poonepalli A, Leong SH, Ong CE, Yu Y, Tan PB, See SJ, Ng WH, Ng I, Hande MP, Kon OL, Ang BT, Tang C (2009) Cryopreservation of neurospheres derived from human glioblastoma multiforme. Stem Cells 27(1):29–39
50. Karimi-Busheri F, Zadorozhny V, Carrier E, Fakhrai H (2013) Molecular integrity and global gene expression of breast and lung cancer stem cells under long-term storage and recovery. Cell Tissue Bank 14(2):175–186
51. Fujita J (1999) Cold shock response in mammalian cells. J Mol Microbiol Biotechnol 1(2):243–255
52. Walters EM, Benson JD, Woods EJ, Critser JK (2009) The history of sperm cryopreservation. In: Pacey AA, Tomlinson MJ (eds) Sperm banking: theory and practice. Cambridge University Press, Cambridge, pp 1–10

53. Asghar W, El Assal R, Shafiee H, Anchan RM, Demirci U (2014) Preserving human cells for regenerative, reproductive, and transfusion medicine. Biotechnol J 9(7):895–903
54. Pegg DE (2002) The history and principles of cryopreservation. Semin Reprod Med 20(1):5–13
55. Manipalviratn S, Huang A, Decherney AH (2006) Future of infertility treatment. Int Surg 91(5 Suppl):S3–S14
56. Woods EJ, Benson JD, Agca Y, Critser JK (2004) Fundamental cryobiology of reproductive cells and tissues. Cryobiology 48(2):146–156
57. Huang H, Choi JK, Rao W, Zhao S, Agarwal P, Zhao G, He X (2015) Alginate hydrogel microencapsulation inhibits devitrification and enables large-volume Low-CPA cell vitrification. Adv Funct Mater 25(44):6939–6850
58. Marquez-Curtis LA, Janowska-Wieczorek A, McGann LE, Elliott JA (2015) Mesenchymal stromal cells derived from various tissues: biological, clinical and cryopreservation aspects. Cryobiology 71(2):181–197
59. Tavukcuoglu S, Al-Azawi T, Afshin Khaki A, Al-Hasani S (2012) Is vitrification standard method of cryopreservation. Middle East Fertil Soc J 17(3):152–156
60. Moon JH, Lee JR, Jee BC, Suh CS, Kim SH, Lim HJ, Kim HK (2008) Successful vitrification of human amnion-derived mesenchymal stem cells. Hum Reprod 23(8):1760–1770
61. Lahmy R, Bolyukh VF, Castilla SM, Laurent LC, Katkov II, Itkin-Ansari P (2015) Tolerance of human embryonic stem cell derived islet progenitor cells to vitrification-relevant solutions. Cryobiology 70(3):283–286
62. Coopman K (2011) Large-scale compatible methods for the preservation of human embryonic stem cells: current perspectives. Biotechnol Prog 27(6):1511–1521
63. Veneziani BM, De Placido S (2013) Issues of banking breast cancer cells to generate mammospheres. Cell Tissue Bank 14(2):153–8
64. Suzuki T, Kiyokawa N, Taguchi T, Sekino T, Katagiri YU, Fujimoto J (2001) CD24 induces poptosis in human B cells via the glycolipid-enriched membrane domains/rafts-mediated signaling ystem. J Immunol 166(9):5567–5577
65. Horton SJ, Huntly BJ (2012) Recent advances in acute myeloid leukemia stem cell biology. Haematologica 97(7):966–974
66. van der Gun BT, Melchers LJ, Ruiters MH, de Leij LF, McLaughlin PM, Rots MG (2010) EpCAM in carcinogenesis: the good, the bad or the ugly. Carcinogenesis 31(11):1913–1921
67. Visvader JE, Lindeman GJ (2008) Cancer stem cells in solid tumours: accumulating evidence and unresolved questions. Nat Rev Cancer 8(10):755–768
68. Bi Y, Meng Y, Wu H, Cui Q, Luo Y, Xue X (2016) Expression of the potential cancer stem cell markers CD133 and CD44 in medullary thyroid carcinoma: a ten-year follow-up and prognostic analysis. J of Surg Oncol 113(2):144–151
69. Cherciu I, Bărbălan A, Pirici D, Mărgăritescu C, Săftoiu A (2014) Stem cells, colorectal cancer and cancer stem cell markers correlations. Curr Health Sci J 40(3):153–161
70. Lan J, Huang B, Liu R, Ju X, Zhou Y, Jiang J, Liang W, Shen Y, Li F, Pang L (2015) Expression of cancer stem cell markers and their correlation with pathogenesis in vascular tumors. Int J Clin Exp Pathol 8(10):12621–12633
71. Honing J, Pavlov KV, Mul VE, Karrenbeld A, Meijer C, Faiz Z, Smit JK, Hospers GA, Burgerhof JG, Kruyt FA, Kleibeuker JH, Plukker JT (2015) CD44, SHH and SOX2 as novel biomarkers in esophageal cancer patients treated with neoadjuvant chemoradiotherapy. Radiother Oncol 117(1):152–158
72. Chitteti BR, Bethel M, Kacena MA, Srour EF (2013) CD166 and regulation of hematopoiesis. Curr Opin Hematol 20(4):273–280
73. Podergajs N, Motaln H, Rajčević U, Verbovšek U, Koršič M, Obad N, Espedal H, Vittori M, Herold-Mende C, Miletic H, Bjerkvig R, Turnšek TL (2016) Transmembrane protein CD9 is glioblastoma biomarker, relevant for maintenance of glioblastoma stem cells. Oncotarget 7(1):593–609
74. Nishida H, Yamazaki H, Yamada T, Iwata S, Dang NH, Inukai T, Sugita K, Ikeda Y, Morimoto C (2009) CD9 correlates with cancer stem cell potentials in human B-acute lymphoblastic leukemia cells. Biochem Biophys Res Commun 382(1):57–62
75. Fang D, Nguyen TK, Leishear K, Finko R, Kulp AN, Hotz S, Van Belle PA, Xu X, Elder DE, Herlyn M (2005) A tumorigenic subpopulation with stem cell properties in melanomas. Cancer Res 65(20):9328–9337
76. Haraguchi N, Ishii H, Mimori K, Tanaka F, Ohkuma M, Kim HM, Akita H, Takiuchi D, Hatano H, Nagano H, Barnard GF, Doki Y, Mori M (2010) CD13 is a therapeutic target in human liver cancer stem cells. J Clin Invest 120(9):3326–3339
77. Gemei M, Mirabelli P, Di Noto R, Corbo C, Iaccarino A, Zamboli A, Troncone G, Galizia G, Lieto E, Del Vecchio L, Salvatore F (2013) CD66c is a novel marker for colorectal cancer stem cell isolation, and its silencing halts tumor growth in vivo. Cancer 119(4):729–738
78. Zhang R, Zhang P, Wang H, Hou D, Li W, Xiao G, Li C (2015) Inhibitory effects of metformin at low concentration on epithelial-mesenchymal transition of CD44(+)CD117(+) ovarian cancer stem cells. Stem Cell Res Ther 6:262
79. Walter RB, Appelbaum FR, Estey EH, Bernstein ID (2014) The role of CD33 as therapeutic target in acute myeloid leukemia. Expert Opin Ther Targets 18(7):715–718
80. Mardiros A, Forman SJ, Budde LE (2015) T cells expressing CD123 chimeric antigen receptors for

treatment of acute myeloid leukemia. Curr Opin Hematol 22(6):484–488
81. Taussig DC, Pearce DJ, Simpson C, Rohatiner AZ, Lister TA, Kelly G, Luongo JL, Danet-Desnoyers GA, Bonnet D (2005) Hematopoietic stem cells express multiple myeloid markers: implications for the origin and targeted therapy of acute myeloid leukemia. Blood 106(13):4086–4092
82. Yin C, Xie WF (2014) Hepatocellular carcinoma: basic and transitional research. Gastrointest Tumors 1(2):76–83
83. Hosen N, Park CY, Tatsumi N, Oji Y, Sugiyama H, Gramatzki M, Krensky AM, Weissman IL (2007) CD96 is a leukemic stem cell-specific marker in human acute myeloid leukemia. Proc Natl Acad Sci U S A 104(26):11008–11013
84. Du W, Hu Y, Lu C, Li J, Liu W, He Y, Wang P, Cheng C, Hu YU, Huang S, Yao J, Zheng J (2015) Cluster of differentiation 96 as a leukemia stem cell-specific marker and a factor for prognosis evaluation in leukemia. Mol Clin Oncol 3(4):833–838
85. Weissman IL (2007) CD96 is a leukemic stem cell-specific marker in human acute myeloid leukemia. Proc Natl Acad Sci U S A 104(26):11008–11013
86. Fuchs A, Cella M, Giurisato E, Shaw AS, Colonna M (2004) Cutting edge: CD96 (tactile) promotes NK cell-target cell adhesion by interacting with the poliovirus receptor (CD155). J Immunol 172(7):3994–3998
87. Majeti R, Chao MP, Alizadeh AA, Pang WW, Jaiswal S, Gibbs KD Jr, van Rooijen N, Weissman IL (2009) CD47 is an adverse prognostic factor and therapeutic antibody target on human acute myeloid leukemia stem cells. Cell 138(2):286–299
88. Hart SP, Alexander KM, Dransfield I (2004) Immune complexes bind preferentially to Fc gamma RIIA (CD32) on apoptotic neutrophils, leading to augmented phagocytosis by macrophages and release of proinflammatory cytokines. J Immunol 172(3):1882–1887
89. Saito Y, Kitamura H, Hijikata A, Tomizawa-Murasawa M, Tanaka S, Takagi S, Uchida N, Suzuki N, Sone A, Najima Y, Ozawa H, Wake A, Taniguchi S, Shultz LD, Ohara O, Ishikawa F (2010) Identification of therapeutic targets for quiescent, chemotherapy-resistant human leukemia stem cells. Sci Transl Med 2(17):17ra9
90. Amu S, Brisslert M (2011) Phenotype and function of CD25-expressing B lymphocytes isolated from human umbilical cord blood. Clin Dev Immunol 2011:481948
91. Yang ZF, Ho DW, Ng MN, Lau CK, Yu WC, Ngai P, Chu PW, Lam CT, Poon RT, Fan ST (2008) Significance of CD90+ cancer stem cells in human liver cancer. Cancer Cell 13(2):153–166
92. Valent P, Sadovnik I, Ráčil Z, Herrmann H, Blatt K, Cerny-Reiterer S, Eisenwort G, Lion T, Holyoake T, Mayer J (2014) DPPIV (CD26) as a novel stem cell marker in Ph+chronic myeloid leukaemia. Eur J Clin Invest 44(12):1239–1245
93. Pruszak J, Ludwig W, Blak A, Alavian K, Isacson O (2009) CD15, CD24, and CD29 define a surface biomarker code for neural lineage differentiation of stem cells. Stem Cells 27(12):2928–2940
94. Zola H, Swart B, Nicholson I, Aasted B et al (2005) CD molecules 2005: human cell differentiation molecules. Blood 106(9):3123–3126
95. Uniport website: http://www.uniprot.org/docs/cdlist. Accessed 13 Dec 2016
96. Ahn SM, Goode RJ, Simpson RJ (2008) Stem cell markers: insights from membrane proteomics? Proteomics 8(23–24):4946–4957
97. Alberts B, Johnson A, Lewis J, Raff M, Roberts K, Walter P (2002) Molecular biology of the cell, 4th edn. Garland Science, New York
98. Kim YY, Ku SY, Liu HC, Cho HJ, Oh SK, Moon SY, Choi YM (2011) Cryopreservation of human embryonic stem cells derived-cardiomyocytes induced by BMP2 in serum-free condition. Reprod Sci 18(3):252–260
99. Gordiyenko OI, Anikieieva MO, Rozanova SL, Kovalenko SY, Kovalenkol IF, Gordiyenko EO (2015) Development of a model to investigate red blood cell surface characteristics after cryopreservation. Cryo Lett 36(3):221–226
100. Munévar JC, Gutiérrez N, Jiménez NT, Lafaurie GI (2015) Evaluation of two human dental pulp stem cell cryopreservation methods. Acta Odontol Latinoam 28(2):114–121
101. Paynter SJ (2008) Principles and practical issues for cryopreservation of nerve cells. Brain Res Bull 75(1):1–14
102. Shima T, Iwasaki H, Yamauchi T, Kadowaki M, Kiyosuke M, Mochimaru T, Takenaka K, Miyamoto T, Akashi K, Teshima T (2015) Preserved in vivo reconstitution ability of PBSCs cryopreserved for a decade at −80°C. Bone Marrow Transplant 50(9):1195–1200
103. Mirzayans R, Andrais B, Scott A, Murray D (2012) New insights into p53 signaling and cancer cell response to DNA damage: implications for cancer therapy. J Biomed Biotechnol 2012:170325
104. Campisi J, d'Adda di Fagagna F (2007) Cellular senescence: when bad things happen to good cells. Nat Rev Mol Cell Biol 8(9):729–740
105. Pérez-Mancera PA, Young AR, Narita M (2014) Inside and out: the activities of senescence in cancer. Nat Rev Cancer 14(8):547–558
106. Pollock K, Sumstad D, Kadidlo D, McKenna DH, Hubel A (2015) Clinical mesenchymal stromal cell products undergo functional changes in response to freezing. Cytotherapy 17(1):38–45
107. Choudhery MS, Harris DT (2014) Cryopreservation can be used as an anti-aging strategy. Cytotherapy 16(12):1771–1773
108. Karimi-Busheri F, Rasouli-Nia A, Mackey JR, Weinfeld M (2010) Senescence evasion by MCF-7 human breast tumor-initiating cells. Breast Cancer Res 12(3):R31

Cryopreservation: Evolution of Molecular Based Strategies

2

John M. Baust, William Corwin, Kristi K. Snyder, Robert Van Buskirk, and John G. Baust

Abstract

Cryopreservation (CP) is an enabling process providing for on-demand access to biological material (cells and tissues) which serve as a starting, intermediate or even final product. While a critical tool, CP protocols, approaches and technologies have evolved little over the last several decades. A lack of conversion of discoveries from the CP sciences into mainstream utilization has resulted in a bottleneck in technological progression in areas such as stem cell research and cell therapy. While the adoption has been slow, discoveries including molecular control and buffering of cell stress response to CP as well as the development of new devices for improved sample freezing and thawing are providing for improved CP from both the processing and sample quality perspectives. Numerous studies have described the impact, mechanisms and points of control of cryopreservation-induced delayed-onset cell death (CIDOCD). In an effort to limit CIDOCD, efforts have focused on CP agent and freeze

J.M. Baust (✉) • K.K. Snyder
CPSI Biotech, 2 Court St, Owego, NY 13827, USA

Institute of Biomedical Technology, State University of New York at Binghamton, 4400 Vestal Parkway East, Binghamton, NY 13902, USA
e-mail: jmbaust@cpsibiotech.com; ksnyder@cpsibiotech.com

W. Corwin
Department of Immunology and Carole and Ray Neag Comprehensive Cancer Center, University of Connecticut School of Medicine, Farmington, CT 06030, USA
e-mail: corwin@uchc.edu

R. Van Buskirk
CPSI Biotech, 2 Court St, Owego, NY 13827, USA

Institute of Biomedical Technology, State University of New York at Binghamton, 4400 Vestal Parkway East, Binghamton, NY 13902, USA

Department of Biological Sciences, Binghamton University, 4400 Vestal Parkway East, Binghamton, NY 13902, USA
e-mail: rvanbus@binghamton.edu

J.G. Baust
Institute of Biomedical Technology, State University of New York at Binghamton, 4400 Vestal Parkway East, Binghamton, NY 13902, USA

Department of Biological Sciences, Binghamton University, 4400 Vestal Parkway East, Binghamton, NY 13902, USA
e-mail: baustcryo@aol.com

F. Karimi-Busheri, M. Weinfeld (eds.), *Biobanking and Cryopreservation of Stem Cells*, Advances in Experimental Medicine and Biology 951, DOI 10.1007/978-3-319-45457-3_2

media formulation to provide a solution path and have yielded improvements in survival over traditional approaches. Importantly, each of these areas, new technologies and cell stress modulation, both individually and in combination, are now providing a new foundation to accelerate new research, technology and product development for which CP serves as an integral component. This chapter provides an overview of the molecular stress responses of cells to cryopreservation, the impact of the hypothermic and cell death continuums and the targeted modulation of common and/or cell specific responses to CP in providing a path to improving cell quality.

Keywords
Cryopreservation • Apoptosis • Molecular control • Biopreservation • Thawing • Cell Storage • Cryopreservation induced cell death • Improved survival • Necroptosis

Abbreviations

CCM	Complex cryopreservation media
CIDOCD	Cryopreservation induced delayed onset cell death
CP	Cryopreservation
CPA	Cryoprotective agent
DMSO	Dimethyl sulfoxide
ER	Endoplasmic reticulum
RNAi	RNA interference
ROCK	Rho-associated kinase
TAC	Target apoptotic control
Tg	Glass transition temperature
TNF-α	Tumor necrosis factor alpha
UPR	Unfolded protein response

2.1 Introduction

Biopreservation is a multi-disciplinary scientific endeavor which focuses on the development of methods for the storage of cells, tissues and organs and their subsequent return to pre-storage functionality [1]. As a field, biopreservation incorporates a number of areas including cryobiology, bioengineering, computer sciences, molecular biology, and cellular biology among others [2]. Currently, biopreservation is experiencing unprecedented growth and development as advances in areas such as cell therapy, stem cell research, personalized medicine, cell banking and cancer research drive the need for optimal processing and storage of the unique biologics (cells and tissues) for downstream utilization [3]. Even as advances continue to improve outcomes, there remain significant issues with current preservation practices including: continued cell death, loss of function, use of animal based products in storage media, activation of molecular stress response pathways in 'surviving' cells and downstream effects on epigenetic control of gene expression and protein levels [4, 5]. This last point highlights the profound change in our understanding of the consequences of low temperature, as the focus has moved from the physical/chemical changes (i.e. osmolarity, membrane disruption, etc.) to the complex molecular and stress pathway associated alterations over the last few decades.

Typically, the utilization of low temperatures is the primary method for the preservation of biologics with the goal to extend the utility and deliver them "on demand". Biopreservation can be divided into two branches, hypothermic storage and cryopreservation. Hypothermic storage is the use of non-frozen temperatures to store biologics, from below normothermic temperatures (32–37 °C) to usually no lower than 0 °C. A temperature range of 4–10 °C is most commonly utilized for this regime. In contrast, cryopreservation is the utilization of ultra-low temperatures, typically at or below −80 °C, and usually below −140 °C, as this is below the reported nominal glass transition temperature (Tg) of water (an important point which will be addressed later) to achieve long term storage. While this approach has demonstrated successful outcomes, advancements in fields such as stem cell research, regenerative medicine and tissue engineering are placing increasing demands on current protocols revealing the limitations of current cryopreservation strategies [6].

As such, a molecular-based focus has evolved, which is not only identifying stress pathway responses subsequent to storage (i.e. apoptosis), but is also guiding the development of next generation methodologies focused on stress response mitigation in an effort to improve outcome. This molecular focus has impacted both the area of cryopreservation as well as hypothermic storage. This chapter focuses primarily on the molecular evolution in cryopreservation while touching upon hypothermic storage to provide perspective into discipline overlap.

2.2 Hypothermic Preservation

Hypothermic storage is built upon the well documented fact that the use of cold temperature has a protective effect on biologic materials. While the benefits of hypothermia have been known for centuries, our understanding of the molecular level effects on biologics has only recently evolved. This "modern era" of low temperature cell preservation began in the 1950s [5]. This was led by Carrel's investigations on normothermic perfusion of organs prior to transplantation in which he described the fundamental characteristics of a perfusion medium leading to the advancement of the cold perfusion technique [7–9].

Both hypothermic and cryopreservation techniques were developed during this "modern era". Each provide a means of effectively storing biologics for various intervals while limiting the negative consequence associated with removal of cells and tissues from their normal, in vivo setting including ischemia and hypoxia [10–13]. The ability of cold temperatures to slow metabolic activity within cells in a reversible manner is the driving principle behind these techniques. This reduction in activity decreases cellular metabolic needs as well as the accumulation of damaging products from the ischemic/hypoxic environment, such as free radicals [14], thereby allowing biologics to be held in a state of "suspended animation" [1, 5]. Further, the use of ultra-low temperatures in cryopreservation can effectively halt molecular motion [3, 15], theoretically allowing for indefinite storage. However, the effectiveness of current cryopreservation techniques is restricted to cell suspensions or simple tissues. Cryopreservation of complex, whole organ systems results in significant cryoinjury as current protocols result in incomplete and uneven cooling and freezing, followed by cell death and in turn, loss of higher order functions [16]. Vitrification, the process of adding high molar concentrations of cryoprotectant in a stepwise fashion during the cooling interval allowing for cryopreservation in an "ice-free" environment, is another approach to tissue and organ cryopreservation. While this technique has demonstrated potential for organ cryopreservation [17, 18] it has limited usefulness and introduces other challenges, such as the cytotoxic 'solution effects' resulting from the inherent high cryoprotectant concentration exposure during the transient hypothermic state [19]. As such, hypothermic storage currently serves as the most effective paradigm for organ preservation [20]. While this approach has proven to be successful in increasing the effective storage interval for organs compared to warm perfusion and storage [21], it is essentially a tradeoff of more profound and fast acting stresses for less severe, but nonetheless significant, stresses associated with cold exposure. In an effort to understand the complex changes associated with hypothermic storage, numerous reports have investigated the metabolic, biochemical and physical characteristics of cells in their normothermic environment and how they change in response to exposure to cold [22–27]. Concurrent advancements in cell biology have furthered our understanding of many of the intricate normal cellular functions, from biochemical processes to cell survival pathway activation. Through a more complete and thorough understanding of how a cell normally functions, it has become possible to establish a basis for comparison when examining cells exposed to cold. Further, this deeper understanding has allowed for a targeted and intelligent approach to improve storage outcome; a fact illustrated by the advances made in storage solution design over the last few decades.

As our understanding of cellular processes has progressed, including how cells regulate the flow of ions to create gradients and electron potential across membranes to drive reactions or how specific biochemical processes such as the citric acid cycle creates energy for the cell, it has allowed for calculated improvements focused on the formulation of the storage solutions. A major evolution was achieved with the development of an 'intracellular-like' preservation media, ViaSpan® (University of Wisconsin solution), which was formulated in the 1980s to mitigate many of the known causes of cell death at that time and became the 'gold standard' for organ transplantation [20, 28]. In the decades that followed, additional preservation solutions have been developed including Celsior, HTK-Custodial, HypoThermosol, Unisol, etc. [5, 26, 29]. This shift in media design has allowed for more successful short-term storage of cells, tissues and organs at hypothermic temperatures, as well as built the foundation upon which many recent advances in cryopreservation solution design have been based (discussed below) [2]. These advances include the addition of compounds, such as antioxidants and inhibitors, in a targeted molecular-based approach to improve storage outcome [3, 5]. As such, now recognized as essential to optimization of the cryopreservation process is the maintenance of proper cold-dependent ion ratios, control of pH at lowered temperature, prevention of the formation of free radicals, osmotic balance and the supply of energy substitutes [30, 31]. Traditional media fall short in addressing changes in solution pH, free radical production, energy deprivation, among others. Accordingly, the basal properties of these historical preservation media often do not provide for protection at the cellular level [32]. While this approach has yielded significant improvements, there remains a considerable void that solution design alone cannot bridge. Moving forward, the field of biopreservation will need to continue to elucidate the complex molecular changes associated with preservation stress pathway activation. This will likely result in a process which carefully controls biologic response through all stages of the storage process (before, during and after), to be optimized in a cell/tissue specific manner to achieve protocols better suited for today's demands.

2.3 Hypothermic Continuum

Medical literature references four stages or intensities of hypothermia, including mild (32–35 °C), moderate (27–32 °C), deep or profound (10–27 °C) and ultraprofound (<10 °C) [33]. Classically, there has been a relative isolation between hypothermic storage and cryopreservation research, likely explained by a disparity in conventional aims between these two techniques. While hypothermic storage research has made considerable advancements through a deeper understanding of complex cellular processes, cryopreservation initially focused on the physical facets associated with the phase change and osmotic flux during the freezing process. Considering the significant overlap within these two fields, this divide has delayed potential gains that an integrated approach could have achieved; particularly given the fact that the cryopreservation process involves transitioning biologics through a deepening state of hypothermia ending when the sample transitions through the glass transition temperature (Tg, nominally −140 °C). This is again repeated during the warming phase.

As stated previously, while the use of hypothermic temperatures increases the time that biologics can remain viable through decreasing metabolism thus reducing oxygen and nutrient demand, there are a number of negative consequences associated with cold exposure. A central problem revolves around the depletion of energy

(ATP) within the cell as the temperature drops resulting in a decrease of kinetic energy necessary to drive the biochemical reactions within the cell [34, 35]. As a result, a number of essential cellular functions become compromised, including membrane mediated transport [36]. The combination of decreased membrane mediated transport along with the structural changes in the membrane as it changes from liquid-crystalline to solid gel-like state in the cold temperature results in pronounced ionic imbalances within the cell. Specifically, a cell experiences increases in intracellular sodium and calcium concentrations coupled with decreased potassium and intracellular acidosis approaching a pH of 4 [23, 26, 37]. Concurrent with these membrane-related events, a number of other deleterious effects occur within a cell including generation of free radicals, disruption of cytoskeletal components, leakage of hydrolases, and mitochondrial-related events leading to apoptotic activation [3, 36, 38–40]. Taken together, these accumulated stresses trigger molecular-based stress and death responses and illustrate the importance of managing these effects as future developments are made in biopreservation.

2.4 Evolution of Cryopreservation Strategies

Cryopreservation is the use of ultra-low temperatures, usually at or below −80 °C, to preserve biologics for extended periods of time (days to years). Cryopreservation studies initially focused on efforts to reduce intracellular ice formation through the use of cryoprotective agents (CPAs). In 1949 Polge, Parks and Smith [41] reported on the "chance" discovery of glycerol's cryoprotective function during their efforts to preserve avian spermatozoa in the frozen state. In the following year, Smith [42] extended these observations by successfully cryopreserving human red blood cells (RBCs) in glycerol. These two reports identified key elements that would play a crucial role in the evolution of the field of cryopreservation including the need for a CPA, the process by which cells could successfully be exposed to penetrating CPA and the manner of freezing and thawing. In 1959, Lovelock and Bishop [43] first described the use of dimethyl sulfoxide as a CPA with its advantage of enhanced permeability versus glycerol for many cell types. In the following decades, incremental advances were made focusing on changes in and the study of carrier media containing the CPAs as well as the mechanisms of cell cryoinjury and cryopreservation [1, 44]. Most notable was a report by Mazur et al. [45] in 1972 which put forth the "Two-factor Hypothesis" to describe the inter-related relationships between cooling rates and survival as influenced by either toxic "solution effects" experienced at sub-optimal slow cooling rates or lethal intracellular ice present at high cooling rates. Specifically, it was determined that a faster than optimal cooling rate results in an incomplete dehydration of cells and thus increases the likelihood of intracellular ice formation [45–48]; conversely, if cooling is done too slowly, cells become exposed to higher concentration of solutes, a condition known as "solution effects", for extended periods of time and as a result experience toxicity [19, 45]. From this it was determined that for mammalian cell systems a cooling rate of −1 °C per minute provided the optimal rate for maintaining viability following cryopreservation. Further, a rapid thawing process was also shown to improve outcomes by limiting the time cell systems are exposed to these damaging temperatures and solutes following the storage stage [49]. In effect, these studies established the biophysical foundation upon which cryopreservation research has rested for nearly half a century.

2.4.1 Cryopreservation Process

Cryopreservation protocols begin with hypothermic exposure which persist through the period of active extracellular ice growth until equilibrium is reached in the glassy-state (vitrified). As previously discussed, this journey of deepening hypothermic stress experienced by a cell has been termed the hypothermic continuum [1]. CPA exposure represents the second step in the preservation process introducing a diversity of penetrating (membrane permeable) and non-penetrating agents contained within a carrier media to the hypothermic cell [32]. Incubation in the cryoprotective cocktail lasts between 10 and 30 min at 4 °C followed by cooling at a nominal ("optimal") cooling rate (ranging from −1 to −10 °C · min^{-1} is common for many mammalian cells). Seeding (ice nucleation) at a temperature close to the equilibrium freezing point of the cryoprotective medium (−2 to −6 °C) supports the controlled growth of extracellular ice and prevents "supercooling" and resultant "flash freezing" of the system. Seeding supports the osmotic efflux of water from the cell and the equilibration of cryoprotective agents across the cell membrane [50]. As extracellular ice formation continues, the availability of freezable water in the cell decreases, while freeze concentration of solutes increases the intracellular viscosity. When cooling rates are too rapid, cellular dehydration is inadequate, increasing the probability of lethal intracellular ice formation [47]. Non-optimal freezing effects are recognized by increased cell rupture and early stage necrosis occurring over the first few hours post-thaw [48, 51, 52]. If cooling rates are too slow, prolonged exposure to multimolar levels of the freeze concentrated solutes results in cell toxicity (solution effects) [45, 53]. An indication of "solution effect" toxicity is the appearance of delayed necrosis peaking 6–12 h post-thaw as well as apoptosis 12–36 h post-thaw (timing is cell type dependent) [54–56].

Controlled rate cooling (CRC) is accomplished by devices which support the controlled injection of liquid nitrogen to achieve active controlled rate cooling, or by passive methods often using insulated alcohol baths placed in a −80 °C freezer. Active CRC devices monitor a representative sample vial, straw or bag, and follow a pre-established program to achieve a desired cooling profile. Profiles are typically set to maintain a standard rate of cooling (e.g. −1 °C/min) over a prescribed temperature range and include a "seeding" event, nucleation through a thermal shock administered by a surge in cryogen to "flatten" the temperature rebound resulting from the latent heat of fusion of ice formation. Passive CRC devices contain the sample surrounded by, but isolated from, an alcohol bath or a thermal-insulation material, and when placed in a freezer (−80 °C), a curvilinear rate (e.g. Approximately −1 °C/min for a given cell type) is achieved in the samples. In passive CRC process seeding is often accomplished via mechanical agitation to create a nucleation event at a prescribed time during the cooling period. After the nucleation point, sample cooling continues at the controlled rate until a user-defined temperature is achieved, usually −40 to −80 °C. Once this target temperature is reached, samples are typically transferred to an ultra-low temperature environment, such as liquid nitrogen immersion, liquid nitrogen vapor phase, or mechanical storage of <−135 °C. While not all cryopreservation storage is conducted at ultra-low temperatures, with some instead choosing to maintain samples at −80 °C, ultra-low temperatures are critical to reach given that they fall below the reported glass transition temperature (Tg) for water [56]. This threshold is important as once crossed all molecular translational movement ceases and thus theoretically stops all biochemical activity, including the accumulation of damaging products, such as free radicals [15]. The transition through Tg also signifies an end to the hypothermic continuum as cells enter a vitrified (glassy) state surrounded by extracellular ice. It is at these temperatures that an indefinite storage period is theoretically possible given this state of 'suspended animation' in which cells are held. Retrieving samples from storage requires rapid thawing often accomplished by placing the sample in a 37–40 °C stirred water bath or dry thawing system (SmartThaw, ThawStar or other device) until most of the ice melts. The rapid thaw limits the time biologics spend in the damaging post-storage hypothermic conditions (between Tg and 0 °C, nominally). Once the ice has dispersed, elution of the cryoprotectant cocktail with cell culture media in a single-step or a step-wise (for high CPA concentrations) dilution process is used. Step-wise elution minimizes the volume excursions of the cell thereby preventing mechanical damage to the cell membrane and rupture.

Overall the cryopreservation process has advanced from these initial findings and others relating to physical parameters of the freezing/thawing process to become an "optimized" technique which is used today with varying degrees of success on more complex cell systems [24, 57]. However, even as these techniques have improved, there continues to be limitations in effectiveness when applied to more complex and sensitive cell systems.

2.4.2 Cryopreservation Induced Delayed Onset Cell Death

Despite the improvements and an increased understanding of cryopreservation, there remain significant challenges which need to be overcome. As "successful" methods of cryopreservation are applied to more sensitive and complex cell systems, less than ideal results are obtained. Often cells appear healthy and viable when examined immediately after thawing; however when examined 24–48 h later, a significant portion (30–70 %) of these cells are lost due to cryopreservation-induced delayed-onset cell death (CIDOCD) [30, 48]. This phenomenon has been reported in many cell systems and demonstrates that while the cryopreservation process may be optimized to protect cells structurally, it fails to adequately manage the other stresses associated with this process. In sensitive biologics, such as stem cells, CIDOCD can account for a significant loss (>50 %) of the total population [58–60] and result in high system failure, compromise in functionality, spontaneous differentiation and alterations in gene expression following what was initially deemed a 'successful' cryopreservation event [3]. As researchers began investigating this phenomenon, it became apparent that cell stress response and the activation of apoptosis played a critical role in CIDOCD. Given all that is now known about both apoptotic induction and cryopreservation related stresses, a connection is now clear. Following apoptosis identification, subsequent research has revealed how complex the cell death process is and the myriad of ways that molecular pathways within a cell can be activated following thawing. It is precisely this shift towards a deeper understanding of the molecular stress responses of a cell that continue to drive the field of cryopreservation to meet the ever-changing demands on it, such as the preservation of stem cells.

2.5 Modes of Cell Death

It is well understood that multiple modes of cell death contribute to cryopreservation failure. The two main molecular paths of cell death associated with cryopreservation failure include necrosis and apoptosis. Cellular rupture due to intracellular ice formation is a critical third mode, however, this is primarily controlled through CPA incorporation. While ice related cell rupture has been extensively researched [61] resulting in significant improvements since cryopreservation's inception, there remains significant issues to address with the other modes of cell death that contribute to preservation failure, particularly in complex, sensitive systems.

2.5.1 Ice Rupture

As ice forms outside the cell during the freezing process, solutes are excluded resulting in their concentration. Freeze concentration of solutes results in an increase in sample osmolality from approximately 350 mosmol to upwards of 10,000 mosmol [45, 61]. Cells exposed to these conditions shrink severely, but do not necessarily experience a lethal event. During thawing, many cells will be subject to significant cell membrane damage resulting in rupture while other cells may experience membrane damage that is repairable. The majority of membrane rupture occurs within minutes after thawing. Those cells rupturing 1 or more hours post-thaw experience non-repairable membrane damage and typically die through necrosis.

2.5.2 Necrosis

Necrotic cell death has been observed in numerous cases of cryopreservation failure [51, 62, 63]. Necrosis is an energy independent form of cell death characterized by the swelling of a cell and its constituent organelles, loss of membrane integrity, lysosomal rupture, random DNA fragmentation by endonucleases and ultimately cell lysis [64–67]. As a result of cell rupture and the associated release of cytokines, there is typically an activation of an immune and inflammatory response in vivo [64, 65, 67]. The progression of necrosis often occurs rapidly in a matter of minutes to hours. Induction is typically seen in response to severe cellular stress and results in the activation of detrimental intracellular signaling cascades. Necrotic cell death has been reported to be activated by stressors such as ischemia, osmotic shock, severe thermal stress, ionic dysregulation and toxic agents. Interestingly, many of these necrotic activating stressors are involved in or associated with cryopreservation.

2.5.3 Apoptosis

Apoptosis is a form of gene regulated cell death often referred to as programmed cell death. It is an energy-dependent process characterized by cell shrinkage, chromatin condensation, intact membranes but with phosphatidyl serine inversion, non-random DNA cleavage, and the formation of organelle containing "blebs" [64–69]. Unlike necrosis, apoptosis does not elicit an immune response in vivo but instead the cell bleb which are then recycled through phagocytosis. Apoptosis is induced by a number of different stressors which can activate responses in the mitochondria, the plasma membrane or the nucleus [68–70]. Apoptosis can be induced by starvation (nutrient deprivation), temperature changes, viral infection, hypoxia, radiation, toxic compounds, osmotic stress and many other stresses. The two canonical "branches" of apoptosis which have been identified in cryopreservation failure are the extrinsic (membrane-mediated) and the intrinsic (mitochondrial-mediated) pathways. Additionally, studies show that cross-talk, feedback and amplification pathways exist [71–73]. The identification of a third, nuclear-mediated apoptotic pathway further complicates a complete delineation of the cell death process following cryopreservation.

2.5.4 Necroptosis

As ongoing cell death research continues to elucidate the specific biochemical mechanisms that trigger and propagate programmed cell death, an alternative form of cell death has been identified [74]. This recently identified type of cell death has been shown to result in a necrotic-like execution with classical hallmarks such as cell swelling and membrane lysis while remaining highly regulated, distinguishing it from conventional necrosis. As such it has been termed necroptosis. Research has now begun to reveal the distinct mechanism of action responsible for the activation of this pathway. Specifically, it has been shown that this mode of programmed necrosis is triggered through the signaling of death receptors, such as tumor necrosis factor receptor 1 [75]. The binding of the respective ligand (TNF-α) to the death receptor, similar to membrane-mediated apoptosis, results in the recruitment of intracellular signaling proteins and the formation of an active complex responsible for downstream effects. Central to this necrotic complex is the kinase activity of receptor interacting proteins 1 and 3 (RIP1 and RIP3, respectively) and their substrate, the pseudokinase mixed lineage kinase domain-like protein (MLKL) as the core machinery for execution [76, 77]. Continued efforts are necessary to elucidate the specific signaling cascade of necroptosis and how the apoptotic/necroptotic balance is controlled during programmed death. However, the role of necroptosis in cryopreservation related stress conditions such as ischemia/reperfusion injury is becoming more evident [78, 79]. As such, efforts to understand the complex cell death interplay at the molecular level will be paramount for improving future cryopreservation approaches.

2.5.5 Transitional Cell Death Continuum

The original perception of molecular-based cell death was that there were two clear and distinct modes: apoptosis and necrosis [64]. Apoptosis is characterized by a particular set of molecular events which are highly organized and represent a true molecular response while necrosis was viewed as an uncontrolled process resulting from extreme stresses that involve random molecular changes [67, 80]. As continued research has further elucidated our understanding of the cell death process, it has become clear that necrosis and apoptosis do not exist as separate entities but comprise opposite ends of a 'cell death continuum' [81].

Reports have describe instances in which a multitude of cell types have demonstrated a cell death that did not follow a classical route of demise, but instead switches between necrosis and apoptosis and in some case display characteristics of both in a single cell [82–84]. To this end, Leist et al. [73] demonstrated that Jurkat cells can be shunted to a necrotic cell death following apoptotic induction through the removal of energy (ATP) from the system. Further, they demonstrated that returning ATP resulted in a reversion to more apoptotic-like death and that there is a point in the process when reversal can no longer be achieved. More recent reports have detailed two distinct forms of cell death that involve this shifting from apoptosis to necrosis, namely necroptosis and secondary necrosis [83, 84]. Secondary necrosis describes circumstances for when a cell has committed to cell death through apoptosis but experiences a depletion of ATP before completing the process and therefore shunts to necrosis to complete cellular demise. In contrast, necroptosis is the term given to a distinct, regulated form of necrosis which results in a cell death with necrotic hallmarks such as cell and organelle swelling, membrane rupture and the absence of chromatic condensation. To further complicate the picture, the literature also further divides 'programmed cell death' into distinct categories with apoptosis as merely one of a host of different programmed pathways a cell can utilize to achieve cell death. Other pathways such as autophagy, caspase-independent apoptosis, necroptosis, anoikis, cornification and exocitotoxicity demonstrate the breadth of programmed cell death.

Cryopreservation most likely triggers an array of stress factors and cell death mechanisms. As research on cell death continues to expand our understanding of the complexities involved in cellular demise, it further complicates the effective management in cryopreservation regimes. This needs to be more fully understood so that more effective modulation strategies can be developed to improve outcome.

2.6 Preservation-Induced Stress-Dependent Cell Death

The identification of programmed cell death involvement following cryopreservation, while a relatively recent discovery, has led to a number of improvements and preservation techniques in an effort to meet the demands of newer and more complex biologics. Given what is now known about both programmed cell death and low temperature induces cell stresses, it seems logical that a multitude of cell death processes, including apoptosis, are triggered in cells during cryopreservation. Since initial reports on the observation of apoptosis following cryopreservation, numerous reports have documented this phenomenon occurring post-thaw in a wide variety of cell types [25, 85–92]. Further, a myriad of reports also demonstrated its involvement in response to hypothermic storage and other related stresses, such as ischemia and hypoxia [11, 32, 79, 93–96]. As such, it is now recognized that molecular-based cell death has a significant role and effect on storage outcomes that is seemingly independent of cell type.

The temporal aspect is one of the most important components to understand the impact of molecular-based cell death on a cell system. While many assume that once a cell emerges from the cryopreservation process *intact*, it has "survived", remaining a viable cell. However, studies have shown that accumulated stresses during the preservation event can trigger molecular-based cell death pathways within the cell hours to days following thawing resulting in delayed cell death. Cryopreservation-Induced, Delayed-Onset Cell Death (CIDOCD), a term coined by Baust et al. [48], describes the phenomenon of a delayed peak in both necrosis and apoptosis many hours following cryopreservation. It is this delayed manifestation which demonstrates why timing is vital when assessing cell survival as well as investigating CIDOCD. If the assessment of cell death is conducted too early, while there may be an observable apoptotic and necrotic population, a significant portion may be missed as death continues to occur over the 24–48 h post-thaw, thus resulting in inflated cell survival estimates. On the other hand, if assessment is too late, the observed cell survival may be closer to the actual survival, but the contributions of apoptosis or necrosis may be missed analytically, leading to the incorrect conclusion that they did not play a role. Continued research has further proven this occurrence through more in-depth analyses showing the up-regulation of pro-apoptotic genes and proteins and activation (cleavage) of caspases and other pro-apoptotic signaling molecules [4, 30, 88].

While assessing viability is an important indicator of preservation efficacy, particularly when performed at the proper time post-thaw, functionality is an equally important post-preservation consideration that is often overlooked. While a cell may emerge from the cryopreservation process "viable" and have avoided CIDOCD, its functionality may have been compromised by the process. As improved techniques continue to enhance our ability to maintain viability in more complex and sensitive biologics, it has become clear that "cell in does not equal cell out". Studies have demonstrated changes to cells following the preservation process such as diminished functions, loss of surface markers, and molecular differences such as changes in gene expression [54, 56, 97–103]. A potential concern is that the preservation process will select for cells with a molecular constitution with increased resistance to preservation stresses. Whether differences existed prior to preservation or as a result of preservation based selection, if the technique is selective for cells which have increased thermal or stress tolerance, it can be a cause for concern. The alteration of cell survival signaling pathways is a hallmark of cancer formation and progression. Changes in cell signaling related proteins, such as an over-expression of Bcl-2 or modification of P53, have been shown experimentally to be involved with cancer formation [104–107]. In this respect, preservation techniques must be designed to prevent changes in a cell's molecular constitution that would otherwise limit downstream utilization.

2.7 Targeted Control of Molecular-Based Death

The identification of a molecular-based cell death component to cryopreservation failure has served to complicate our understanding of the complex post-storage response. Despite the complications revealed by our furthered understanding of the entire post-storage cellular response, this also identifies a wide array of targets whose modulation may improve preservation outcome. Recent advances have revolved around changes to preservation solutions as well as the addition of chemical modulators to help control the molecular response following storage.

In cryopreservation, typical cryopreservation solutions are prepared by supplementing a cell's basal culture media with serum proteins and DMSO. While this formulation provides adequate physical protection and control of intracellular ice formation, it provides only low level protection against many of the other stresses a cell encounters during the freeze/thaw process. To address this, the development and utilization of more "intracellular-like" solutions was undertaken in an effort to further buffer against changes that an "extracellular-like" solution (high Na+, low K+), such as basal culture media, has trouble managing. Studies have shown improved outcomes for cell systems cryopreserved using "intracellular-like" solutions supplemented with CPA's as compared to standard culture media based cryopreservation solution [31, 60, 103,

108–110]. These advanced cryopreservation media are often referred to as Complex Cryopreservation Media (CCM) as they are designed to balance ionic needs of cells at low temperatures, buffer pH, provide energy substrates, osmotic stabilization, protection against free radical damage, among others. Typically, the use of CCM's results in improved sample quality post-thaw including increased viability, functionality, and regrowth which are usually not evident until hours after thawing. These improvements by CCM's have been shown to be a result of decreases in molecular based cell death during the recovery period [32].

More recently, advances in cryopreservation have been achieved through the incorporation of compounds into the cryopreservation media with the aim to control cell death through the targeting of molecular stress response pathways. This strategy is often referred to as Targeted Apoptotic Control (TAC). In this regard, the use of chemical modulators such as free radical scavengers, protease inhibitors and ion chelators have demonstrated the ability to improve cryopreservation and hypothermic storage outcomes [39, 111–113]. For instance, Mathew et al. [40], demonstrated the beneficial effect of the addition of vitamin E, a potent antioxidant, during hypothermic preservation. Other reports from this group described the beneficial effect of EDTA addition and demonstrated implications of apoptotic involvement [113]. In the area of cryopreservation, early reports by Baust et al. [51], Yagi et al. [114], and others [115] demonstrated the benefit of caspase inhibition in improving cryopreservation outcome. These initial reports were important first steps in demonstrating the ability to control cell death and ultimately improve post-preservation outcomes in a number of different cell systems.

More recent attempts at targeted molecular control of cell death have focused on targeting apoptosis. Through the use of caspase inhibitors during the cryopreservation, reports have demonstrated significant improvements in outcome through the blocking of apoptosis activated by the preservation process [51, 86, 92, 114, 115]. Interestingly, studies have shown that the utilization of caspase inhibitors has a beneficial effect in multiple cell types through a decrease in both apoptosis and necrosis, further strengthening the linkage of the cell death continuum [30, 90, 95]. Another central finding of this research was the demonstration of the temporal aspect of the molecular-based cell death following cryopreservation. Continued research in this area has suggested that the post-storage addition of agents such as caspase inhibitors, also has a beneficial effect on outcome [92, 115]. To this end, a combinatorial approach of targeting stress pathways both during and after cryopreservation may result in improved outcomes for biologics.

As research continues to reveal the multitude of different pathways and molecular agents that play a role in the balance of a cell's fate following cryopreservation, a number of new targets and strategies have been developed to modulate cellular control. One such area of focus has centered on apoptotic control through the management of Rho-associated kinase (ROCK) through the use of ROCK inhibitor [116, 117]. ROCK has an important role in the morphological changes associated with apoptotic activation as a direct target of active caspase 3. Specifically, research has shown beneficial effects of ROCK inhibitor on stem-cell cryopreservation when incorporated during and post-preservation, including increased survival and decreased spontaneous differentiation [118]. Interestingly, studies have shown that utilization of ROCK inhibitor prior to single-cell dissociation of pluripotent stem cells, a necessary step for cryopreservation and a host of other applications, blocks the dissociation based apoptotic activation in these cells and thus improves preservation outcomes. This important finding again illustrates the impact that the entire preservation process, before, during and after, has on the overall success.

As advances in molecular biology techniques continue to improve their efficacy, and in turn increase their utilization, new strategies for improving storage outcomes become more feasible. Specifically, the ability to modulate post-thaw cell stress response at the level of transcriptional and translational events may hold the key for future improvements. Typically, improved outcomes resultant from targeted approaches act through a change to cellular function at the proteomic level to achieve their effect. However, newer molecular tools such as RNAi, gene manipulation and epigenetic control are demonstrating the potential for alterations at the level of DNA and RNA as novel approaches for improved cryopreservation. RNA interference (RNAi) in particular has become an important and heavily researched cellular system since its initial discovery in 1998 and may offer potential in improving cryopreservation outcome [119]. RNAi works through innate cellular mechanisms that degrade double-stranded RNA in both an effort to protect the cell from foreign genetic material (i.e. retro-virus) and as a form of post-transcriptional gene silencing to regulate protein synthesis. Through the use of microRNA and small interfering RNA, it is possible to insert a complimentary RNA sequence to mRNA transcribed by the cell thus forming a double-stranded RNA segment, which is in-turn, degraded by the cell. The implication for cryopreservation then would be to utilize this technology to insert complimentary sequences to the mRNA of pro-death signaling molecules, such as caspases and pro-apoptotic Bcl-2 family (e.g. Bax, Bid, Bcl-XL, etc.), in order to transiently "knock-down" their expression, thus effectively increasing pro-survival signaling.

Similarly, the ability to manipulate genes at the level of DNA transcription through gene insertion or deletion has the potential to improve cryopreservation outcome. Through these techniques it may be possible to control levels of both pro-death and pro-survival proteins in an effort to increase survival through a targeted approach. However, there still remain concerns when manipulating genes, especially in a cell therapy application, because of the potential for unintended consequences such as cancer development. To the end, research is continuing to develop safer methods of gene manipulation as well as elucidate and expand on our understanding of the related cellular events of epigenetic control. Reports have begun to show a connection between thermal tolerance and epigenetic changes [120], demonstrating the potential for epigenetic manipulations as a modulator to improve cryopreservation outcome. Epigenetics could potentially allow for the targeted control of gene expression in a reversible manner that may not alter a cell's DNA sequence, limiting the latent negative consequences of gene manipulation.

Lastly, while our understanding of the roles that organelles such as the cell membrane, nucleus and mitochondria play in the cell death cascade continue to evolve, research has revealed other organelles may contribute to a cell's fate as well. Specifically, the lysosomes, Golgi apparatus and endoplasmic reticulum (ER) have been shown to play a role in cellular stress monitoring with the ability to trigger a molecular-based response if irregularities are detected [121]. The ER in particular has become a major area of research in recent years as reports have revealed its role in cellular homeostasis and its subsequent ability to trigger an apoptotic response in the event of profound cellular distress [122–124]. Specifically, the mechanism through which the ER conducts these functions is known as the Unfolded Protein Response (UPR) [125–129]. As our understanding of the role and impact these areas have on cryopreservation outcome continues to evolve, they too may provide additional targets for modulation to improve post-thaw outcome (survival, function and genomic stability).

2.8 Summary

As cell based research and development continues to advance at an exponential rate, it has become clear that cryopreservation protocols have become outdated and not kept pace with advances in other related fields. While today's standard of practice for cryopreservation still focuses primarily on the control of osmotic flux, ice formation and associated stressors, numerous reports have emerged demonstrating that the combined targeting of physical, biochemical and molecular responses to the freezing/thaw process can greatly improve outcome. The impact of this combined approach extends well beyond influencing cell death but also has a long term impact on biochemical pathways and cellular functionality. The results of these practices are, of course, only realized upon sample thawing and utilization. Given the complexity of events which occur during the seemingly simple and routine process of freezing, it is important to recognize and understand the impact of sample preparation, cooling, storage temperature, storage time, and warming have on overall sample quality. As such, the ability to successfully cryopreserve samples depends on the integration of the molecular strategy to obtain the highest quality cell product post-thaw. This strategic shift includes a) structural methodologies which are reasonably effective ("optimized") in preventing ice-related damage, b) the use of CCM based cryopreservation media and c) integration of modulations strategies to reduce CIDOCD. While this new approach is well supported, there remains a continuing need to decipher the cell's responsiveness to the stress events associated with cryopreservation to overcome the significant cell death after thawing [3, 48]. As such, investigations into the complex molecular changes associated with cold exposure are necessary to elucidate additional targets for improving storage outcomes. Particularly important is the integrated study of cell response to hypothermic and sub-freezing temperatures as they are experienced by cells during both freezing and thawing processes associated with cryopreservation. Through a more complete understanding of low temperature stress response and how associated changes alter a cell's typical, normothermic baseline of function, significant improvements in cryopreservation strategies can be achieved in turn yielding improved biologics for downstream utilization.

References

1. Baust JG (2007) Concepts in biopreservation. In: Baust JG, Baust JM (eds) Advances in biopreservation. CRC Press, Boca Raton, pp 1–14
2. Baust JM, Snyder KK, Van Buskirk RG, Baust JG (2009) Changing paradigms in biopreservation. Biopreservation Biobanking 7(1):3–12
3. Baust JM (2007) Properties of cells and tissues influencing preservation outcome: molecular basis of preservation-induced cell death. In: Baust JG, Baust JM (eds) Advances in biopreservation. CRC Press, Boca Raton, pp 63–87
4. Fuller BJ (2003) Gene expression in response to low temperatures in mammalian cells: a review of current ideas. Cryo Lett 24(2):95–102
5. Taylor MJ (2007) Biology of cell survival in the cold: the basis for biopreservation of tissues and organs. In: Baust JG, Baust JM (eds) Advances in biopreservation. CRC Press, Boca Raton, pp 15–62
6. Fuller BJ, Lane N, Benson EE (eds) (2004) Life in the frozen state. CRC Press, Boca Raton
7. Carrel A, Lindbergh CA (1935) The culture of whole organs. Science 81(2112):621–623
8. Carrel A (1912) On the permanent life of tissue outside of the organism. J Exp Med 15:516–528
9. Carrel A (1938) Perfusing media. In: Carrel A, Lindbergh CA (eds) The culture of organs. Paul B. Hoeber, New York, pp 55–74
10. White BC, Wiegenstein JG, Winegar CD (1984) Brain ischemic anoxia. Mechanisms of injury. JAMA 251(12):1586–1590
11. Saikumar P, Dong Z, Weinberg JM, Venkatachalam MA (1998) Mechanisms of cell death in hypoxia/reoxygenation injury. Oncogene 17(25):3341–3349
12. Jennings RB, Ganote CE (1976) Mitochondrial structure and function in acute myocardial ischemic injury. Circ Res 38(5 Suppl 1):I80–I91
13. Neely JR, Grotyohann LW (1984) Role of glycolytic products in damage to ischemic myocardium. Dissociation of adenosine triphosphate levels and recovery of function of reperfused ischemic hearts. Circ Res 55(6):816–824
14. Khar A, Pardhasaradhi BV, Ali AM, Kumari AL (2003) Protection conferred by Bcl-2 expression involves reduced oxidative stress and increased glutathione production during hypothermia-induced apoptosis in AK-5 tumor cells. Free Radic Biol Med 35(8):949–957
15. Rall WF, Mazur P, McGrath JJ (1983) Depression of the ice-nucleation temperature of rapidly cooled mouse embryos by glycerol and dimethyl sulfoxide. Biophys J 41(1):1–12
16. Fahy GM, DaMouta C, Tsonev L, Khirabadi BJ, Mehl P, Meryman HT (1995) Cellular injury associated with organ cryopreservation: chemical toxicity and cooling injury. In: Lemasters JJ, Oliver C (eds) Cell biology of trauma. CRC Press, Boca Raton, pp 333–356
17. Fahy GM, MacFarlane DR, Angell CA, Meryman HT (1984) Vitrification as an approach to cryopreservation. Cryobiology 21(4):407–426
18. Taylor MJ, Song YC, Brockbank KGM (2004) Vitrification in tissue preservation. In: Fuller BJ, Lane N, Benson EE (eds) Life in the frozen state. Academic, Boca Raton, pp 603–692
19. Fahy GM (1986) The relevance of cryoprotectant "toxicity" to cryobiology. Cryobiology 23(1):1–13

20. Southard JH, Belzer FO (1995) Organ preservation. Annu Rev Med 46:235–247
21. Toledo-Pereyra LH, Hau T, Simmons RL, Najarian JS (1977) Lung preservation techniques. Ann Thorac Surg 23(5):487–494
22. Nejepinska J, Juklova K, Jirsova K (2010) Organ culture, but not hypothermic storage, facilitates the repair of the corneal endothelium following mechanical damage. Acta Ophthalmol 88(4):413–419
23. Renlund DG, Gerstenblith G, Lakatta EG, Jacobus WE, Kallman CH, Weisfeldt ML (1984) Perfusate sodium during ischemia modifies post-ischemic functional and metabolic recovery in the rabbit heart. J Mol Cell Cardiol 16(9):795–801
24. De Loecker P, Fuller BJ, Koptelov VA, Grischenko VI, De Loecker W (1997) Cryopreservation of isolated rat hepatocytes: effects of iron-mediated oxidative stress of metabolic activity. Cryobiology 34(2):150–156
25. Vanhulle VP, Neyrinck AM, Pycke JM, Horsmans Y, Delzenne NM (2006) Role of apoptotic signaling pathway in metabolic disturbances occurring in liver tissue after cryopreservation: study on rat precision-cut liver slices. Life Sci 78(14):1570–1577
26. Baicu SC, Taylor MJ (2002) Acid-base buffering in organ preservation solutions as a function of temperature: new parameters for comparing buffer capacity and efficiency. Cryobiology 45(1):33–48
27. Borderie VM, Laroche L (1999) Ultrastructure of cultured and cryopreserved human corneal keratocytes. Cornea 18(5):589–594
28. Southard JH, van Gulik TM, Ametani MS, Vreugdenhil PK, Lindell SL, Pienaar BL, Belzer FO (1990) Important components of the UW solution. Transplantation 49(2):251–257
29. Van Buskirk RG, Taylor MJ, Shen R, Tang W, Mathew A, Baust JG (1995) Optimization of HypoThermosol using cultured cardiomyocytes and the fluorescent multiple endpoint assay. Cryobiology 32:590
30. Baust JM (2002) Molecular mechanisms of cellular demise associated with cryopreservation failure. Cell Preserv Technol 1(1):17–31
31. Taylor MJ, Campbell LH, Rutledge RN, Brockbank KG (2001) Comparison of Unisol with Euro-Collins solution as a vehicle solution for cryoprotectants. Transplant Proc 33(1–2):677–679. doi:S0041-1345(00)02198-9 [pii]
32. Baust JM (2005) Advances in media for cryopreservation and hypothermic storage. Bioprocess Int 3(Supp 3):46–56
33. Taylor MJ (2000) Encyclopedia of stress. In: (ed) Academic, London, pp 484–496
34. Stubenitsky BM, Ametani M, Danielewicz R, Southard JH, Belzer FO (1995) Regeneration of ATP in kidney slices after warm ischemia and hypothermic preservation. Transpl Int 8(4):293–297
35. Lanir A, Clouse ME, Lee RG (1987) Liver preservation for transplant. Evaluation of hepatic energy metabolism by 31P NMR. Transplantation 43(6):786–790
36. Stefanovich P, Ezzell RM, Sheehan SJ, Tompkins RG, Yarmush ML, Toner M (1995) Effects of hypothermia on the function, membrane integrity, and cytoskeletal structure of hepatocytes. Cryobiology 32(4):389–403
37. Tani M, Neely JR (1989) Role of intracellular Na+ in Ca2+ overload and depressed recovery of ventricular function of reperfused ischemic rat hearts. Possible involvement of H+-Na+ and Na+-Ca2+ exchange. Circ Res 65(4):1045–1056
38. Fuller BJ, Gower JD, Green CJ (1988) Free radical damage and organ preservation: fact or fiction? A review of the interrelationship between oxidative stress and physiological ion disbalance. Cryobiology 25(5):377–393
39. Nagasaki H, Nakano H, Boudjema K, Jaeck D, Alexandre E, Baek Y, Kitamura N, Yamaguchi M, Kumada K (1998) Efficacy of preconditioning with N-acetylcysteine against reperfusion injury after prolonged cold ischaemia in rats liver in which glutathione had been reduced by buthionine sulphoximine. Eur J Surg 164(2):139–146
40. Mathew AJ, Hollister WR, Addona T, Baust JG, Van Buskirk RG (1999) Vitamin E and EDTA improve the efficacy of hypothermosol-implication of apoptosis. In Vitro Mol Toxicol 12(3):163–172
41. Polge C, Smith AU, Parkes AS (1949) Revival of spermatozoa after vitrification and dehydration at low temperatures. Nature 164(4172):666
42. Smith AU (1950) Prevention of haemolysis during freezing and thawing of red blood-cells. Lancet 2(6644):910–911. doi:S0140-6736(50)91861-7 [pii]
43. Lovelock JE, Bishop MW (1959) Prevention of freezing damage to living cells by dimethyl sulphoxide. Nature 183(4672):1394–1395
44. Fuller BJ (2004) Cryoprotectants: the essential antifreezes to protect life in the frozen state. Cryo Lett 25(6):375–388
45. Mazur P, Leibo SP, Chu EH (1972) A two-factor hypothesis of freezing injury. Evidence from Chinese hamster tissue-culture cells. Exp Cell Res 71(2):345–355
46. Kubota S, Graham F, Crabo BG, Lillehei RC, Dietman RH (1976) The effect of freeze rate, duration of phase transition, and warming rate on survival of frozen canine kidneys. Cryobiology 13(4):455–462
47. Mazur P (1977) The role of intracellular freezing in the death of cells cooled at supraoptimal rates. Cryobiology 14(3):251–272. doi:0011-2240(77)90175-4 [pii]
48. Baust JM, Vogel MJ, Van Buskirk R, Baust JG (2001) A molecular basis of cryopreservation failure and its modulation to improve cell survival. Cell Transplant 10(7):561–571
49. Loretz LJ, Li AP, Flye MW, Wilson AG (1989) Optimization of cryopreservation procedures for rat and human hepatocytes. Xenobiotica 19(5):489–498
50. Mazur P, Cole KW (1985) Influence of cell concentration on the contribution of unfrozen fraction and

salt concentration to the survival of slowly frozen human erythrocytes. Cryobiology 22(6):509–536
51. Baust JM, Van Buskirk R, Baust JG (2000) Cell viability improves following inhibition of cryopreservation-induced apoptosis. In Vitro Cell Dev Biol-Anim 36(4):262–270
52. Baust JM, Van Buskirk RG, Baust JG (2002) Gene activation of the apoptotic caspase cascade following cryogenic storage. Cell Preserv Technol 1(1):63–80
53. Lane N (2004) The future of cryobiology. In: Fuller B, Lane N, Benson E (eds) Life in the frozen state. CRC Press, Boca Raton, pp 645–657
54. Corominola H, Mendola J, Esmatjes E, Saenz A, Fernandez-Cruz L, Gomis R (1998) Cryopreservation of pancreatic islets prior to transplantation: a comparison between UW solution and RPMI culture medium. Cryobiology 37(2):110–118
55. Fisher RL, Hasal SJ, Sanuik JT, Hasal KS, Gandolfi AJ, Brendel K (1996) Cold- and cryopreservation of dog liver and kidney slices. Cryobiology 33(1):163–171. doi:S0011-2240(96)90016-4 [pii] 10.1006/cryo.1996.0016
56. Franks F (1982) The properties of aqueous solutions at sub-zero temperature. In: Franks F, Mathias S (eds) Biophysics of water. Wiley, New York, pp 279–294
57. Kurtz J, Seetharaman S, Greco N, Moroff G (2007) Assessment of cord blood hematopoietic cell parameters before and after cryopreservation. Transfusion 47(9):1578–1587. doi:TRF01327 [pii] 10.1111/j.1537-2995.2007.01327.x
58. Abrahamsen JF, Bakken AM, Bruserud O, Gjertsen BT (2002) Flow cytometric measurement of apoptosis and necrosis in cryopreserved PBPC concentrates from patients with malignant diseases. Bone Marrow Transplant 29(2):165–171. doi:10.1038/sj.bmt.1703346
59. Heng BC, Ye CP, Liu H, Toh WS, Rufaihah AJ, Yang Z, Bay BH, Ge Z, Ouyang HW, Lee EH, Cao T (2006) Loss of viability during freeze-thaw of intact and adherent human embryonic stem cells with conventional slow-cooling protocols is predominantly due to apoptosis rather than cellular necrosis. J Biomed Sci 13(3):433–445. doi:10.1007/s11373-005-9051-9
60. Stylianou J, Vowels M, Hadfield K (2005) CryoStor significantly improves cryopreservation of haematopoetic stem cells (HSC). Cryotherapy 7(Suppl 1):117
61. Mazur P (1984) Freezing of living cells: mechanisms and implications. Am J Physiol 247(3 Pt 1):C125–C142
62. Fowke KR, Behnke J, Hanson C, Shea K, Cosentino LM (2000) Apoptosis: a method for evaluating the cryopreservation of whole blood and peripheral blood mononuclear cells. J Immunol Methods 244(1–2):139–144. doi:S0022-1759(00)00263-5 [pii]
63. Martin H, Bournique B, Sarsat JP, Albaladejo V, Lerche-Langrand C (2000) Cryopreserved rat liver slices: a critical evaluation of cell viability, histological integrity, and drug-metabolizing enzymes. Cryobiology 41(2):135–144. doi:10.1006/cryo.2000.2275 S0011-2240(00)92275-2 [pii]
64. Searle J, Kerr JF, Bishop CJ (1982) Necrosis and apoptosis: distinct modes of cell death with fundamentally different significance. Pathol Annu 17(Pt 2):229–259
65. Walker NI, Harmon BV, Gobe GC, Kerr JF (1988) Patterns of cell death. Methods Achiev Exp Pathol 13:18–54
66. Kerr JF (1972) Shrinkage necrosis of adrenal cortical cells. J Pathol 107(3):217–219. doi:10.1002/path.1711070309
67. Columbano A (1995) Cell death: current difficulties in discriminating apoptosis from necrosis in the context of pathological processes in vivo. J Cell Biochem 58(2):181–190. doi:10.1002/jcb.240580207
68. Zimmermann KC, Green DR (2001) How cells die: apoptosis pathways. J Allergy Clin Immunol 108(4 Suppl):S99–S103. doi:a117819 [pii]
69. Thornberry NA, Lazebnik Y (1998) Caspases: enemies within. Science 281(5381):1312–1316
70. Bras M, Queenan B, Susin SA (2005) Programmed cell death via mitochondria: different modes of dying. Biochemistry (Mosc) 70(2):231–239. doi:BCM70020284 [pii]
71. Kroemer G, Dallaporta B, Resche-Rigon M (1998) The mitochondrial death/life regulator in apoptosis and necrosis. Annu Rev Physiol 60:619–642. doi:10.1146/annurev.physiol.60.1.619
72. Melino G, Knight RA, Nicotera P (2005) How many ways to die? How many different models of cell death? Cell Death Differ 12(Suppl 2):1457–1462. doi:4401781 [pii] 10.1038/sj.cdd.4401781
73. Leist M, Single B, Castoldi AF, Kuhnle S, Nicotera P (1997) Intracellular adenosine triphosphate (ATP) concentration: a switch in the decision between apoptosis and necrosis. J Exp Med 185(8):1481–1486
74. Degterev A, Huang Z, Boyce M, Li Y, Jagtap P, Mizushima N, Cuny GD, Mitchison TJ, Moskowitz MA, Yuan J (2005) Chemical inhibitor of nonapoptotic cell death with therapeutic potential for ischemic brain injury. Nat Chem Biol 1(2):112–119. doi:nchembio711 [pii] 10.1038/nchembio711
75. Vandenabeele P, Galluzzi L, Vanden Berghe T, Kroemer G (2010) Molecular mechanisms of necroptosis: an ordered cellular explosion. Nat Rev Mol Cell Biol 11(10):700–714. doi:10.1038/nrm2970 nrm2970 [pii]
76. Sun L, Wang X (2014) A new kind of cell suicide: mechanisms and functions of programmed necrosis. Trends Biochem Sci 39(12):587–593. doi:S0968-0004(14)00185-6 [pii] 10.1016/j.tibs.2014.10.003

77. Linkermann A, Brasen JH, Himmerkus N, Liu S, Huber TB, Kunzendorf U, Krautwald S (2012) Rip1 (receptor-interacting protein kinase 1) mediates necroptosis and contributes to renal ischemia/reperfusion injury. Kidney Int 81(8):751–761. doi:10.1038/ki.2011.450 ki2011450 [pii]
78. Liu CY, Liu YH, Lin SM, Yu CT, Wang CH, Lin HC, Lin CH, Kuo HP (2003) Apoptotic neutrophils undergoing secondary necrosis induce human lung epithelial cell detachment. J Biomed Sci 10(6 Pt 2):746–756. doi:73962 73962 [pii]
79. Jaeschke H, Lemasters JJ (2003) Apoptosis versus oncotic necrosis in hepatic ischemia/reperfusion injury. Gastroenterology 125(4):1246–1257. doi:S0016508503012095 [pii]
80. Farber E (1994) Programmed cell death: necrosis versus apoptosis. Mod Pathol 7(5):605–609
81. Zeiss CJ (2003) The apoptosis-necrosis continuum: insights from genetically altered mice. Vet Pathol 40(5):481–495
82. Henriquez M, Armisen R, Stutzin A, Quest AF (2008) Cell death by necrosis, a regulated way to go. Curr Mol Med 8(3):187–206
83. Christofferson DE, Yuan J (2010) Necroptosis as an alternative form of programmed cell death. Curr Opin Cell Biol 22(2):263–268
84. Wu W, Liu P, Li J (2012) Necroptosis: an emerging form of programmed cell death. Crit Rev Oncol Hematol 82(3):249–258
85. Baust JM, Cosentino M, Meeks E, Baer J, Van Buskirk RG, Baust JG (2005) Apoptotic cell death contributes significantly to peripheral blood mononuclear cells cryopreservation failure. Cryobiology 51(4):354–355
86. Fujita R, Hui T, Chelly M, Demetriou AA (2005) The effect of antioxidants and a caspase inhibitor on cryopreserved rat hepatocytes. Cell Transplant 14(6):391–396
87. Greco NJ, Seetharaman S, Kurtz J, Lee WR, Moroff G (2006) Evaluation of the reactivity of apoptosis markers before and after cryopreservation in cord blood CD34(+) cells. Stem Cells Dev 15(1):124–135
88. Park SY, Kim EY, Cui XS, Tae JC, Lee WD, Kim NH, Park SP, Lim JH (2006) Increase in DNA fragmentation and apoptosis-related gene expression in frozen-thawed bovine blastocysts. Zygote 14(2):125–131
89. Geissler HJ, Fischer UM, Foerster S, Krahwinkel A, Antonyan A, Kroener A, Addicks K, Mehlhorn U, Bloch W (2006) Effect of thawing on nitric oxide synthase III and apoptotic markers in cryopreserved human allografts. Ann Thorac Surg 82(5):1742–1746
90. Cosentino LM, Corwin W, Baust JM, Diaz-Mayoral N, Cooley H, Shao W, Van Buskirk R, Baust JG (2007) Preliminary report: evaluation of storage conditions and cryococktails during peripheral blood mononuclear cell cryopreservation. Cell Preserv Technol 5(4):189–204
91. Chaveiro A, Santos P, da Silva FM (2007) Assessment of sperm apoptosis in cryopreserved bull semen after swim-up treatment: a flow cytometric study. Reprod Domest Anim 42(1):17–21
92. Heng BC, Clement MV, Cao T (2007) Caspase inhibitor Z-VAD-FMK enhances the freeze-thaw survival rate of human embryonic stem cells. Biosci Rep 27(4–5):257–264
93. Hagl C, Tatton NA, Khaladj N, Zhang N, Nandor S, Insolia S, Weisz DJ, Spielvogel D, Griepp RB (2001) Involvement of apoptosis in neurological injury after hypothermic circulatory arrest: a new target for therapeutic intervention? Ann Thorac Surg 72(5):1457–1464
94. Malhotra JD, Miao H, Zhang K, Wolfson A, Pennathur S, Pipe SW, Kaufman RJ (2008) Antioxidants reduce endoplasmic reticulum stress and improve protein secretion. Proc Natl Acad Sci U S A 105(47):18525–18530
95. Mathew AJ, Van Buskirk RG, Baust JG (2002) Improved hypothermic preservation of human renal cells through suppression of both apoptosis and necrosis. Cell Preserv Technol 1(4):239–253
96. Snyder KK, Baust JM, Van Buskirk RG, Baust JG (2005) Enhanced hypothermic storage of neonatal cardiomyocytes. Cell Preserv Technol 3(1):61–74
97. Sosef MN, Baust JM, Sugimachi K, Fowler A, Tompkins RG, Toner M (2005) Cryopreservation of isolated primary rat hepatocytes: enhanced survival and long-term hepatospecific function. Ann Surg 241(1):125–133
98. Januskauskas A, Johannisson A, Rodriguez-Martinez H (2003) Subtle membrane changes in cryopreserved bull semen in relation with sperm viability, chromatin structure, and field fertility. Theriogenology 60(4):743–758
99. de Boer F, Drager AM, Pinedo HM, Kessler FL, Monnee-van Muijen M, Weijers G, Westra G, van der Wall E, Netelenbos T, Oberink JW, Huijgens PC, Schuurhuis GJ (2002) Early apoptosis largely accounts for functional impairment of CD34+ cells in frozen-thawed stem cell grafts. J Hematother Stem Cell Res 11(6):951–963
100. Fuller B, Green C, Grischenko VI (2004) Cooling, cryopreservation and gene expression in mammalian cells. Probl Cryobiol 3:58–71
101. Koenigsmann MP, Koenigsmann M, Notter M, Neuloh M, Mucke C, Thiel E, Berdel WE (1998) Adhesion molecules on peripheral blood-derived CD34+ cells: effects of cryopreservation and short-term ex vivo incubation with serum and cytokines. Bone Marrow Transplant 22(11):1077–1085
102. Matsushita T, Yagi T, Hardin JA, Cragun JD, Crow FW, Bergen HR, Gores GJ, Nyberg SL (2003) Apoptotic cell death and function of cryopreserved porcine hepatocytes in a bioartificial liver. Cell Transplant 12(2):109–121
103. Sugimachi K, Sosef MN, Baust JM, Fowler A, Tompkins RG, Toner M (2004) Long-term function

of cryopreserved rat hepatocytes in a coculture system. Cell Transplant 13(2):187–195

104. Kajiwara T, Takeuchi T, Ueki T, Moriyama N, Ueki K, Kakizoe T, Kawabe K (1999) Effect of Bcl-2 overexpression in human prostate cancer cells in vitro and in vivo. Int J Urol 6(10):520–525
105. Kumar S, Mohan A, Guleria R (2009) Prognostic implications of circulating anti-p53 antibodies in lung cancer – a review. Eur J Cancer Care (Engl) 18(3):248–254
106. Mu Z, Hachem P, Pollack A (2005) Antisense Bcl-2 sensitizes prostate cancer cells to radiation. Prostate 65(4):331–340
107. Yoshino T, Shiina H, Urakami S, Kikuno N, Yoneda T, Shigeno K, Igawa M (2006) Bcl-2 expression as a predictive marker of hormone-refractory prostate cancer treated with taxane-based chemotherapy. Clin Cancer Res 12(20 Pt 1):6116–6124
108. Arikura J, Kobayashi N, Okitsu T, Noguchi H, Totsugawa T, Watanabe T, Matsumura T, Maruyama M, Kosaka Y, Tanaka N, Onodera K, Kasai S (2002) UW solution: a promising tool for cryopreservation of primarily isolated rat hepatocytes. J Hepatobiliary Pancreatol Surg 9(6):742–749
109. Baust JM, Van Buskirk R, Baust JG (1998) Cryopreservation outcome is enhanced by intracellular-type medium and inhibition of apoptosis. Cryobiology 37(4):410–411
110. Snyder KK, Baust JM, Van Buskirk RG, Baust JG (2005) Improved cryopreservation of vascular tissue. Cryobiology 51(3):357–358
111. O'Flaherty C, Beconi M, Beorlegui N (1997) Effect of natural antioxidants, superoxide dismutase and hydrogen peroxide on capacitation of frozen-thawed bull spermatozoa. Andrologia 29(5):269–275
112. Hadj-Aissa A, Ramella-Virieux SG, Steghens JP, Barbieux A, Pozet N (1997) Calcium antagonists improve kidney function in the rat after cold storage in high-Na UW but not in high-K UW solution. Transplant Proc 29(5):2439–2441
113. Mathew A, Baust JG, Van Buskirk RG (1997) Optimization of HypoThermosol® for the hypothermic storage of cardiomyocytes – addition of EDTA. In Vitro Toxicol 10(4):407–415
114. Yagi T, Hardin JA, Valenzuela YM, Miyoshi H, Gores GJ, Nyberg SL (2001) Caspase inhibition reduces apoptotic death of cryopreserved porcine hepatocytes. Hepatology 33(6):1432–1440
115. Peter AT, Linde-Forsberg C (2003) Efficacy of the anticaspase agent zVAD-fmk on post-thaw viability of canine spermatozoa. Theriogenology 59(7):1525–1532
116. Heng BC (2009) Effect of Rho-associated kinase (ROCK) inhibitor Y-27632 on the post-thaw viability of cryopreserved human bone marrow-derived mesenchymal stem cells. Tissue Cell 41(5):376–380
117. Martin-Ibanez R, Unger C, Stromberg A, Baker D, Canals JM, Hovatta O (2008) Novel cryopreservation method for dissociated human embryonic stem cells in the presence of a ROCK inhibitor. Hum Reprod 23(12):2744–2754
118. Li X, Krawetz R, Liu S, Meng G, Rancourt DE (2009) ROCK inhibitor improves survival of cryopreserved serum/feeder-free single human embryonic stem cells. Hum Reprod 24(3):580–589
119. Fire A, Xu S, Montgomery MK, Kostas SA, Driver SE, Mello CC (1998) Potent and specific genetic interference by double-stranded RNA in Caenorhabditis elegans. Nature 391(6669):806–811
120. Johnston JW, Benson EE, Harding K (2009) Cryopreservation induces temporal DNA methylation epigenetic changes and differential transcriptional activity in Ribes germplasm. Plant Physiol Biochem 47(2):123–131
121. Ferri KF, Kroemer G (2001) Organelle-specific initiation of cell death pathways. Nat Cell Biol 3(11):E255–E263
122. Szegezdi E, Macdonald DC, Ni Chonghaile T, Gupta S, Samali A (2009) Bcl-2 family on guard at the ER. Am J Physiol Cell Physiol 296(5):C941–C953
123. Lisbona F, Hetz C (2009) Turning off the unfolded protein response: an interplay between the apoptosis machinery and ER stress signaling. Cell Cycle 8(11):1643–1644
124. Verfaillie T, Garg AD, Agostinis P (2013) Targeting ER stress induced apoptosis and inflammation in cancer. Cancer Lett 332(2):249–264
125. Austin RC (2009) The unfolded protein response in health and disease. Antioxid Redox Signal 11(9):2279–2287. doi:10.1089/ARS.2009.2686
126. Fribley A, Zhang K, Kaufman RJ (2009) Regulation of apoptosis by the unfolded protein response. Methods Mol Biol 559:191–204. doi:10.1007/978-1-60327-017-5_14
127. Malhotra JD, Kaufman RJ (2007) Endoplasmic reticulum stress and oxidative stress: a vicious cycle or a double-edged sword? Antioxid Redox Signal 9(12):2277–2293. doi:10.1089/ars.2007.1782
128. Mao XR, Crowder CM (2010) Protein misfolding induces hypoxic preconditioning via a subset of the unfolded protein response machinery. Mol Cell Biol 30(21):5033–5042. doi:10.1128/MCB.00922-10 MCB.00922-10 [pii]
129. Hetz C (2012) The unfolded protein response: controlling cell fate decisions under ER stress and beyond. Nat Rev Mol Cell Biol 13(2):89–102. doi:10.1038/nrm3270 nrm3270 [pii]

Fundamental Principles of Stem Cell Banking

3

Changbin Sun, Jianhui Yue, Na He, Yaqiong Liu, Xi Zhang, and Yong Zhang

Abstract

Stem cells are highly promising resources for application in cell therapy, regenerative medicine, drug discovery, toxicology and developmental biology research. Stem cell banks have been increasingly established all over the world in order to preserve their cellular characteristics, prevent contamination and deterioration, and facilitate their effective use in basic and translational research, as well as current and future clinical application. Standardization and quality control during banking procedures are essential to allow researchers from different labs to compare their results and to develop safe and effective new therapies. Furthermore, many stem cells come from once-in-a-life time tissues. Cord blood for example, thrown away in the past, can be used to treat many diseases such as blood cancers nowadays. Meanwhile, these cells stored and often banked for long periods can be immediately available for treatment when needed and early treatment can minimize disease progression. This paper provides an overview of the fundamental principles of stem cell banking, including: (i) a general introduction of the construction and architecture commonly used for stem cell banks; (ii) a detailed section on current quality management practices; (iii) a summary of questions we should consider for long-term storage, such as how long stem cells can be stored stably, how to prevent contamination during long term storage, etc.; (iv) the prospects for stem cell banking.

C. Sun • J. Yue • N. He • Y. Liu • X. Zhang
China National GeneBank-Shenzhen, BGI-Shenzhen, Yantian District, Shenzhen 518083, China
e-mail: sunchangbin@genomics.cn; yuejianhui@genomics.cn; hena@genomics.cn; liuyaqiong@genomics.cn; zhangxi1@genomics.cn

Y. Zhang (✉)
Huahan Gene Research Institute (HHGene), Futian District, Shenzhen 518083, China
e-mail: zhangy@hhgene.cn

F. Karimi-Busheri, M. Weinfeld (eds.), *Biobanking and Cryopreservation of Stem Cells*, Advances in Experimental Medicine and Biology 951, DOI 10.1007/978-3-319-45457-3_3

Keywords
Stem cells • Biobanks • Architecture • Management • Standardization • Quality management • File management • Long-term maintenance • Application • GMP

Abbreviations

AMI	Acute myocardial infarction
CFC	Colony-forming cells
CGMP	Current good manufacturing practice
CJD	Creutzfeldt-Jakob disease
CMV	Cytomegalovirus
DNA	Deoxyribonucleic acid
EBV	Epstein-Barr virus
EMA	European Medicines Agency
ESCs	Embryonic stem cells
FDA	Food and Drug Administration
GMP	Good manufacturing practice
GvHD	Graft versus host disease
HBV	Hepatitis B virus
HCV	Hepatitis C virus
HIV	Human immunodeficiency virus
HSCs	Hematopoietic stem cells
HTLV	Human T-cell leukemia virus
ICH	International conference on harmonization
iPSCs	Induced pluripotent stem cells
LAL	Limulus amebocyte lysate test method
MCB	Master cell bank
MSCs	Mesenchymal stem cells
NRA	National regulatory authority
PCR	Polymerase chain reaction
QA	Quality assurance
QC	Quality control
SNP	Single nucleotide polymorphism
SPCs	Spermatogonial stem cells
STR	Short tandem repeat
UCB	Umbilical cord blood
WCB	Working cell bank
WHO	World Health Organization

3.1 Introduction

Stem cells are a class of cells that have the capability to indefinitely self-renew and differentiate to special functional cell types under the right conditions or given the right signals [1]. According to their potency, i.e. the range of cell types to which they can differentiate, stem cells commonly can be classified into totipotent stem cells, pluripotent stem cells, multipotent stem cells and unipotent stem cells, such as zygote, embryonic stem cells (ESCs) and induced pluripotent stem cells (iPSCs), hematopoietic stem cells (HSCs) and mesenchymal stem cells (MSCs), and spermatogonial stem cells (SPCs) respectively. Stem cells with properties of self-renewal and potency to generate different cell types are proving to be very promising resources for application in cell therapy, regenerative medicine, drug discovery, toxicology and developmental biology research [2–6]. They hold great potential to revolutionize healthcare and to supply new methods and tools for effective treatment of many diseases [7–10]. In the process, stem cell banks play a pivotal role in stem cell basic and translational research, as well as current and future clinical applications.

Banking stem cell is an extension of biobanking. The concept of biobanking has varied in history and there is no consensus [11]. The practice of "biobanking" has been broadly construed to involve storing health and genetic information and/or various types of biological materials in banks, repositories, or collections [12]. The practice of banking somatic cells, as well as primary cell lines and immortalized cancer lines, has been a vital research resource for many years [12, 20]. In 1988, the first successful related umbilical cord blood (UCB) transplant was conducted in Paris, France on a 6-year-old male patient suffering from a blood disorder called Fanconi's Anemia [3]. In 1998, hESC isolation and culture methods were developed by James Thomson and his research group [5]. In recent years, many kinds of MSCs derived from a large number of different tissues show promising therapy potential [13–16]. In 2007, human iPSCs derived from human somatic cells were shown to have similar properties to hESCs and hold significant potential in disease modeling, drug discovery and selection [4, 6]. All these important events in the field of stem cells have sparked considerable public interest in stem cells banking.

Stem cell banks have been established in many countries in order to preserve stem cells and facilitate their research and clinical application [17, 18]. Many benefits accrue from stem cell banking, but standardization and quality control during banking procedures are essential to allow researchers from different labs to compare their results and to develop safe and effective new therapies. Furthermore, many stem cells come from once-in-a-life time tissues, UCB for example, thrown away in the past, can be used to treat many diseases such as blood cancers nowadays. Over the last several decades, the field of UCB banking and transplantation has grown exponentially. Over 600,000 UCB units have been stored for transplantation worldwide, and more than 30,000 UCB transplantations have been performed [19]. Meanwhile, these stem cells stored in banks are immediately available for treatment when needed and early treatment can minimize disease progression [20].

Before establishment of a stem cell bank, there are many factors that need to be taken into consideration including new technological developments and a deeper understanding of stem cells. For example, pluripotent stem cell culture has clearly improved in the past decade, from dependence on feeder cells to culture on synthetic matrices [21]. Thus, stem cell banks may need to change their SOPs and adopt new ones to increase productivity and stability etc. Although different stem banks may store different typed of stem cells, the major issues are common, such as the design of the bank, how to guarantee storage cell quality, and how do maintain and manage the bank over the long term? These are the main questions that will be addressed in this article, as well as the prospects for stem cell banking.

3.2 Construction and Design

3.2.1 Functional Areas and Equipment

Stem cell banking is a procedure wherein the cells isolated from various sources are collected, proliferated, stored and preserved for their future use [7]. Establishing a well-functioning biobank for stem cells provides the foundation for successful storage, quality control and assessment, and long-term maintenance and management.

The basic issues pertaining to the design and equipping of a stem cell biobank are very similar to those of a regular stem cell laboratory [22], but more functional areas and equipment should be considered and coordinated, as well as more rigorous manufacturing practices. The key considerations when setting up the bank include (1) defining the scope of the work, which includes the numbers and types of cell lines to be banked, and (2) determining the number of people who will work in the bank and their specific tasks. If more than one cell type will be stored in the same bank, what safeguards will be put into place to prevent cross-contamination? Do different cell types need separated areas to quarantine these cell types [22]? Furthermore, the processing of stem cells, including isolation, culture and collection, requires a physical environment in which air quality (i.e., the number of airborne particles) is controlled to minimize risk of contamination. The processing facility should be constructed and operated to minimize the introduction, generation and retention of particles and microorganisms. A formal program of environmental monitoring should be maintained in each stem cell bank to specify and assess key factors and their influence on the microbiological quality of the process and product. This program should assure the manipulation of cells involved in the derivation of stem cell lines and their culture under established limits for airborne particles and for microbial contamination of the air and surfaces, especially for stem cell lines that will be applied to human therapy [23].

When designing and building a well-functioning stem cell bank, the design team usually includes architects, contractors, builders, tradespeople and a laboratory director or manager. One needs to consider budget, space, the biosafety level of the laboratory, major equipment, number of people, major functional areas, freezer rooms, storage areas in and adjacent to the laboratory, and databases for information storage and extraction etc. To avoid cross-contamination of different functional areas, carrying out cell isolation, culture and quality control in different laboratories is recommended. Figure 3.1 outlines the major functional areas commonly needed in a stem cell bank. The major function of the preparation areas focuses on cell isolation and expansion, while the function of the quality assessment area is to test the cell characteristics and insure the cells generated meet the requirement of good manufacturing practice (GMP) or good laboratory practice (GLP) or similar quality requirements according to the local government.

3.2.2 File Management

Figure 3.2 shows the workflow for stem cell banking under good manufacturing practice conditions, which can be partitioned into three components: manufacturing, quality control and file management. Following cell isolation, expansion, characterization, storage, and quality control and assessment, many record files will be formed and archived, including data files, informed consent, personal information from donors, SOPs for practice etc. All should be well managed to insure standardization, efficiency and privacy protection, and should guarantee cell quality. Ideally file management of stem cell banks should conform good manufacturing practice (GMP) or good laboratory practice (GLP) or similar quality requirements mandated by local government.

Fig. 3.1 Major functional area in a stem cell bank

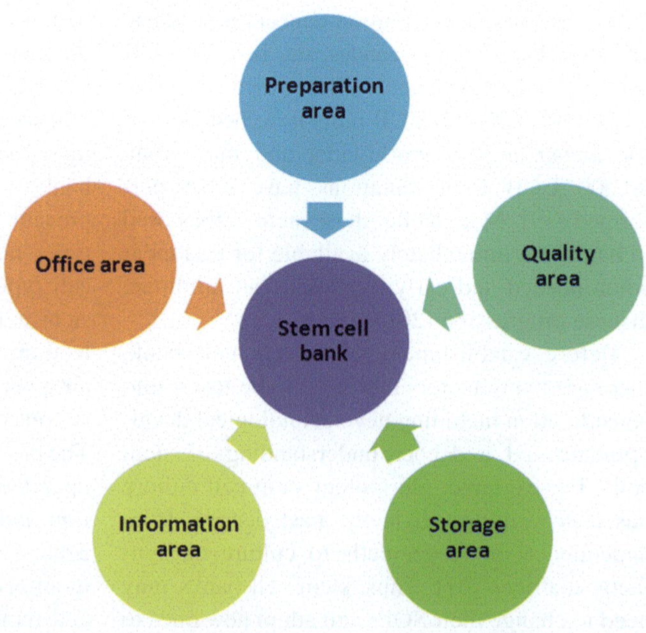

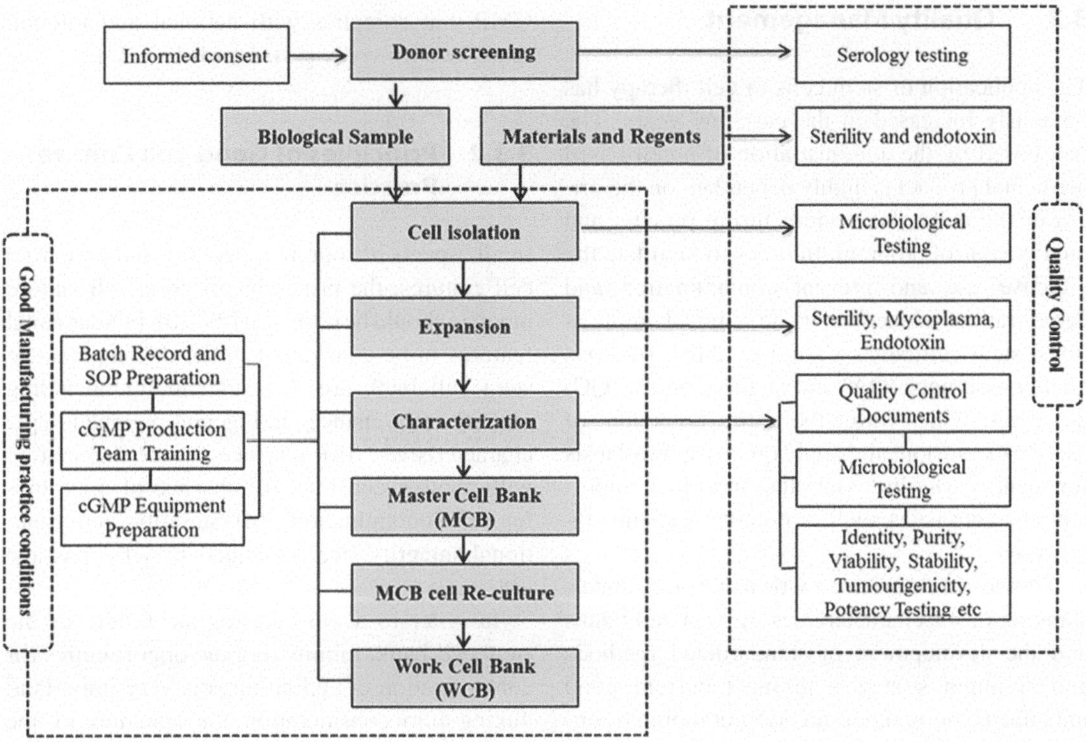

Fig. 3.2 Workflow for stem cells banking under GMP conditions

3.2.3 Master Cell Banks and Working Cell Banks

When a cell line is to be used over many manufacturing cycles and may be distributed to others, a two-tiered cell banking system consisting of a master cell bank (MCB) and a working cell bank (WCB) is recommended [24]. Upon arrival at a banking facility, tissue and newly acquired cells lines should be quarantined until shown to be clear of contamination (such as mycoplasma, HIV etc.), and then clean tissues and cells can be used to create the MCB [12]. This MCB must be characterized and extensively tested for contaminants such as bacteria, fungi, and mycoplasma after isolation and expansion before freezing. Most cell lines also require sterility and PCR testing for viruses. The analysis process starts with the isolation of a sufficient number of clones to adequately sample the population of cells in the culture. The characterization of single clone isolates insures that there are no significant numbers of cells with missing or mutated genetic elements present in the cell bank.

A WCB can be prepared from live cultures or a single frozen vial from the MCB. Cells are grown for several passages, allowing for minimal necessary characterization and scale-up prior to aliquoting into vials and freezing. Cells from the WCB are normally used for distribution to researchers. Thus, creation of a two-tiered frozen cell bank enables a supply of reliable quality-controlled cells at the same passage over a long period of time, which is the principal requirement for biomedical research and product development [12, 24]. Furthermore, WCBs are best maintained at low passage to avoid phenotypic and genetic variation. In order to minimize the risk of complete loss of a cell line due to, for example, contamination or equipment or facility failure, it is critical that frozen stocks are also stored at a second site [25].

3.3 Quality Management

The application of stem cells in cell therapy has obviously increased in the past few years. The risk posed by the administration of a cell-based medicinal product is highly dependent on the origin of the cells, the manufacturing process and quality control program. In order to facilitate the effective use and prevent contamination and deterioration of stem cells in clinical applications, stem cell banks should establish rigorous quality assurance (QA) and quality control (QC) processes, which cover the entire procedure of ex vivo expansion, and apply quality control tests for identity, sterility, viability, and other important characteristics such as potency, genetic stability, etc.

This section aims to provide a thorough understanding of the characteristics of stem cell banks and the development of standardized methods and common strategies to meet current good manufacturing practice (cGMP) compliant conditions, including tissue sourcing, manufacturing, testing, and storage.

3.3.1 Current Good Manufacturing Practices

The goal of banking stem cells is to preserve their characteristics, prevent contamination and deterioration, and facilitate their effective use in research and clinical applications. National and international regulatory authorities such as the World Health Organization (WHO), European Medicines Agency (EMA) and Food and Drug Administration (FDA) recommend that the general principles of GMP should be applied from the first stage of cell banking onwards [26–28]. Many important guidelines on good cell culture practice [29] and cell banking [30, 31] have been developed. Briefly, the fundamental regulatory principles of stem cell banking include the creation of quality controls for master cell bank and working cell banks, supported by traceability of procured cell lines and thorough documentation of the whole stem cell banking process. Ideally banks should operate minimally in "the spirit" of GMP that complies with national and international quality systems standards [32, 33].

3.3.2 Principles of Good Cell Culture Practice

In all aspects of sourcing, banking and preparing cell cultures, the principles of good cell culture practice should be observed [29, 30]. Fundamental features to be considered in the development of stem cell bank are: (a) authenticity, including identity, provenance and genotypic/phenotypic characteristics; (b) absence of contamination with another cell line; (c) absence of microbiological contamination; (d) stability and functional integrity over extended in vitro passage [28].

In order to avoid catastrophic failure of the stem cell bank, minimizing the opportunities for contamination of cell cultures is very important. Taking into consideration the specifics of the manufacturing process, the steps of cell manipulation and open processing should be minimized. It is critical to employ a rigorous aseptic technique and appropriate environmental controls for cell culture processing and the preparation of growth media [28]. Meanwhile, the variation of physical culture parameters such as pH, temperature, humidity, gas composition also can significantly affect the viability of cells and should be specified within established standard operating procedures. In summary, a consistent process should be demonstrated.

Although antibiotic and antifungal agents may be required for primary cell cultures, the using of antibiotics in a stem cell bank is discouraged. When antibiotics have been used, the potential inhibitory effects of the antibiotics on contaminating organisms should be considered in sterility-testing.

3.3.3 Selection of Donors, Materials and Reagents

All materials should be subjected to risk assessment and testing when necessary because raw materials can be a primary source for the introduction of adventitious agents into the stem cell bank [28]. Careful attention should be paid to sourcing, production, handling, testing and quality control. According to international guidelines [28, 34, 35], the sections below detail the information on manufacturing components and materials that should be considered:

- **Donor screening and testing**: Donor screening and testing is performed to determine donor eligibility. Details of the requirements for screening and testing donors of human cells and tissues are described in FDA and EMA documents [26, 28, 34, 35]. Briefly, the donors of stem cells must be screened and tested for HIV-1, HIV-2, hepatitis B virus (HBV), hepatitis C virus (HCV), Treponema pallidum (syphilis), and CJD. Donors of viable leukocyte-rich cells or tissues should be screened and tested for human T-lymphotropic virus types 1 and 2 (HTLV-1, HTLV-2) and CMV.
- **Quality control of reagents**: Reagents which are used for cellular growth, differentiation, selection, purification, or other critical manufacturing process can affect the safety, potency, and purity of the stem cells, especially by introducing adventitious agents such as bacteria, fungi, mycoplasmas and viruses [26, 28]. For example, trypsin and bovine serum have been known to contain mycoplasma species and sometimes more than one viral contaminant. The common reagents include fetal bovine serum, trypsin, digestion enzymes, growth factors, cytokines, monoclonal antibodies, antibiotics, cell separation devices, media components and so on. Documentation should be gathered on their manufacturing history, production, quality control in order to ensure the reagents' quality and safety [26–28]. The reagents should be subjected to strict tests for quality and freedom from contamination by adventitious agents to assure their acceptability for use in cell culture. Meanwhile, the reduction and elimination from the manufacturing process of reagents and materials derived from humans and animals is encouraged, where feasible.

3.4 Quality Control Testing of the Stem Cell Bank

The recommendations for characterization testing for cells in the stem cell bank include, but are not limited to, identity, purity, viability, microbiological testing (including sterility, mycoplasma, and adventitious viral agent testing), tumorigenicity, stability (including stability during cryostorage and genetic stability), potency etc.

3.4.1 Identity

Cell line misidentification, contamination and poor annotation affect scientific reproducibility [36–38]. Cell banks should be authenticated by a cell identification method approved by the NRA in order to confirm that no switching or cross-contamination has occurred [28]. We recommend that the identity of the cells in the MCB and WCB should be characterized by phenotypic and/or genotypic profiles including cell morphology, genetic tests (STR, SNP), specific surface markers and gene expression products.

For adherent cells, morphology may be a useful method in conjunction with other tests. For human stem cells, genetic tests such as short tandem repeat (STR) analysis are most commonly used to confirm cell line identity [39–42], and single nucleotide polymorphism (SNP) genotyping is also used for identification of cells as well as for monitoring their genomic stability. We also recommend additional identification based on cell surface markers or genetic polymorphisms to distinguish between the multiple cell lines.

In summary, the combined use of variety methods for cell identification depending on the different cell types and tissues can achieve more rigorous identification.

3.4.2 Purity

The cellular population of interest usual contains other cells, which belong to different lineages or differentiation stages that may be unrelated to the intended population. The unwanted cells should be defined and their amount in the final cell bank should be controlled by appropriate specifications, Acceptance criteria for the amounts of contaminating cells should be set. Cell surface markers, genetic polymorphisms and specific biological activities are recommended methods to test the cell purity [43].

3.4.3 Viability

A high level of viability of cell lines derived from stem cell banks is important for efficient and reliable cell-based products [28]. Documents of the China FDA recommend that different testing methods (such as cell counting, cell doubling time, cell cycle, cloning efficiency, telomerase activity) can be used to determine cell viability and growth status [43]. The recommendations of WHO state that thawed cells should typically have viability levels in excess of 80 %, though this is not always achieved and may depend on the cell line [28]. Lower viabilities may still result in suitable growth recovery and in acceptable product qualities. Meanwhile, for somatic cellular therapies, the minimum acceptable viability specification recommended by the FDA is generally set at 70 % [27].

Cell viability can be easily assessed in culture by employing widely applied assays. However, in some cases, such as pre-apoptotic cells excluding trypan blue, viability assays may give misleading results. So we suggest that when a membrane-integrity test is used for cell viability, additional cell markers such as indicators of apoptosis should be studied in order to avoid significant overestimation of viability.

3.4.4 Microbiological Testing

3.4.4.1 Sterility Tests

It is important in risk evaluation for the manufacturer to bear in mind that standard compendial tests for "sterility" are intended to give an indication of the effectiveness of aseptic processing in preventing general bacterial or fungal contamination. For suitable sterility tests one should refer to certified documentation and Pharmacopoeias. If antibiotics were used in manufacturing, documentation should be provided indicating that the antibiotics were removed prior to sterility testing.

3.4.4.2 Mycoplasma

Animal serum products used in culture and the culture facility environment are the two main sources of mycoplasma contamination. Testing for mycoplasma contamination should be conducted on both cells and supernatant. Due to the limited dating period of many cellular products, it is frequently not feasible for a sponsor to perform the recommended culture-based assay. The EMA and FDA recommend the use of polymerase chain reaction (PCR)-based mycoplasma assays or another rapid detection assay during cell manufacturing [26, 44].

3.4.4.3 Pyrogenicity/Endotoxins

The rabbit pyrogen test is the currently required method for testing biological products for pyrogenic substances. Although the pyrogenicity test is required, there may be specific cases where this test method cannot be performed due to properties of the cellular product. Another test such as the Limulus Amebocyte Lysate test (LAL) may be used as an alternative method [27]. For any parenteral drug, the FDA recommends that the upper limit of acceptance criterion for endotoxin be 5 EU/kg body weight/hour. For intrathecally-administered drugs, an upper limit of acceptance criterion is 0.2 EU/kg body weight/hour [27].

3.4.4.4 Adventitious Agent Testing

As appropriate, adventitious agent testing should include in vitro viral testing, in vivo viral testing and selected species-specific testing for adventitious viruses.

In vitro viral testing should be conducted by inoculating the test sample (MCB) onto various susceptible indicator cell lines such as the human cell line MRC-5 or Vero cells. We recommend that in vivo viral assays be carried out by inoculating the test sample (MCB) into animals such as adult and suckling mice and embryonated hen eggs. Finally, assays for the presence of species-specific viruses also should be performed especially when the cell lines are used as a therapeutic product. We recommend that testing for human pathogens (CMV, HIV-1 & 2, HTLV-1 & 2, EBV, HBV, HCV, B19), and other human viral agents, for which PCR-based assays are available, should be performed [44, 45].

3.4.5 Tumorigenicity

Testing for tumorigenicity of cells derived from cell culture and banked will be required with a test approved by the NRA. Several in vitro systems, such as cell growth in soft agar [46] and muscle organ culture [47], have been explored as alternatives to in vivo tests for tumorigenicity. The test should involve a comparison between the cell line and a suitable positive reference preparation (e.g. HeLa cells) and a standardized procedure for evaluating results.

Other assays can also be considered, such as proliferative capacity, dependence on exogenous stimuli, response to apoptosis stimuli and genomic modification, in order to evaluate the risk of cellular transformation and subsequent potential for tumorigenicity.

3.4.6 Stability

The stability of cell banks during cryostorage, and the genetic stability of cell lines are key elements in a successful cell bank programme.

3.4.6.1 Stability During Cryostorage

Data should be generated to support the stability or suitability of the cryopreserved cell banks during storage. Continuous monitoring records and successful periodic production runs could be useful strategies to monitor the stability of the cell bank.

3.4.6.2 Genetic Stability

Cell genome stability is the key quality and safety indicator of stem cell banks. Genomic instability refers to an increased tendency to generate alterations in the genome during the life cycle of cells. It is main driving force for tumorigenesis [40]. The WHO has pointed out that any features of the cell lines that might affect quality should be discussed with the NRA to ensure that tests used by the manufacturer to monitor genetic stability are adequate [28].

However, there is no standard method to evaluate the genetic stability of cell lines in MCB and WCB at present. The common methods such as doubling time of cell lines and karyotyping already cannot meet the comprehensive assessment of genomic stability. Here we recommend high-throughput sequencing which has better resolution and accuracy to assess genetic stability. Other approaches include examining telomerase activity, chromosomal integrity, expression of proto oncogenes, etc.

3.4.7 Potency

According to ICH guideline 6QB, potency is the quantitative measure of biological activity [48], but in addition, qualitative biological assay may be included.

Basically, potency assays can be performed in vitro using cell systems and in vivo using animal models [48, 49]. As stem cells have a tendency to spontaneously differentiate in culture, the stem cell bank should ensure that pluripotency is not lost during in vitro expansion and cryopreservation. As the function of cells such as viability, self-renewal, differentiation and apoptosis are critical to their quality, potency assays may need to be evaluated during manufacturing using appropriate markers and technology, such as gene expression profiles, flow cytometry, cell cloning, PCR, etc. [50]. In vivo assays for potency may also be useful especially using available animal models. As described in the referred guideline, the combination of multiple methods may be better to adequately evaluate the potency of the cells in the stem cell bank.

3.5 Long -Term Maintenance and Management

Stem cells have been applied in research and clinical use for many years, and appropriate cryopreservation protocols for long-term storage is essential [51]. Standardization of processes and the implementation of quality control programmes should be established during the banking process in order to prevent contamination and deterioration and maintain post-thaw quality, pluripotency and genetic stability of recovered cells [17, 24]. Moreover, viability tests, functional assays, stability measurements, authentication and documentation have to be carried out [52, 53].

3.5.1 Quality Control During Stem Cell Cryopreservation Process

3.5.1.1 Storage Temperature

Storage temperature is a key factor for long-term storage of stem cells in a stable state. Lower temperatures can reduce metabolism of cells and efficiently reduce the rate of degradation [53]. In general, the lower the storage temperature, the more stable the cells remain and the longer they can be preserved [54, 55]. The traditional method for long-term cryopreservation of stem cell is storage in liquid nitrogen, which can reach a terminal temperature of −196 °C [52].

Quality control methods should be applied to keep a stable temperature environment in long-term storage vessels as stored cells may be subjected to temperature changes during the maintenance of storage vessels. For example, storage vessels are opened frequently to gain access to stored vials, intermittent warming may occur to cryovials. Furthermore, some storage racking material can act as good heat conductors and may promote warming cycles and temperature gradients within the storage vessel. Finally, there may be failures in the liquid nitrogen filling process [25]. A number of straightforward procedures can be taken to keep a stable temperature environment in stem cell storage vessels during the long-term storage process: (1) choose a suitable terminal temperature for the long-term storage of stem cells; (2) use appropriate facilities to cryopreserve samples and manual or automatic continuous temperature monitoring to detect and provide useful monitoring data of temperatures; (3) set up an automatic liquid nitrogen filling system; (4) keep vessel vacuum checks, electrical testing, and other maintenance procedures regularly and periodically audited [25, 53].

3.5.1.2 Contamination

It is crucial to prevent microbial/cross contamination during the long-term storage of stem cells. The risk of stem cell products is mainly associated with contaminated liquid nitrogen during long-term storage [51] as microbial flora can easily accumulate within ice sludge debris [25]. Quality control methods should be taken to prevent contamination of cell samples in these long-term storage vessels [56]. It has been shown that vapour phase storage can largely overcome sample contamination as it can prevent the liquid nitrogen permeating into cryovials thereby avoiding contamination by contaminated liquid [57]. However, all long-term storage samples run the risk of cross contamination risk when they are stored in the same cryotank, rewarmed in the same water bath, and because of other potential environmental sources.

Quality control methods can be taken as follows to reduce the contamination of stem cells during long-term storage: (1) choose a vapour phase liquid nitrogen containment vessel; (2) set security caps when sample have to be submersed in liquid nitrogen; (3) employ internal threads and double bagging to store cell samples; (4) maintain the storage container closure integrity and within the product shelf life; (5) set up a quarantined storage area to control the spread of pathogens and infectious agents; (6) keep storage containers in a clean and controllable environment to reduce the contamination risk from the surrounding area; (7) make sure the source of liquid nitrogen is clean [53].

3.5.2 Quality Validation for Recovered Stem Cell

Stem cells should have the ability to survive long-term storage and keep their functions after recovery. Post-thaw cell viability testing is vital to evaluate the effects of cryopreservation on stem cells. Quality assessment including viability tests, functional assays, stability measurements and authentication should be applied to analyze the post-thaw stem cells metabolic functions, genetic integrity and totipotency after recovery [25, 58].

These quality control measures have been shown to be necessary in various studies. For example, the numbers of colony-forming cells (CFC) and CD34 cells, which can effectively predict engraftment kinetics and hematopoietic recovery, were measured to evaluate the function of 9.5-year cryopreserved hematopoietic stem cells [59], and the differentiation potential of long-term cryopreserved human adipose-derived stem cells was measured by using different induction medium to test totipotency after rewarming [60].

Trypan Blue exclusion assay is used to determine the post-thaw cell viability of stem cells, and a light microscope is used to count total number of dead cells and live cells [60]. Stability assessment for stem cells includes: evaluation of totipotency by cell differentiation potential assays and morphological characteristics; studies of phenotypes; cytological and karyotype testing; confirmation of functional biochemistry (metabolites/proteins, proteins/enzymes); omics analyses; biomarker studies;, investigations of genome structure, DNA damage and repair, chromatin analysis, analysis of DNA adducts (methylation, oxidation) and confirmation of genetic modifications [61]. Furthermore, since cells that have been held in long-term storage may have been cross-contaminated by other samples, authentication is particularly important and quality control needs to be performed to insure sample identity following the cells retrieval from storage [53].

3.6 Prospects for Stem Cell Banking

Many countries have now established stem cell banks. Table 3.1 shows the major stem cell banks in the world. Stem cells stored in banks play an important role in promoting biological development and providing medical science with a powerful tool. Several types of conditions are currently being treated with stem cell-based therapies, including autoimmune diseases, neurological disorders, cancers, and infertility [62]. To date, several stem cell therapeutic products have been approved in different countries (Table 3.2). Especially worthy of mention is umbilical cord blood. Since the first successful transplantation, the field of UCB banking and transplantation has grown exponentially. Over 600,000 UCB units have been stored for transplantation worldwide, and more than 30,000 UCB transplantations have been performed [19].

Treatments in regular clinical use are limited to adult stem cells. Many stem cell therapies are based on the regenerative capacities of stem cells to produce a variety of tissues, either in the patient's body or in vitro [62]. And other therapies depend on the transplanted stem cells to provide signals that regulate the activities of nearby cells [63]. Although unanticipated challenges in safety or efficacy still exist, we can optimistically believe that products and medicine based on stem cells will be widely used in the future.

3.7 Summary

Since significant progress has been made in stem cell research and application over the last few decades, stem cell banks have become very important platforms to preserve their characteristics, prevent contamination and deterioration and facilitate their effective use in baicl and translational research, as well as clinical application all over the world [17, 18]. A number of stem cell banks have been established in many countries

Table 3.1 Example of stem cell banks in the word

Name of bank	Stem cell type	Country/region	Website and/or reference
UK Stem Cell Bank	Comprehensive	UK	http://www.nibsc.org
National Centre for Cell Science Cell Repository	Comprehensive	US	http://www.nccs.res.in
WiCell International Stem Cell Bank	Comprehensive	US	http://www.nationalstemcellbank.org
Cord Blood Registry	Cord blood/cord tissue stem cells	US	http://www.cordblood.com/benefits-cord-blood
Australian cord blood bank	Cord blood	Australian	http://www.stemlife.com.au
BioEden	Child milk tooth stem cells	US	https://us.bioeden.com
Human Pluripotent Stem Cell Registry	Human pluripotent stem cells	Global registry	http://hpscreg.eu
Swiss Stem Cell Bank	Cord blood	Swiss	http://parentsguidecordblood.org/en/bank/swiss-stem-cell-bank
Stem Med	Haemopoietic stem cells/ mesenchymal stem cells	Singapore	http://stem-med.sg/about-stem-med
Dental stem cell banking	Dental stem cells	US	http://www.store-a-tooth.com
Cell Center of China National Genebank	Comprehensive	China	http://stemcell.cngb.org
The International Placenta Stem Cell Society	Placenta stem cells	International	http://www.iplass.net/ii

Table 3.2 Example of stem cell therapeutic products approved in the world

Country	Year	Product/company	Source	Adaptation disease
Europe – EMA	2009	Chondro Celect/TiGenix	Autologous chondrocytes	Knee cartilage defects
USA – Food and Drug Administration (FDA)	2009	Prochymal/Osiris	Allogeneic bone marrow mesenchymal stem cells	GvHD and Crohn disease
Australia-Therapeutic Goods Administration (TGA)	2010	Mesenchymal Preursor/ Mesoblast	Autologous mesenchymal progenitor cells	Bone repair
Korea – Food and Drug Administration (FDA)	2011	Hearticellgram-AMI/ FCB-Pharmicell	Autologous bone marrow mesenchymal stem cells	Acute myocardial infarction(AMI)
USA – FDA biologics license	2011	Hemacord/New York Blood Centre	Umbilical cord blood hematopoietic progenitor cells	Genetic acquired hematopoietic system disease
Korea – Food and Drug Administration (FDA)	2012	Cartistem/Medi-post	Mesenchymal stem cells from umbilical cord blood	Degenerative arthritis and knee joint cartilage injury
Korea – Food and Drug Administration (FDA)	2012	Cuepistem/Anterogen	Autologous fat source of mesenchymal stem cells	Complexity Crohn's disease complicated by anal fistula
Canada	2012	Prochymal/Osiris	Bone marrow stem cell	Children graft versus host disease(GvHD)

and regions, and the number will increase. There are many factors that need to be carefully considered before establishing a stem cell bank. New technology development and knowledge will undoubtedly lead to improvements in stem cell bank maintenance and management. For example, at present most banks banking MSCs adhere to the minimal criteria proposed by the International Society for Cellular Therapy (ISCT), i.e. (1) MSCs must be plastic-adherent when maintained in standard culture conditions; (2) MSCs must express CD105, CD73 and CD90, and lack expression of CD45, CD34, CD14 or CD11b, CD79alpha or CD19 and HLA-DR surface molecules. (3) MSCs must differentiate to osteoblasts, adipocytes and chondroblasts in vitro [64]. However, many investigations reported recently have shown that MSCs isolated utilizing these criteria are heterogeneous and different MSCs subtypes may have discordant functional properties, implying their applications need further elaboration [65, 66], and as a result banking SOPs need corresponding modification. Thus, the practice of stem cell banking is a dynamic process, which will advance with the times.

Acknowledgements Preparation of this report was supported by National Natural Science Foundation of China (No. 31401259) and the Shenzhen Municipal Government of China (NO. JCYJ20140418203036946).

References

1. National Institute of Health web site http://stemcells.nih.gov/info/scireport/pages/chapter1.aspx. Accessed Feb 2016
2. Baksh D, Song L, Tuan RS (2004) Adult mesenchymal stem cells: characterization, differentiation, and application in cell and gene therapy. J Cell Mol Med 8:301–316
3. Gluckman E, Broxmeyer HA, Auerbach AD et al (1989) Hematopoietic reconstitution in a patient with Fanconi's anemia by means of umbilical-cord blood from an HLA-identical sibling. N Engl J Med 321:1174–1178
4. Takahashi K, Tanabe K, Ohnuki M et al (2007) Induction of pluripotent stem cells from adult human fibroblasts by defined factors. Cell 131:861–872

5. Thomson JA, Itskovitz-Eldor J, Shapiro SS et al (1998) Embryonic stem cell lines derived from human blastocysts. Science 282:1145–1147
6. Yu J, Vodyanik MA, Smuga-Otto K et al (2007) Induced pluripotent stem cell lines derived from human somatic cells. Science 318:1917–1920
7. Deshpande HA, Akkala S, Rasquinha S et al (2015) Stem cell banking: an update on current scenario. Indian J Res Pharm Biotechnol 3:392
8. Khanna A, Shin S, Rao MS (2008) Stem cells for the treatment of neurological disorders. CNS Neurol Disord Drug Targets 7:98–109
9. Le Blanc K, Frassoni F, Ball L et al (2008) Mesenchymal stem cells for treatment of steroid-resistant, severe, acute graft-versus-host disease: a phase II study. Lancet 371:1579–1586
10. Resnick IB, Shapira MY, Slavin S (2005) Nonmyeloablative stem cell transplantation and cell therapy for malignant and non-malignant diseases. Transpl Immunol 14:207–219
11. M.F.A O, D N, M.J.A S (2010) Biobanks information paper. J Law Inf Sci 20:97–227
12. Efthymiou AG, Rao M, Lowenthal J (2014) Banking of pluripotent stem cells: issues and opportunities from the NIH perspective. In: Stem cell banking. Springer, New York, pp 77–93
13. Caplan AI (1991) Mesenchymal stem cells. J Orthop Res 9:641–650
14. Gimble J, Guilak F (2003) Adipose-derived adult stem cells: isolation, characterization, and differentiation potential. Cytotherapy 5:362–369
15. Romanov YA, Svintsitskaya VA, Smirnov VN (2003) Searching for alternative sources of postnatal human mesenchymal stem cells: candidate MSC-like cells from umbilical cord. Stem Cells 21:105–110
16. Shi S, Gronthos S (2003) Perivascular niche of postnatal mesenchymal stem cells in human bone marrow and dental pulp. J Bone Mineral Res 18:696–704
17. Luong MX, Smith KP, Crook JM et al (2012) Biobanks for pluripotent stem cells. In: Loring JF, Peterson SE (eds) Human stem cell manual: a laboratory guide. Elsevier Inc, Amsterdam, pp. 105–125
18. Mirabet V, Solves P (2013) Cryopreservation of hematopoietic stem cells from umbilical cord blood for transplantation. In: Stem cells and cancer stem cells, vol 9. Springer, dordrecht, pp 3–11
19. Ballen KK, Gluckman E, Broxmeyer HE (2013) Umbilical cord blood transplantation: the first 25 years and beyond. Blood 122:491–498
20. Stewart JA (2005) Cancer management: a multidisciplinary approach. Oncology:1272–1272
21. Qian X, Villa-Diaz LG, Krebsbach PH (2013) Advances in culture and manipulation of human pluripotent stem cells. J Dent Res 92:956–962
22. Wesselschmidt RL, Schwartz PH (2011) The stem cell laboratory: design, equipment, and oversight. Methods Mol Biol 767:3–13
23. Cobo F, Stacey GN, Cortes JL et al (2006) Environmental monitoring in stem cell banks. Appl Microbiol Biotechnol 70:651–662
24. Kiatpongsan S, Wacharaprechanont T, Tannirandorn Y (2006) Embryonic stem cell bank. J Med Assoc Thai = Chotmaihet Thangphaet 89:1055–1063
25. Stacey G (2004) Fundamental issues for cell-line banks in biotechnology and regulatory affairs. In: Fuller BJ, Lane N, Benson EE (eds) Life in the frozen state. CRC Press, Boca Raton, pp 437–452
26. European Medicines Agency (Emea) (2008) Guideline on human cell-based medical products
27. FDA (2008) Content and review of Chemistry, Manufacturing, and Control (CMC) Information for human somatic cell therapy Investigational New Drug Applications (INDs)
28. WHO (2010) Recommendations for the evaluation of animal cell cultures as substrates for the manufacture of biological medicinal products and for the characterization of cell banks
29. Coecke S, Balls M, Bowe G et al (2005) Guidance on good cell culture practice. a report of the second ECVAM task force on good cell culture practice. Altern Lab Anim: ATLA 33:261–287
30. International Stem Cell Banking I (2009) Consensus guidance for banking and supply of human embryonic stem cell lines for research purposes. Stem Cell Rev 5:301–314
31. OECD (2007) OECD best practice guidelines for biological resource centres. Website http://www.oecd.org/sti/biotech/oecdbestpracticeguidelinesforbiologicalresourcecentres.htm. Accessed Mar 2016
32. FDA (2010) Good laboratory practice for nonclinical laboratory. Website http://www.reginfo.gov/public/do/eAgendaViewRule?pubId=201004&RIN=0910-AG47. Accessed Jan 2016
33. Niazi SK EU Guidelines to good manufacturing practice: medicinal products for human and veterinary use
34. FDA (2007) Guidance for industry: eligibility determination for donors of human cells, tissues, and cellular and tissue-based products (HCT/Ps)
35. Food, Drug Administration HHS (2004) Eligibility determination for donors of human cells, tissues, and cellular and tissue-based products. Final rule. Fed Regist 69:29785–29834
36. Anonymous (2009) Identity crisis. Nature 457:935–936
37. Asn- ATCCSDOW (2010) Cell line misidentification: the beginning of the end. Nat Rev Cancer 10:441–448
38. Lorsch JR, Collins FS, Lippincott-Schwartz J (2014) Cell biology. Fixing problems with cell lines. Science 346:1452–1453
39. Barallon R, Bauer SR, Butler J et al (2010) Recommendation of short tandem repeat profiling for authenticating human cell lines, stem cells, and tissues. In Vitro Cell Develop Biol Animal 46:727–732
40. Butler JM (2006) Genetics and genomics of core short tandem repeat loci used in human identity testing. J Forensic Sci 51:253–265
41. Dirks WG, Drexler HG (2013) STR DNA typing of human cell lines: detection of intra-and interspecies

cross-contamination. In: Basic cell culture protocols. Springer, Dordrecht, pp 27–38
42. Yu M, Selvaraj SK, Liang-Chu MM et al (2015) A resource for cell line authentication, annotation and quality control. Nature 520:307–311
43. China Food and Drug Administration (CFDA). http://eng.sfda.gov.cn/WS03/CL0755/. Accessed Jan 2016
44. FDA (1993) Points to consider in the characterization of cell lines used to produce biologicals. Center for Biologics Evaluation and Research, Rockville
45. FDA (1998) Guidance for industry Q5A viral safety evaluation of biotechnology products derived from cell lines of human or animal origin. U.S. Department of Health and Human Services, Food and Drug Administration, Center for Drug Evaluation and Research, Rockville
46. Macpherson I, Montagnier L (1964) Agar suspension culture for the selective assay of cells transformed by polyoma virus. Virology 23:291–294
47. Petricciani JC, Levenbook I, Locke R (1983) Human muscle: a model for the study of human neoplasia. Invest New Drugs 1:297–302
48. Ema ICH Topic Q 6 B specifications: test procedures and acceptance criteria for biotechnological/biological products
49. Emea/Chmp (2007) guideline on potency testing of cell-based immunotherapy medicinal products for the treatment of cancer
50. FDA (2011) Guidance for industry: potency tests for cellular and gene therapy products. U.S. Department of Health and Human Services, Food and Drug Administration, Center for Biologics Evaluation and Research, Rockville
51. Hunt CJ (2011) Cryopreservation of human stem cells for clinical application: a review. Transfusion Med Hemother 38:107–123
52. Andrews PW, Baker D, Benvinisty N et al (2015) Points to consider in the development of seed stocks of pluripotent stem cells for clinical applications: International Stem Cell Banking Initiative (ISCBI). Regen Med 10:1–44
53. Benson EE, Betsou F, Fuller BJ et al (2013) Translating cryobiology principles into trans-disciplinary storage guidelines for biorepositories and biobanks: a concept paper. CryoLetters 34:277–312
54. Ballesteros D, Estrelles E, Walters C et al (2011) Effect of storage temperature on green spore longevity for the ferns Equisetum ramosissimum and Osmunda regalis. CryoLetters 32:89–98
55. Baust J, Gage A, Klossner D et al (2007) Issues critical to the successful application of cryosurgical ablation of the prostate. Technol Cancer Res Treat 6:97–109
56. Tedder R, Zuckerman M, Brink N et al (1995) Hepatitis B transmission from contaminated cryopreservation tank. Lancet 346:137–140
57. Kyuwa S, Nishikawa T, Kaneko T et al (2003) Experimental evaluation of cross-contamination between cryotubes containing mouse 2-cell embryos and murine pathogens in liquid nitrogen tanks. Exp Anim 52:67–70
58. Day JG, Stacey G (2007) Cryopreservation and freeze-drying protocols. Springer, Dordrecht
59. Winter JM, Jacobson P, Bullough B et al (2014) Long-term effects of cryopreservation on clinically prepared hematopoietic progenitor cell products. Cytotherapy 16:965–975
60. Yong KW, Pingguan-Murphy B, Xu F et al (2015) Phenotypic and functional characterization of long-term cryopreserved human adipose-derived stem cells. Scientific reports
61. Chen YK, Liu QH, Li J et al (2010) Effect of long-term cryopreservation on physiological characteristics, antioxidant activities and lipid peroxidation of red seabream (Pagrus major) sperm. Cryobiology 61:189–193
62. Daley GQ, Scadden DT (2008) Prospects for stem cell-based therapy. Cell 132:544–548
63. ClinicalTrials.gov https://clinicaltrials.gov/ct2/results?term=ips+cell. Accessed Feb 2016
64. Dominici M, Le Blanc K, Mueller I et al (2006) Minimal criteria for defining multipotent mesenchymal stromal cells. The International Society for Cellular Therapy position statement. Cytotherapy 8:315–317
65. Freeman BT, Jung JP, Ogle BM (2015) Single-cell RNA-Seq of bone marrow-derived mesenchymal stem cells reveals unique profiles of lineage priming. PLoS One 10:e0136199
66. Samsonraj RM, Rai B, Sathiyanathan P et al (2015) Establishing criteria for human mesenchymal stem cell potency. Stem Cells 33:1878–1891

Biobanking: An Important Resource for Precision Medicine in Glioblastoma

4

Si Yan Melanie Tan, Edwin Sandanaraj,
Carol Tang*, and Beng Ti Christopher Ang*

Abstract

The Cancer Genome Atlas effort has generated significant interest in a new paradigm shift in tumor tissue analysis, patient diagnosis and subsequent treatment decision. Findings have highlighted the limitation of sole reliance on histology, which can be confounded by inter-observer variability. Such studies demonstrate that histologically similar grade IV brain tumors can be divided into four molecular subtypes based on gene expression, with each subtype demonstrating unique genomic aberrations and clinical outcome. These advances indicate that curative therapeutic strategies must now take into account the molecular information in tumor tissue, with the goal of

*Carol Tang and Beng Ti Christopher Ang are Co-corresponding authors for this chapter.

S.Y.M. Tan
Department of Research, National Neuroscience Institute, Singapore 308433, Singapore

School of Biological Sciences, Nanyang Technological University, Singapore 637551, Singapore
e-mail: melanie_sy_tan@nni.com.sg

E. Sandanaraj
Department of Research, National Neuroscience Institute, Singapore 308433, Singapore

School of Biological Sciences, Nanyang Technological University, Singapore 637551, Singapore

Singapore Institute for Clinical Sciences, Agency for Science, Technology and Research (A*STAR), Singapore 117609, Singapore
e-mail: Sandanaraj_Edwin@nni.com.sg

C. Tang (✉)
Department of Research, National Neuroscience Institute, Singapore 308433, Singapore

Humphrey Oei Institute of Cancer Research, National Cancer Centre, Singapore 169610, Singapore

Duke-National University of Singapore Medical School, Singapore 169857, Singapore
e-mail: carol_tang@nni.com.sg

B.T.C. Ang
Singapore Institute for Clinical Sciences, Agency for Science, Technology and Research (A*STAR), Singapore 117609, Singapore

Duke-National University of Singapore Medical School, Singapore 169857, Singapore

Department of Neurosurgery, National Neuroscience Institute, Singapore 308433, Singapore

Department of Physiology, Yong Loo Lin School of Medicine, National University of Singapore, Singapore 117593, Singapore
e-mail: ang.beng.ti@singhealth.com

F. Karimi-Busheri, M. Weinfeld (eds.), *Biobanking and Cryopreservation of Stem Cells*, Advances in Experimental Medicine and Biology 951, DOI 10.1007/978-3-319-45457-3_4

identifying molecularly stratified patients that will most likely to receive treatment benefit from targeted therapy. This in turn spares non-responders from chemotherapeutic side effects and financial costs. In advancing clinical stage drug candidates, the banking of brain tumor tissue necessitates the acquisition of not just tumor tissue with clinical history and robust follow-up, but also high quality molecular information such as somatic mutation, transcriptomic and DNA methylation profiles which have been shown to predict patient survival independent of current clinical indicators. Additionally, the derivation of cell lines from such tumor tissue facilitates the development of clinically relevant patient-derived xenograft mouse models that can prospectively reform the tumor for further studies, yet have retrospective clinical history to associate bench and *in vivo* findings with clinical data. This represents a core capability of Precision Medicine where the focus is on understanding inter- and intra-tumor heterogeneity so as to best tailor therapies that will result in improved treatment outcomes.

Keywords
Glioblastoma • Histology • Glioma-propagating cells • TCGA • Patient-derived xenografts • Tumor resource • Precision medicine • Connectivity map • Patient stratification • Bioinformatics

Abbreviations

bFGF	Basic fibroblast growth factor
CMAP	Connectivity map
CNS	Central nervous system
EGF	Epidermal growth factor
GBM	Glioblastoma multiforme
GEMM	Genetically engineered mouse model
GPCs	Glioma-propagating cells
MRI	Magnetic resonance imaging
NIH	National Institutes of Health
PDX	Patient-derived xenograft
TCGA	The Cancer Genome Atlas

4.1 Introduction

Glioblastoma multiforme (GBM) remains a cancer with the worst prognosis. Patients often show a median survival period of 15 months, even with advanced surgical intervention, chemotherapy and radiation treatment [1]. Temozolomide, the standard of care drug in the clinic, has annual global sales of US$ 1 billion, yet it merely extends survival by 3 months. Among the reasons for the highly invasive and recurrent nature of the disease lies in the cellular and molecular heterogeneity of GBM. Recent deep molecular profiling efforts such as The Cancer Genome Atlas (TCGA) demonstrated that histologically identical GBM tumors are molecularly heterogeneous, further suggesting that their regulatory pathway networks determine each tumor's sensitivity to targeted therapeutic approaches [2, 3]. While these computational analyses reveal the putative mechanisms underlying tumor resistance and recurrence, biological or functional validation in preclinical animal models is lacking. In the past decade, the use of mouse xenograft models created from commercially procured serum-grown glioma cells has been challenged by studies demonstrating that such xenografted tumors fail to recapitulate the patient's original tumor morphology and transcriptomic profiles [4]. In addition, such *in vitro* serially passaged serum-grown cells often contain genomic aberrations not found in the original primary tumor [5, 6]. Indeed, the widely used US National Cancer Institute NCI-60 panel of human cancer cell lines commonly passaged in serum-containing media will soon be decommissioned, with an aim to launch a rejuvenated repository of cancer models that are derived from fresh patient samples and tagged with details about their clinical past [7].

Glioma-propagating cells (GPCs) have been isolated from malignant brain tumors, and cultivated as spheroid structures in serum-free media supplemented with growth factors [8]. This media composition is similar to that used to passage neural stem cells, and helps promote self-renewal and tumorigenicity of GPCs [9]. In contrast, the addition of serum encourages differentiation of GPCs, resulting in cessation of cell proliferation with subsequent involution of tumor growth. We and others previously demonstrated the preservation of karyotypic hallmarks in these cells, similar to the original tumors [6, 10]. We developed a method to passage these spheroid structures by mechanical trituration, thus avoiding constant use of harsh enzymatic solutions that have been shown to alter karyotypic patterns in human embryonic stem cells grown as embryoid bodies [11]. We discuss vitrification, a technique adapted from *in vitro* fertilization procedures, and emphasize the importance of preserving these spheroid structures with reduced water content to prevent damage from ice crystals during the thawing process. Collectively, these methodologies preserve the integrity of GPCs and their ability to establish patient-derived xenograft (PDX) tumors that faithfully phenocopy the original tumor pathophysiology, cytogenetic and transcriptomic profiles. This capability reflects the importance of establishing a cell line and tumor tissue bank that presents the most clinically relevant resource to test and validate computational predictions generated from large patient glioma databases.

Crucial to establishing clinical relevance of our brain tumor resource, we discuss computational platforms such as the Connectivity Map to link molecular data acquired from *in vitro* and animal studies with multi-dimensional clinical information such as the patient's age, tumor grade, molecular information, Karnofsky score and magnetic resonance imaging (MRI) scans available in large clinical databases such as TCGA [12]. These computational advances have expanded the scope of studies made possible with our brain tumor resource [13–17]. Importantly, we now have a brain tumor biobank that facilitates studies in Precision Medicine where biological validation of patient-centric predictions is achievable.

4.2 Brain Tumor Resource

With the advance of TCGA efforts, several brain tumor banks containing low-passage cell lines have been established with a focus on acquisition of clinical history with deep content molecular information (genotype-phenotype databases). The goal of such tumor banks is to facilitate studies requiring capability to remodel the disease accurately so as to prospectively test computational predictions based on patient information [18]. Central to this biobank creation, we previously demonstrated three important criteria: (1) Growth of patient-derived glioma cells in serum-free media supplemented with growth factors, (2) Vitrification as a cryopreservation method, and (3) Ability to establish orthotopic PDX models that recapitulate the patient's original tumor pathophysiology. As such, genotype–phenotype databases previously needed only simple computing technologies, including very basic data fields relating to pathogenicity, but did not capture the process of pathogenicity interpretation. Going forward, this approach will have to change, especially if we wish to deliver truly Precision Medicine-based findings, which will require mechanistic in addition to probabilistic modeling, and hence even more sophisticated sources of input information and tools for the recording of results.

GPCs can be maintained and propagated as tumor neurosphere cultures in defined serum-free condition supplemented with epidermal growth factor (EGF) and basic fibroblast growth factor (bFGF), a paradigm that is adopted from the traditional neurosphere culture [19, 20]. Furthermore, Lee and colleagues have shown that tumor stem-like cells grown in serum-free condition closely mimic the genotype, transcriptomic profile and morphological features of their parental tumors [6]. Thus, the establishment of a tumor neurosphere repository with preservation of essential features of tumor heterogeneity would provide a clinically relevant resource to investigate the disease. Such a method would also allow us to return to the same experimental cell line passages to reduce variability in experimental replication. Recent efforts by the Roadmap Epigenomics Project spearheaded by the National Institutes of Health (NIH) highlighted epigenetic silencing differences of *in vitro* cultured cells when compared to similar cell lineages in primary tissue, thus underscoring the importance of cryopreserving low-passage GPCs with retention of its molecular fingerprint [21]. In many studies involving the prospective isolation of tumor-propagating cells, only small amounts of clinical material are available, and this limitation is compounded by lack of appropriate methods to preserve such cells at convenient time points. Although *in vivo* serial passage of GPC-derived tumors has been described as the most reliable method to preserve the cells, practically, it is not always feasible to have access to suitably-aged immunocompromised mice [8]. We developed mechanical trituration as a method to passage three-dimensional spheroid cultures for the reason that acute dissociation of such structures into single cells promotes cellular senescence over an extended period [9]. In addition, induction of cellular differentiation, such as by the presence of serum, results in loss of tumorigenicity and consequent involution of tumor growth [22].

We adapted a method used in cryopreservation of embryos from *in vitro* fertilization procedures. Vitrification is a process of glass-like solidification in which an aqueous solution is prevented from crystallization by rapid cooling [23]. This method has been commonly used for the cryopreservation of embryos at different developmental stages from various species such as murine, rabbit, sheep and bovine [24–29]. Furthermore, human and mouse multi-cell embryos have been successfully cryopreserved using this strategy [30]. This highlights the feasibility of cryopreserving cellular aggregates. In addition, it has been demonstrated that vitrified embryonic stem cells retained their pluripotency, cytogenetic profile and viability upon thawing [31]. Taken together, vitrification could provide an effective means of storage of brain tumor-propagating cells cultured as spherical structures. Although adherent GPC cultures using laminin have been proposed, these growth conditions resulted in transcriptomic shift of the cells [32]. In support, Dirks and colleagues showed that a chemical genetics screen utilizing GPC spheroid cultures identified small molecules affecting neurotransmission in the adult central nervous system (CNS), thus suggesting that clinically approved neuromodulators may remodel the mature CNS and find application in the treatment of brain cancer [33].

To facilitate testing of patient stratification methods and identification of molecular mechanisms contributing to disease progression, a genomic roadmap is created to characterize the cells, tumor xenografts and primary tumors. The pivotal goal of molecular profiling of patient-derived cells aims to stratify patient cohorts and identify amenable therapeutic strategies. The content-rich patient tumor molecular profiles need to be systematically archived for efficient data mining to evaluate proof-of-concept studies. Initial steps at identifying potential cancer-specific biomarkers require patient-centric bioinformatics interrogation, information from which can then be used for the analysis of samples stored in tissue banks. In TCGA, the quality of samples acquired was assessed from several participating tissue banks that surprisingly showed only one percent of the samples being reliable for downstream molecular data acquisition [2, 34, 35]. Sample quality and the associated clinical information are important factors in tissue banking. The importance in having expertise at tissue sampling and culturing of patient-derived cells, with preservation of cytogenetic and transcriptomic hallmarks found in the original primary tumor has previously been reported by our colleagues [6, 10]. Precision Medicine-driven studies, dependent on this molecular heterogeneity, warrant a preclinical mouse model that recapitulates the patient's original tumor pathophysiology and aids at advancing chemotherapeutic candidates into clinical trials.

4.3 Animal Model Established from the Biobank: An Informative Preclinical Mouse Model

Modeling brain tumors in animals reveals genetic events and molecular mechanisms that contribute to oncogenesis. The mouse shares extensive molecular and physiological similarities to human beings and is a powerful tool for studying cancer [36]. Unlike invertebrate model systems, tumor development in mice is accompanied by other complex processes such as angiogenesis and metastasis, similar to those in human cancer [37]. More importantly, mouse tumor models provide a temporal perspective and genetically-controlled system for studying the tumorigenic process, as well as response to specific therapies.

Genetically engineered mouse models (GEMMs) are a popular model to study tumor biology [37]. There are several limitations to using GEMMs as more than one driver mutation is required to initiate tumorigenesis [38, 39]. The expression of the transgene is often elevated to levels that exceed those in patients. Tumors that arise in this model are often sporadic, resulting in difficulty of study designs that require significant animal numbers for reproducibility. TCGA efforts have also demonstrated that the spectrum of driver mutations differs significantly among patients, thus the relevance of a particular GEMM may be limited [2, 40]. Nevertheless, GEMMs will likely provide useful insight into the tumor cell-of-origin and initiating events.

Xenograft models established from commercially procured serum-grown cell lines date back to the late 1960s. However, several studies demonstrated that such xenografted tumors exhibit significant morphological and molecular features not found in the original primary tissue [4, 6]. Such issues have subsequently been overcome through the development of PDX models. Several studies revealed that the orthotopic xenograft model established from patient-derived glioma cell lines or tumor explants bear more clinical relevance [41–43]. This is due to the presence of the microenvironment provided by the normal brain parenchyma, where better measurement of drug delivery and clearance kinetics can be evaluated. The orthotopic PDX model is useful as tumor formation with high incidence and the ability to generate large cohorts of animals in preclinical studies are attainable. Several investigators have provided evidence that PDX tumors phenocopy the pathophysiology and molecular features of their parental tumor [6, 44]. Importantly, this model allows assessment of therapeutic responses using stratified clinical material, a core capability of Precision Medicine. We and others previously integrated the use of such an in-house brain tumor resource to interrogate lab findings in clinical glioma databases [7, 10, 14–17]. We showed that transcriptomic patterns derived from *in vitro* drug-treated or genetically manipulated cells mapped to patient clinical databases, and dictated primary tumor phenotype.

Despite the frequent use of the PDX model in studying drug therapy response, we recognize that immunogenic and microenvironmental factors may not be fully represented. Additionally, engraftment inefficiency can be as high as 90 %, depending on the type of cancer [45]. Such limitations suggest that the use of mouse models should be carefully considered to provide maximum information about the study question.

4.4 Enabling Precision Medicine-Based Studies: Highlighting the Importance of a Biobank

A brain tumor biobank that merges multi-platform data from patient material (cells, xenograft and primary tumors) and seeks to address clinical and imaging phenotypes with molecular data will advance studies in stratified medicine. The development and preclinical validation of novel anti-cancer drugs require low-passage cell lines that are representative, scalable and reproducible in experimental models. In the absence of an integrated human brain biobank, research findings from *in vitro* and *in vivo* models of neurological disorders cannot be functionally validated in the actual disease context. TCGA efforts revealed that gene expression drives GBM disease progression and survival outcome [3, 46–48]. Although an important prognostic factor of glioma progression relies on the World Health Organization (WHO) grading scheme, the wide differences in treatment response and survival suggest that the aggressiveness of treatment cannot be decided just by histology alone. These findings underscore the limitation of relying solely on morphological criteria to diagnose patients. Future brain tumor classification must now include molecular information which in turn guides diagnosis and subsequent treatment decision [49]. Development and maintenance of biobanks as an international resource for the study of human diseases provides the scientific community with well-characterized cells and rich phenotypic data. Such resources facilitate prospective remodeling of the disease in mouse models, with retrospective clinical information to evaluate correlation patterns and directly validate the mechanism.

Targeted therapy is an attractive approach to overcome the highly infiltrative and recurrent nature of GBM. Recent characterization of the epigenome, somatic mutation profile and transcriptome of tumor tissue has now provided a deeper understanding of the alterations underlying the disease phenotype [2, 3, 50, 51]. Tumor cells are assessed for the underlying pattern using unsupervised computational approaches to discern their molecular heterogeneity. An important evaluation is to computationally identify regulatory pathways that can be targeted with small molecule drugs. These predictions are then functionally validated in PDX models. Transcriptomic resources of xenograft and primary tumors are scrutinized for highly variable genes which can reflect a common bias present among primary tumors and xenografts established from GPCs. This common and systematic bias can be controlled and nullified by statistical algorithms such as Anova-based batch effect removal and principal component based analysis [52, 53]. A scatter plot accounting for major principal components across matched primary and orthotopic tumors demonstrates close proximity of matched samples, indicating that GPCs can recreate the original tumor molecular profile *in vivo* (Fig. 4.1).

Computational evaluation of matched molecular data from biobanked patient tumors and GPCs requires systematic validation of molecular profiles using bioinformatics approaches. A computational pipeline is required to interrogate gene signatures (transcriptomic classifiers) derived from GPCs and a large set of independent predictive database collection from patients' molecular data [12, 54, 55]. Molecular perturbation experiments on GPCs have demonstrated promising evidence in conferring the patient's prognosis in predictive databases and *in vivo* experiments [13, 15]. The most comprehensive glioma patient's database was established by TCGA, where key components of clinical phenotype and genomics information were catalogued into a multi-tier organizational structure for 33 different tumor types [56]. Each tier is confined with a data structure from genomic platforms including somatic mutation, copy number, methylation, transcriptomic and proteomic technologies. Importantly, we have merged our cell and xenograft tumor

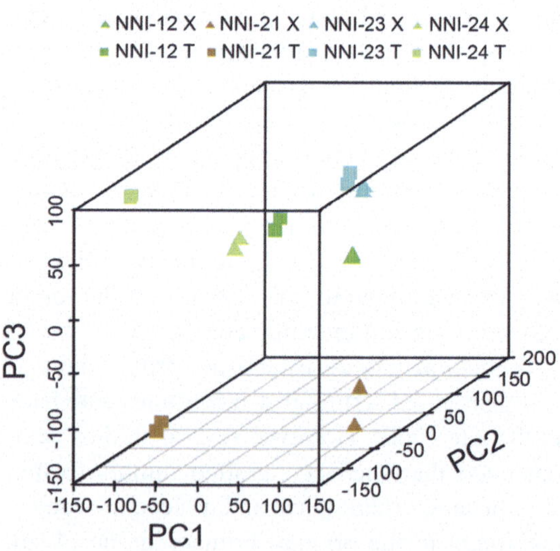

Fig. 4.1 Recapturing molecular portrait of primary tumors in orthotopic xenograft tumors derived from GPCs. Principal Component Analysis (PCA) map demonstrates similar transcriptomic profiles between matched xenograft and primary tumors. Each *color* denotes similar patient material with corresponding xenograft tumor. *Triangle*, xenograft tumor; *square*, patient tumor

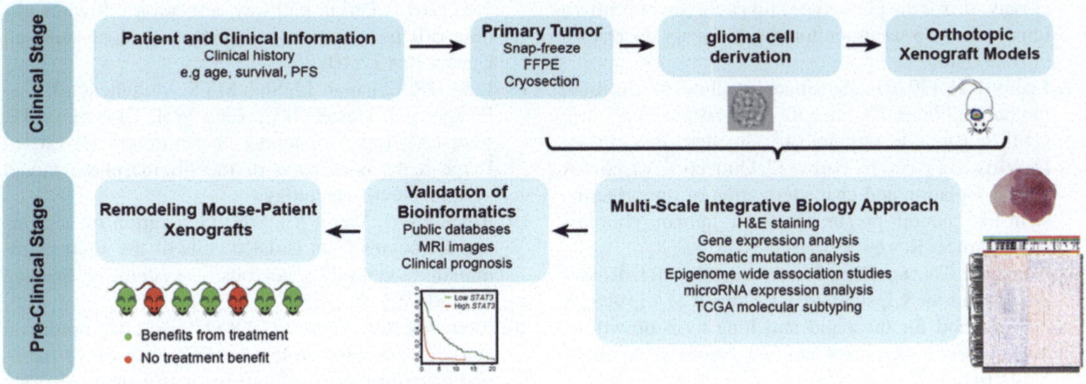

Fig. 4.2 Graphical summary for brain tumor resource. A brain tumor resource allows orthotopic xenograft tumors to be re-established from clinical material, thus facilitating future testing of small molecules identified through molecular stratification approaches

molecular patterns with international collections by adapting the statistical framework to control for systematic batch effects across different collections [52, 53]. This integrated evaluation also confirms the consistency across tumor cell types and provides greater statistical power [57]. We have successfully adapted a qualitative enrichment pipeline, the Connectivity Map (CMAP) to interrogate active pathway programs coded as gene signatures in our biobanked cells with our patients' transcriptomic patterns [12, 13, 17]. These patients' prognoses can then be retrospectively predicted by mapping survival and clinical response parameters with enrichment scores from the CMAP pipeline. Thus, molecular perturbation experiments tapping into our brain tumor biobank serve as effective tools to biologically validate these patient-centric computational predictions.

4.5 Conclusion

Biobanking coupled with deep molecular characterization is a core capability for Precision Medicine-based studies (Fig. 4.2). The ability to remodel brain tumors in mice facilitates biological and functional validation of computationally predicted pathway networks that should be therapeutically targeted.

References

1. Louis DN, Ohgaki H, Wiestler OD, Cavenee WK, Burger PC, Jouvet A, Scheithauer BW, Kleihues P (2007) The 2007 WHO classification of tumours of the central nervous system. Acta Neuropathol 114(2):97–109
2. Cancer Genome Atlas Research N (2008) Comprehensive genomic characterization defines human glioblastoma genes and core pathways. Nature 455(7216):1061–1068
3. Verhaak RG, Hoadley KA, Purdom E, Wang V, Qi Y, Wilkerson MD, Miller CR, Ding L, Golub T, Mesirov JP, Alexe G, Lawrence M, O'Kelly M, Tamayo P, Weir BA, Gabriel S, Winckler W, Gupta S, Jakkula L, Feiler HS, Hodgson JG, James CD, Sarkaria JN, Brennan C, Kahn A, Spellman PT, Wilson RK, Speed TP, Gray JW, Meyerson M, Getz G, Perou CM, Hayes DN, Cancer Genome Atlas Research N (2010) Integrated genomic analysis identifies clinically relevant subtypes of glioblastoma characterized by abnormalities in PDGFRA, IDH1, EGFR, and NF1. Cancer Cell 17(1):98–110
4. Hodgson JG, Yeh RF, Ray A, Wang NJ, Smirnov I, Yu M, Hariono S, Silber J, Feiler HS, Gray JW, Spellman PT, Vandenberg SR, Berger MS, James CD (2009) Comparative analyses of gene copy number and mRNA expression in glioblastoma multiforme tumors and xenografts. Neuro Oncol 11(5):477–487
5. Clark MJ, Homer N, O'Connor BD, Chen Z, Eskin A, Lee H, Merriman B, Nelson SF (2010) U87MG decoded: the genomic sequence of a cytogenetically aberrant human cancer cell line. PLoS Genet 6(1):e1000832
6. Lee J, Kotliarova S, Kotliarov Y, Li A, Su Q, Donin NM, Pastorino S, Purow BW, Christopher N, Zhang W, Park JK, Fine HA (2006) Tumor stem cells derived from glioblastomas cultured in bFGF and EGF more

closely mirror the phenotype and genotype of primary tumors than do serum-cultured cell lines. Cancer Cell 9(5):391–403
7. Ledford H (2016) US cancer institute to overhaul tumour cell lines. Nature 530(7591):391
8. Galli R, Binda E, Orfanelli U, Cipelletti B, Gritti A, De Vitis S, Fiocco R, Foroni C, Dimeco F, Vescovi A (2004) Isolation and characterization of tumorigenic, stem-like neural precursors from human glioblastoma. Cancer Res 64(19):7011–7021
9. Svendsen CN, ter Borg MG, Armstrong RJ, Rosser AE, Chandran S, Ostenfeld T, Caldwell MA (1998) A new method for the rapid and long term growth of human neural precursor cells. J Neurosci Methods 85(2):141–152
10. Chong YK, Toh TB, Zaiden N, Poonepalli A, Leong SH, Ong CE, Yu Y, Tan PB, See SJ, Ng WH, Ng I, Hande MP, Kon OL, Ang BT, Tang C (2009) Cryopreservation of neurospheres derived from human glioblastoma multiforme. Stem Cells 27(1):29–39
11. Pera MF (2004) Unnatural selection of cultured human ES cells? Nat Biotechnol 22(1):42–43
12. Lamb J, Crawford ED, Peck D, Modell JW, Blat IC, Wrobel MJ, Lerner J, Brunet JP, Subramanian A, Ross KN, Reich M, Hieronymus H, Wei G, Armstrong SA, Haggarty SJ, Clemons PA, Wei R, Carr SA, Lander ES, Golub TR (2006) The connectivity map: using gene-expression signatures to connect small molecules, genes, and disease. Science 313(5795):1929–1935
13. Chong YK, Sandanaraj E, Koh LW, Thangaveloo M, Tan MS, Koh GR, Toh TB, Lim GG, Holbrook JD, Kon OL, Nadarajah M, Ng I, Ng WH, Tan NS, Lim KL, Tang C, Ang BT (2016) ST3GAL1-associated transcriptomic program in glioblastoma tumor growth, invasion, and prognosis. J Natl Cancer Inst 108(2)
14. Choudhury Y, Tay FC, Lam DH, Sandanaraj E, Tang C, Ang B-T, Wang S (2012) Attenuated adenosine-to-inosine editing of microRNA-376a* promotes invasiveness of glioblastoma cells. J Clin Invest 122(11):4059–4076
15. Koh LW, Koh GR, Ng FS, Toh TB, Sandanaraj E, Chong YK, Phong M, Tucker-Kellogg G, Kon OL, Ng WH, Ng IH, Clement MV, Pervaiz S, Ang BT, Tang CS (2013) A distinct reactive oxygen species profile confers chemoresistance in glioma-propagating cells and associates with patient survival outcome. Antioxid Redox Signal 19(18):2261–2279
16. Ng FS, Toh TB, Ting EH, Koh GR, Sandanaraj E, Phong M, Wong SS, Leong SH, Kon OL, Tucker-Kellogg G, Ng WH, Ng I, Tang C, Ang BT (2012) Progenitor-like traits contribute to patient survival and prognosis in oligodendroglial tumors. Clin Cancer Res 18(15):4122–4135
17. Yeo CWS, Ng FSL, Chai C, Tan JMM, Koh GRH, Chong YK, Koh LWH, Foong CSF, Sandanaraj E, Holbrook JD, Ang B-T, Takahashi R, Tang C, Lim K-L (2012) Parkin pathway activation mitigates glioma cell proliferation and predicts patient survival. Cancer Res 72(10):2543–2553
18. Laks DR, Crisman TJ, Shih MYS, Mottahedeh J, Gao F, Sperry J, Garrett MC, Yong WH, Cloughesy TF, Liau LM, Lai A, Coppola G, Kornblum HI (2016) Large-scale assessment of the gliomasphere model system. Neuro-Oncology
19. Reynolds B, Weiss S (1992) Generation of neurons and astrocytes from isolated cells of the adult mammalian central nervous system. Science 255(5052):1707–1710
20. Reynolds BA, Weiss S (1996) Clonal and population analyses demonstrate that an EGF-responsive mammalian embryonic CNS precursor is a stem cell. Dev Biol 175(1):1–13
21. Roadmap Epigenomics C, Kundaje A, Meuleman W, Ernst J, Bilenky M, Yen A, Heravi-Moussavi A, Kheradpour P, Zhang Z, Wang J, Ziller MJ, Amin V, Whitaker JW, Schultz MD, Ward LD, Sarkar A, Quon G, Sandstrom RS, Eaton ML, Wu YC, Pfenning AR, Wang X, Claussnitzer M, Liu Y, Coarfa C, Harris RA, Shoresh N, Epstein CB, Gjoneska E, Leung D, Xie W, Hawkins RD, Lister R, Hong C, Gascard P, Mungall AJ, Moore R, Chuah E, Tam A, Canfield TK, Hansen RS, Kaul R, Sabo PJ, Bansal MS, Carles A, Dixon JR, Farh KH, Feizi S, Karlic R, Kim AR, Kulkarni A, Li D, Lowdon R, Elliott G, Mercer TR, Neph SJ, Onuchic V, Polak P, Rajagopal N, Ray P, Sallari RC, Siebenthall KT, Sinnott-Armstrong NA, Stevens M, Thurman RE, Wu J, Zhang B, Zhou X, Beaudet AE, Boyer LA, De Jager PL, Farnham PJ, Fisher SJ, Haussler D, Jones SJ, Li W, Marra MA, McManus MT, Sunyaev S, Thomson JA, Tlsty TD, Tsai LH, Wang W, Waterland RA, Zhang MQ, Chadwick LH, Bernstein BE, Costello JF, Ecker JR, Hirst M, Meissner A, Milosavljevic A, Ren B, Stamatoyannopoulos JA, Wang T, Kellis M (2015) Integrative analysis of 111 reference human epigenomes. Nature 518(7539):317–330
22. Piccirillo SGM, Reynolds BA, Zanetti N, Lamorte G, Binda E, Broggi G, Brem H, Olivi A, Dimeco F, Vescovi AL (2006) Bone morphogenetic proteins inhibit the tumorigenic potential of human brain tumour-initiating cells. Nature 444(7120):761–765
23. Rall WF, Wood MJ, Kirby C, Whittingham DG (1987) Development of mouse embryos cryopreserved by vitrification. J Reprod Fertil 80(2):499–504
24. Ali J, Shelton JN (1993) Design of vitrification solutions for the cryopreservation of embryos. J Reprod Fertil 99(2):471–477
25. Kasai M, Komi JH, Takakamo A, Tsudera H, Sakurai T, Machida T (1990) A simple method for mouse embryo cryopreservation in a low toxicity vitrification solution, without appreciable loss of viability. J Reprod Fertil 89(1):91–97
26. Kasai M, Nishimori M, Zhu SE, Sakurai T, Machida T (1992) Survival of mouse morulae vitrified in an ethylene glycol-based solution after exposure to the solu-

tion at various temperatures. Biol Reprod 47(6):1134–1139
27. Saha S, Otoi T, Takagi M, Boediono A, Sumantri C, Suzuki T (1996) Normal calves obtained after direct transfer of vitrified bovine embryos using ethylene glycol, trehalose, and polyvinylpyrrolidone. Cryobiology 33(3):291–299
28. Ali J, Shelton JN (1993) Successful vitrification of day-6 sheep embryos. J Reprod Fertil 99(1):65–70
29. Kasai M, Hamaguchi Y, Zhu SE, Miyake T, Sakurai T, Machida T (1992) High survival of rabbit morulae after vitrification in an ethylene glycol-based solution by a simple method. Biol Reprod 46(6):1042–1046
30. Mukaida T, Wada S, Takahashi K, Pedro PB, An TZ, Kasai M (1998) Vitrification of human embryos based on the assessment of suitable conditions for 8-cell mouse embryos. Hum Reprod 13(10):2874–2879
31. Reubinoff BE, Pera MF, Vajta G, Trounson AO (2001) Effective cryopreservation of human embryonic stem cells by the open pulled straw vitrification method. Hum Reprod 16(10):2187–2194
32. Pollard SM, Yoshikawa K, Clarke ID, Danovi D, Stricker S, Russell R, Bayani J, Head R, Lee M, Bernstein M, Squire JA, Smith A, Dirks P (2009) Glioma stem cell lines expanded in adherent culture have tumor-specific phenotypes and are suitable for chemical and genetic screens. Cell Stem Cell 4(6):568–580
33. Diamandis P, Wildenhain J, Clarke ID, Sacher AG, Graham J, Bellows DS, Ling EKM, Ward RJ, Jamieson LG, Tyers M, Dirks PB (2007) Chemical genetics reveals a complex functional ground state of neural stem cells. Nat Chem Biol 3(5):268–273
34. Scott CT, Caulfield T, Borgelt E, Illes J (2012) Personal medicine -the new banking crisis. Nat Biotech 30(2):141–147
35. Suh KS, Sarojini S, Youssif M, Nalley K, Milinovikj N, Elloumi F, Russell S, Pecora A, Schecter E, Goy A (2013) Tissue banking, bioinformatics, and electronic medical records: the front-end requirements for personalized medicine. J Oncol 2013:368751
36. Rosenthal N, Brown S (2007) The mouse ascending: perspectives for human-disease models. Nat Cell Biol 9(9):993–999
37. Wee B, Charles N, Holland EC (2011) Animal models to study cancer-initiating cells from glioblastoma. Front Biosci (Landmark Ed) 16:2243–2258
38. Alcantara Llaguno S, Chen J, Kwon C-H, Jackson EL, Li Y, Burns DK, Alvarez-Buylla A, Parada LF (2009) Malignant astrocytomas originate from neural stem/progenitor cells in a somatic tumor suppressor mouse model. Cancer Cell 15(1):45–56
39. Zheng H, Ying H, Yan H, Kimmelman AC, Hiller DJ, Chen AJ, Perry SR, Tonon G, Chu GC, Ding Z, Stommel JM, Dunn KL, Wiedemeyer R, You MJ, Brennan C, Wang YA, Ligon KL, Wong WH, Chin L, DePinho RA (2008) p53 and Pten control neural and glioma stem/progenitor cell renewal and differentiation. Nature 455(7216):1129–1133
40. Network TC (2013) Corrigendum: comprehensive genomic characterization defines human glioblastoma genes and core pathways. Nature 494(7438):506
41. Shapiro WR, Basler GA, Chernik NL, Posner JB (1979) Human brain tumor transplantation into nude mice. J Natl Cancer Inst 62(3):447–453
42. Kaye AH, Morstyn G, Gardner I, Pyke K (1986) Development of a xenograft glioma model in mouse brain. Cancer Res 46(3):1367–1373
43. Horten BC, Basler GA, Shapiro WR (1981) Xenograft of human malignant glial tumors into brains of nude mice. A histopatholgical study. J Neuropathol Exp Neurol 40(5):493–511
44. Joo KM, Kim J, Jin J, Kim M, Seol HJ, Muradov J, Yang H, Choi YL, Park WY, Kong DS, Lee JI, Ko YH, Woo HG, Lee J, Kim S, Nam DH (2013) Patient-specific orthotopic glioblastoma xenograft models recapitulate the histopathology and biology of human glioblastomas in situ. Cell Rep 3(1):260–273
45. Herter-Sprie GS, Kung AL, Wong KK (2013) New cast for a new era: preclinical cancer drug development revisited. J Clin Invest 123(9):3639–3645
46. Gravendeel LA, Kouwenhoven MC, Gevaert O, de Rooi JJ, Stubbs AP, Duijm JE, Daemen A, Bleeker FE, Bralten LB, Kloosterhof NK, De Moor B, Eilers PH, van der Spek PJ, Kros JM, Sillevis Smitt PA, van den Bent MJ, French PJ (2009) Intrinsic gene expression profiles of gliomas are a better predictor of survival than histology. Cancer Res 69(23):9065–9072
47. Li A, Walling J, Ahn S, Kotliarov Y, Su Q, Quezado M, Oberholtzer JC, Park J, Zenklusen JC, Fine HA (2009) Unsupervised analysis of transcriptomic profiles reveals six glioma subtypes. Cancer Res 69(5):2091–2099
48. Phillips HS, Kharbanda S, Chen R, Forrest WF, Soriano RH, Wu TD, Misra A, Nigro JM, Colman H, Soroceanu L, Williams PM, Modrusan Z, Feuerstein BG, Aldape K (2006) Molecular subclasses of high-grade glioma predict prognosis, delineate a pattern of disease progression, and resemble stages in neurogenesis. Cancer Cell 9(3):157–173
49. Louis DN, Perry A, Burger P, Ellison DW, Reifenberger G, von Deimling A, Aldape K, Brat D, Collins VP, Eberhart C, Figarella-Branger D, Fuller GN, Giangaspero F, Giannini C, Hawkins C, Kleihues P, Korshunov A, Kros JM, Beatriz Lopes M, Ng HK, Ohgaki H, Paulus W, Pietsch T, Rosenblum M, Rushing E, Soylemezoglu F, Wiestler O, Wesseling P, International Society of N-H (2014) International Society of Neuropathology – Haarlem consensus guidelines for nervous system tumor classification and grading. Brain Pathol 24(5):429–435
50. Brennan CW, Verhaak RG, McKenna A, Campos B, Noushmehr H, Salama SR, Zheng S, Chakravarty D, Sanborn JZ, Berman SH, Beroukhim R, Bernard B, Wu CJ, Genovese G, Shmulevich I, Barnholtz-Sloan J, Zou L, Vegesna R, Shukla SA, Ciriello G, Yung WK, Zhang W, Sougnez C, Mikkelsen T, Aldape K, Bigner DD, Van Meir EG, Prados M, Sloan A, Black

KL, Eschbacher J, Finocchiaro G, Friedman W, Andrews DW, Guha A, Iacocca M, O'Neill BP, Foltz G, Myers J, Weisenberger DJ, Penny R, Kucherlapati R, Perou CM, Hayes DN, Gibbs R, Marra M, Mills GB, Lander E, Spellman P, Wilson R, Sander C, Weinstein J, Meyerson M, Gabriel S, Laird PW, Haussler D, Getz G, Chin L, Network TR (2013) The somatic genomic landscape of glioblastoma. Cell 155(2):462–477

51. Nevo I, Woolard K, Cam M, Li A, Webster JD, Kotliarov Y, Kim HS, Ahn S, Walling J, Kotliarova S, Belova G, Song H, Bailey R, Zhang W, Fine HA (2014) Identification of molecular pathways facilitating glioma cell invasion in situ. PLoS One 9(11):e111783
52. Johnson WE, Li C, Rabinovic A (2007) Adjusting batch effects in microarray expression data using empirical Bayes methods. Biostatistics 8(1):118–127
53. Reese SE, Archer KJ, Therneau TM, Atkinson EJ, Vachon CM, de Andrade M, Kocher JP, Eckel-Passow JE (2013) A new statistic for identifying batch effects in high-throughput genomic data that uses guided principal component analysis. Bioinformatics 29(22):2877–2883
54. Ritchie ME, Phipson B, Wu D, Hu Y, Law CW, Shi W, Smyth GK (2015) Limma powers differential expression analyses for RNA-sequencing and microarray studies. Nucleic Acids Res 43(7):e47
55. Tibshirani RJ, Efron B (2002) Pre-validation and inference in microarrays. Stat Appl Genet Mol Biol 1:Article 1. doi:10.2202/1544-6115.1000
56. Marx V (2013) Drilling into big cancer-genome data. Nat Meth 10(4):293–297
57. Lottaz C, Beier D, Meyer K, Kumar P, Hermann A, Schwarz J, Junker M, Oefner PJ, Bogdahn U, Wischhusen J, Spang R, Storch A, Beier CP (2010) Transcriptional profiles of CD133+ and CD133- glioblastoma-derived cancer stem cell lines suggest different cells of origin. Cancer Res 70(5):2030–2040

Slow Cooling Cryopreservation Optimized to Human Pluripotent Stem Cells

5

Takamichi Miyazaki and Hirofumi Suemori

Abstract

Human pluripotent stem cells (hPSCs) have the potential for unlimited expansion and differentiation into cells that form all three germ layers. Cryopreservation is one of the key processes for successful applications of hPSCs, because it allows semi-permanent preservation of cells and their easy transportation. Most animal cell lines, including mouse embryonic stem cells, are standardly cryopreserved by slow cooling; however, hPSCs have been difficult to preserve and their cell viability has been extremely low whenever cryopreservation has been attempted.

Here, we investigate the reasons for failure of slow cooling in hPSC cryopreservation. Cryopreservation involves a series of steps and is not a straightforward process. Cells may die due to various reasons during cryopreservation. Indeed, hPSCs preserved by traditional methods often suffer necrosis during the freeze-thawing stages, and the colony state of hPSCs prior to cryopreservation is a major factor contributing to cell death.

It has now become possible to cryopreserve hPSCs using conventional cryopreservation methods without any specific equipment. This review summarizes the advances in this area and discusses the optimization of slow cooling cryopreservation for hPSC storage.

Keywords

Human pluripotent stem cell • Human induced pluripotent stem cells • Human embryonic stem cell • Cryopreservation • Slow-cooling • Vitrification

T. Miyazaki
Institute for Integrated Cell-Material Sciences, Kyoto University, Kyoto, Japan
e-mail: tmiyazaki@frontier.kyoto-u.ac.jp

H. Suemori (✉)
Institute for Frontier Medical Sciences, Kyoto University, Kyoto, Japan
e-mail: hsuemori@frontier.kyoto-u.ac.jp

F. Karimi-Busheri, M. Weinfeld (eds.), *Biobanking and Cryopreservation of Stem Cells*, Advances in Experimental Medicine and Biology 951, DOI 10.1007/978-3-319-45457-3_5

Abbreviation

hPSC	Human pluripotent stem cell
hESC	Human embryonic stem cell
hiPSC	Human induced pluripotent stem cell
DMSO	Dimethyl sulfoxide
ROCK	Rho-associated coiled-coil forming kinase
MEF	Mouse embryonic fibroblast
EG	Ethylene glycol
KSR	Knockout serum replacement
FCS	Fetal calf serum

5.1 Introduction

Human pluripotent stem cells (hPSCs), including human embryonic stem cells (hESCs) [1] and human induced pluripotent stem cells (hiPSCs) [2], have an infinite proliferative potential and capacity for differentiation into all cells of the three germ layers. Their unique properties make them promising sources of cells for transplantation therapy, drug discovery and investigations of human congenital abnormalities. For successful applications of hPSCs in regenerative medicine, it is necessary to prepare large numbers of hPSCs with known genetic backgrounds [3] and establish stable cell storage systems (cell banking). To this end, cryopreservation is a key tool that enables long-term preservation of cell stocks. Additionally, cryopreservation is important for easy transportation of cells, enabling the use of hPSCs as desired, and minimizing the loss of their specific cellular potentials. Methodologies for effective hPSC cryopreservation have been actively pursued, in parallel with investigations to improve hPSC yield. However, the progress of hPSC cryopreservation has been ambiguous, and it has also been unclear which factors affect hPSC survival in the cryopreservation process.

Here, we review the cellular characteristics of hPSC cells and the history of hPSC cryopreservation, together with the difficulties that have been encountered with hPSC cryopreservation to date. Finally, we describe an optimized method for hPSC cryopreservation.

5.2 Slow Cooling in hPSC Cryopreservation

In general, cryopreservation is a process of storing cells, tissues, and organs to maintain their properties and viability [4]. Biological material is cooled to freezing temperatures, typically to lower than −140 °C. For cell survival following freezing and thawing, it is necessary to avoid ice crystal formation at icing temperature, which would otherwise cause cell death due to stress from excess intracellular water volume. "Slow Cooling" is one of the cryopreservation methods used to avoid ice crystal formation. The cooling rate is maintained at approximately −1 °C/min (ranging from −1 to −10 °C) in freezing medium that typically contains 10 % (ranging from 2 to 20 %) dimethyl sulfoxide (DMSO) as a cryoprotectant. The cryoprotectant enters the cytoplasm, and dehydration and condensation of the cytoplasm are promoted as the freezing medium cools, inhibiting the generation of ice crystals [4]. This slow cooling technique is widely applied to different cell types, including mouse embryonic stem cells [5]. However, it has been difficult to apply slow cooling cryopreservation to hPSCs. Standard slow cooling resulted in less than 10 % survivability of hPSCs after cryopreservation (as measured 24 h after cryopreservation) [6]. The use of a modified cryoprotectant that contained DMSO, ethylene glycol (EG) and fetal calf serum (FCS) increased hPSC viability to 30 % (as measured by counting cell colonies after 10 days) [7]; while use of propylene glycol or glycerol cryoprotectant solutions resulted in 20–60 % of cells recovering [8]. Most of the improvements to the slow cooling method have focused on modifying the cryoprotectant. The rationale for cryoprotectant modification approaches may come from the impression that hPSCs are fragile and intolerant to the freeze-thawing process; however, the actual reason for the failure of cryopreservation had not been clarified. To better appreciate the underlying reason for the resistance of hPSCs to cryopreservation using the slow cooling method; it is important to understand the specific cellular characteristics of hPSCs.

5.3 Cellular Characteristics of hPSCs

The difficulty with cryopreserving hPSCs might arise from the colony formation pattern that is characteristic of hPSCs. When hPSCs proliferate, they form tightly-packed colonies that are difficult to dissociate without damaging the cells. Cell-to-cell contact and the signals derived from these interactions, such those arising from E-cadherins, are vital for hPSC survival [9]. Indeed, inhibition of cell-to-cell interactions due to E-cadherin antibody leads to death of dissociated hPSCs [10]. To avoid this damage, hPSCs have been cultured in the presence of mild dissociation treatments such as Accutase or Collagenase IV, avoiding the complete single-cell state. Until the discovery of Rho-associated coiled-coil kinase (ROCK) inhibitors' effects on single dissociated hPSC survival [11], sustaining small clusters of cells was essential for hPSC culturing. This approach was applied to the cell detachment process during cryopreservation. Therefore, to cryopreserve hPSCs without single cell dissociation, researchers focused on another cryopreservation manner, vitrification.

5.4 Vitrification for hPSC Cryopreservation

Dramatic improvements in cell survival have been achieved by applying vitrification to hPSC cryopreservation. Vitrification involves flash cooling cells directly in liquid nitrogen. In contrast to slow cooling, vitrification avoids ice crystal formation in both the intracellular and extracellular environments. It is achievable when solutes are sufficiently concentrated, or when they are cooled rapidly, thereby increasing the viscosity [12]. Therefore, vitrification requires a system of efficient thermal conduction and is routinely administered in a narrow vessel such as an open pulled straw [13, 14]. Although most standard cell lines are generally cryopreserved by slow cooling, certain cell types and tissues, including oocytes, fertilized eggs and the embryos of several mammalian species and humans, are often preserved using vitrification [15–18]. Moreover, hESCs have also been successfully cryopreserved using vitrification [13]. In early hESC studies cells were stored in an open pulled straw and the maximum cryopreservation volume was extremely limited. However, an improved vitrification approach was soon developed, that instead used a cryotube and a small volume of freezing medium [19]. In comparison with the low cell survival obtained with slow cooling, vitrification sustained relatively higher cell survival rates after freeze-thawing [14, 20].

However, some problems remain with using vitrification for clinical or industrial hPSC applications. The improved vitrification method still restricts the number of cells that may be treated at any one time, and it is fewer than the number of cells that can be processed using slow cooling. Moreover, the freezing medium that is used for vitrification is significantly more harmful to hPSCs than that used in slow cooling. The small volume of freezing medium easily leads to an increase in temperature when transferring cryotubes at room temperature; consequently, rigorous control of the sample temperature is constantly necessary, and skilled handling is required for reliable vitrification results. Additionally, depending on the container used, there is a risk of contamination with infectious agents via direct contact with the liquid nitrogen. Thus, vitrification is unsuitable for the bulk storage of hPSCs, especially when they are intended for future clinical purposes.

In light of the limitations of applying vitrification for hPSC cryopreservation, it would be ideal if the slow cooling method could be modified to achieve hPSC cryopreservation. To adapt slow cooling to hPSCs, it is first necessary to recognize what factors underlie the difficulty of applying this technique. A review of previous studies shows that slow cooling of hPSCs has been troublesome, as shown in Table 5.1. To clarify the influence of cryopreservation on hPSC survival, a detailed revision of how cell viability is estimated during cryopreservation is warranted.

5.5 The Cryopreservation Process

We often assume that cryopreservation is a simple technique. However, it involves several processing stages, including cell detachment, freezing, storage, thawing, and also 'reseeding'. Significantly, a proportion of cells die as a result of different factors at each processing stage. When the cells are detached there may be a high risk of necrotic cell death due to physical damage from handling. During freezing and thawing, cells may also be at high risk of necrotic cell death due to physical damage; for example, due to the cytotoxicity of cryoprotectant [21], osmotic injury due to excursion of cryoprotectant [22], intracellular ice formation in the cooling process [23], and recrystallization of intracellular ice during warming [24]. On the other hand, at reseeding following freeze-thawing, cells may be at risk of apoptotic cell death due to accumulated damage as a result of freeze-thawing [4] or insufficient adhesion. Together, these possibilities suggest that cell survival during cryopreservation should be individually estimated in a stepwise manner to allow the development of an optimal cryopreservation approach.

5.6 Estimation of hPSC Survival

Before a detailed discussion of cell survival efficiency, we should first review the method by which survival is measured. Development of an efficient methodology is most easily achievable when studies can be compared using a common parameter. However, as shown in Table 5.1, the method for estimating hPSC survival efficiency during cryopreservation is not standardized. In addition, previous studies have had trouble estimating the true survival efficiency. One of the issues in the estimation of hPSC viability is selection of the target cells to be counted. The hPSCs are often counted as colonies grown for several days; however, this is quite inaccurate because it easily leads to an overestimation because of the divisions of the original colonies during the freezing and thawing processes. Therefore, even when hPSCs are seeded as colonies, we should estimate the total number of cells by counting the number of single hPSCs after complete dissociation. The other issue is the timing of the counting and the target cells used for comparison. As also shown in Table 5.1, the time at which cell viability is counted is widely variable between different studies. There is a broad time range between 24 h and 1 week that is used for cell counting, while cell survival rates are estimated by the comparison with the number of living cells prior to freezing. Of course, hPSCs proliferate after seeding; therefore, it is inaccurate to calculate cell survival rates by comparing cell numbers a few days after seeding with those prior to freezing. Additionally, counting cells several days after seeding can mask the reason for reduced proliferation, which may be caused by cell adhesion failure or reduced growth. Thus, cell counting should be performed in a stepwise manner, minimizing the time period after the initial cell count to be used for comparison. Accurate cell estimates will allow the development of reliable improvements in cell survival efficiency using this technology.

5.7 Optimization of hPSC Cryopreservation

Based on the above background, Miyazaki et al. used a stepwise approach to estimate the detailed survival rates of hPSCs at each stage of cryopreservation [25]. Firstly, they estimated survival up until immediately after freeze-thawing using slow cooling. Dispase treatment was used to detach intact hPSCs colonies, after which survival of individual cells was calculated (42.3–59.4 % for each hPSC line). After freeze-thawing, hPSCs from these colonies had low survival rates (15.5–32.5 %). This data shows that most of the colony hPSCs had already died before reseeding.

Table 5.1 Summary of cryopreservation protocols for human pluripotent stem cells

Freezing medium	Cell lines/type of culture	ROCK inhibitor	Cellular state	Cryo-method	Recovery rate	Estimation	Reference
Slow cooling: 90 % FCS, 10 % DMSO, DMEM	hESC(HES-1,HES-2)/MEF feeder	No	Colony	Slow cooling,	Slow cooling: 16 %	Slow cooling: colonies after 2 days, compared with pre-freezing colonies	[13]
Vitrification: 20 % DMSO, 20 % EG, 0.5 mol/L sucrose, DMEM				Vitrification	Vitrification: 100 %	Vitrification: measuring area of colonies	
10 % DMSO, 50 % FBS, 35 mM trehalose, culture medium	hESC(H1,H9)/MEF feeder Or Matrigel	No	Colony	Slow cooling	<25 %	Increase ratio of cells, compared with DMSO alone	[27]
90 % KSR, 10 % DMSO, 0.2 mol/L trehalose	hESC(original)/MEF feeder	No	Colony	Slow cooling	<48 %	Colonies at 7 days after seeding, compared with preseeding colonies	[20]
5 % DMSO, 10 % EG, culture medium	hESC(SNUhES-3)/STO feeder	No	Colony	Slow cooling	<30 %	Colonies at 10 days after seeding	[7]
10 % DMSO, 25 % FBS, culture medium	hESC(H1)/MEF feeder	No	Colony	Slow cooling	<79 %	Area of colonies, relative to non-freezing control	[28]
10 % DMSO, 90 % culture medium	hESC(H9,BG01V/hOG), hiPS	Yes	Single	Slow cooling	<8-folds	Increase ratio compared with cells without ROCK inhibitor at 4 days after seeding	[29]
10 % DMSO, 90 % KSR	hESC(H9)	Yes	Single	Slow cooling	After freezing: <80 %	After thawing: cells by flow cytometry analysis	[30]
					After seeding: <65 %	After reseeding: cells at 24 h after seeding	

(continued)

Table 5.1 (continued)

Freezing medium	Cell lines/type of culture	ROCK inhibitor	Cellular state	Cryo-method	Recovery rate	Estimation	Reference
10 % DMSO, 90 % culture medium	hESC(HS207,HS401)/human foreskin fibroblast feeder	Yes	Single	Slow cooling	<56 %	Cells after thawing by flow cytometry analysis, compared with prefreezing cells	[31]
10 % DMSO, 90 % FCS	hESC(Royan H5,H6), hiPSC(Royan hiPSC1,4)/Matrigel	Yes	Single	Slow cooling	After thawing: <90 %	After thawing: cells, compared with prefreezing cells	[32]
					After seeding: <4–32 %	After seeding: colony formation, compared with initially seeded cells	
10 % DMSO, 90 % KSR	hESC(H9,CHA-hES3)/MEF feeder	NO	Colony	Slow cooling	<51 %	Colonies at 7 days after seeding (relative to non-freezing control)	[33]
10 % EG, 90 % culture medium	hESC(H9), hiPSC/MEF feeder or Matrigel	Yes	Single	Slow cooling	<60 %	Cells at 2 days after seeding	[8]
10 % DMSO, 90 % culture medium	hESC(H9)/MEF feeder	Yes	Colony	Slow cooling	<10 %	Cells at 24 h after seeding, compared with postthawing alive cells	[6]

As described earlier, slow cooling requires sufficient penetration of cryoprotectant and adequate dehydration inside the cells. Therefore, it is thought that hPSCs under a colony state are not tolerate to the process of slow cooling. To overcome this problem, Miyazaki et al. tried changing the state of hPSCs prior to freeze-thawing and estimated hPSC survival immediately after thawing. The hPSCs, dissociated into single cells by treatment with EDTA and TrypLE Select, had much high survival rates before freeze-thawing (68.1–77.7 %). More importantly, hPSCs remained viable even after thawing (59.7–66.1 %, corresponding with 80–90 % survival). This data shows that physical damage during freeze-thawing is minimized when hPSCs are cryopreserved in a single cell state, and suggests that colony morphology enhances direct damage that lead to necrotic cell death at the first step of cryopreservation. Together, these results suggest that the majority of hPSCs that are killed are a result of the process of traditional slow cooling itself, and are already dead before the reseeding stage.

After establishing the importance of the cellular state prior to freeze-thawing, appropriate colony dissociation methods for slow cooling were mostly addressed. The next consideration is avoiding the apoptotic cell death of single hPSCs when they are reseeded. To solve this problem, there are two approaches that may be used to improve the survival of single hPSCs. The first approach is the adjustment of the cell density at reseeding. Adequate cell density is required to allow cell-cell interactions, which promotes hPSC survival [26]. By optimizing the seeding density of freeze-thawed hPSCs, hPSCs showed robust adhesion and increased survival (40–60 %, compared with numbers of seeded alive cells) at 12 h after seeding [25].

The second approach is the use of ROCK inhibitor at cell reseeding, which effectively increased the survival of single hPSCs (55 % on average in comparison with the number of seeded viable cells). Importantly, higher cell density (for example, 3×10^5 cells/cm^2) did not require the use of ROCK inhibitors for hPSC survival, but cells at lower densities (for example, 0.5×10^5 cells/cm^2) benefited from its use. To achieve a density of 0.5×10^5 cells/cm^2, a slightly higher seeding density is required when the cells are subcultured. Therefore, cryopreserved single-hPSCs should be seeded at higher cell densities without the use of ROCK inhibitor, but it should be used for lower seeding densities. The described experiments were carried out using Matrigel-coated culture dishes. However, some culture substrates, such as specific laminin isoform and its fragment, have been shown to enhance the survival of single hPSCs [26]. When these substrates were used to coat culture vessels instead, cryopreserved single hPSCs were efficiently recovered at much lower cell densities (60 % at 12 h, compared against the number of cells seeded). Therefore, following optimization of the culture conditions, hPSCs that have been cryopreserved and thawed can easily and efficiently recover to subculture.

5.8 Conclusion

We have summarized the main difficulties encountered with using the slow cooling method for hPSC cryopreservation. During freeze-thawing, hPSCs that have been dissociated from the culture media and frozen as intact colonies are mostly killed, presumably having suffered necrotic cell death due to insufficient penetration of the cryoprotectant into the colony state. Following thawing, hPSCs survival has also been low, due to cells being seeded at an inadequate density, which is thought to lead to apoptotic cell death in response to earlier damage from the freeze-thawing process. By solving those problems; that is, dissociating hPSCs into a single cell state and optimizing the reseeding density after freeze-thawing, slow cooling becomes an effective method for hPSC cryopreservation. Importantly, because this is a routine cell culture technique, specific equipment is unnecessary.

The use of slow cooling for cryopreservation of hPSCs will be further optimized in the future to suit a wider range of clinical material. In doing so, it is highly recommended that the cellular state of target hPSCs be taken into consideration. If this is done, there should be no significant trouble in the cryopreservation of undifferentiated hPSCs. Where differentiated cells that are derived from hPSCs are to be cryopreserved, the dissociation manner of target cells should be considered. This is because most differentiated cells form robust tissue structures, and these are more sensitive to disruption than those of cells in an undifferentiated state. As long as insufficient penetration of cryoprotectant into tissues remains a problem, it will be necessary to dissociate target cells into a single cell state. Although slow cooling is highly recommended for hPSC cryopreservation, vitrification is recommended instead for certain types of differentiated cells. We conclude that in any situation requiring hPSC cryopreservation it is important to optimize each step of the procedure and consider the cellular state of the hPSCs to be freeze-thawed. This approach is non-specific and may also be applied to the cryopreservation of pluripotent stem cells from other animal species.

References

1. Thomson JA, Itskovitz-Eldor J, Shapiro SS, Waknitz MA, Swiergiel JJ, Marshall VS, Jones JM (1998) Embryonic stem cell lines derived from human blastocysts. Science 282(5391):1145–1147
2. Takahashi K, Tanabe K, Ohnuki M, Narita M, Ichisaka T, Tomoda K, Yamanaka S (2007) Induction of pluripotent stem cells from adult human fibroblasts by defined factors. Cell 131(5):861–872
3. Serra M, Brito C, Correia C, Alves PM (2012) Process engineering of human pluripotent stem cells for clinical application. Trends Biotechnol 30(6):350–359
4. Baust JG, Gao D, Baust JM (2009) Cryopreservation: an emerging paradigm change. Organogenesis 5(3):90–96
5. Hunt CJ (2011) Cryopreservation of human stem cells for clinical application: a review. Transfus Med Hemother 38(2):107–123
6. Wagh V, Meganathan K, Jagtap S, Gaspar JA, Winkler J, Spitkovsky D, Hescheler J, Sachinidis A (2011) Effects of cryopreservation on the transcriptome of human embryonic stem cells after thawing and culturing. Stem Cell Rev 7(3):506–517
7. Ha SY, Jee BC, Suh CS, Kim HS, Oh SK, Kim SH, Moon SY (2005) Cryopreservation of human embryonic stem cells without the use of a programmable freezer. Hum Reprod 20(7):1779–1785
8. Katkov II, Kan NG, Cimadamore F, Nelson B, Snyder EY, Terskikh AV (2011) DMSO-free programmed cryopreservation of fully dissociated and adherent human induced pluripotent stem cells. Stem Cells Int 2011:981606
9. Chen G, Hou Z, Gulbranson DR, Thomson JA (2010) Actin-myosin contractility is responsible for the reduced viability of dissociated human embryonic stem cells. Cell Stem Cell 7(2):240–248
10. Li L, Wang BH, Wang S, Moalim-Nour L, Mohib K, Lohnes D, Wang L (2010) Individual cell movement, asymmetric colony expansion, rho-associated kinase, and E-cadherin impact the clonogenicity of human embryonic stem cells. Biophys J 98(11):2442–2451
11. Watanabe K, Ueno M, Kamiya D, Nishiyama A, Matsumura M, Wataya T, Takahashi JB, Nishikawa S, Muguruma K, Sasai Y (2007) A ROCK inhibitor permits survival of dissociated human embryonic stem cells. Nat Biotechnol 25(6):681–686
12. Pegg DE (2005) The role of vitrification techniques of cryopreservation in reproductive medicine. Hum Fertil (Camb) 8(4):231–239
13. Reubinoff BE, Pera MF, Vajta G, Trounson AO (2001) Effective cryopreservation of human embryonic stem cells by the open pulled straw vitrification method. Hum Reprod 16(10):2187–2194
14. Richards M, Fong CY, Tan S, Chan WK, Bongso A (2004) An efficient and safe xeno-free cryopreservation method for the storage of human embryonic stem cells. Stem Cells 22(5):779–789
15. Vajta G, Holm P, Kuwayama M, Booth PJ, Jacobsen H, Greve T, Callesen H (1998) Open Pulled Straw (OPS) vitrification: a new way to reduce cryoinjuries of bovine ova and embryos. Mol Reprod Dev 51(1):53–58
16. Kuleshova L, Gianaroli L, Magli C, Ferraretti A, Trounson A (1999) Birth following vitrification of a small number of human oocytes: case report. Hum Reprod 14(12):3077–3079
17. Dinnyes A, Dai Y, Jiang S, Yang X (2000) High developmental rates of vitrified bovine oocytes following parthenogenetic activation, in vitro fertilization, and somatic cell nuclear transfer. Biol Reprod 63(2):513–518
18. Dobrinsky JR (2002) Advancements in cryopreservation of domestic animal embryos. Theriogenology 57(1):285–302
19. Suemori H, Yasuchika K, Hasegawa K, Fujioka T, Tsuneyoshi N, Nakatsuji N (2006) Efficient establishment of human embryonic stem cell lines and long-term maintenance with stable karyotype by enzymatic bulk passage. Biochem Biophys Res Commun 345(3):926–932
20. Wu CF, Tsung HC, Zhang WJ, Wang Y, Lu JH, Tang ZY, Kuang YP, Jin W, Cui L, Liu W, Cao YL (2005) Improved cryopreservation of human embryonic stem cells with trehalose. Reprod Biomed Online 11(6):733–739
21. Muldrew K, McGann LE (1994) The osmotic rupture hypothesis of intracellular freezing injury. Biophys J 66(2 Pt 1):532–541

22. Gao DY, Liu J, Liu C, McGann LE, Watson PF, Kleinhans FW, Mazur P, Critser ES, Critser JK (1995) Prevention of osmotic injury to human spermatozoa during addition and removal of glycerol. Hum Reprod 10(5):1109–1122
23. Fujikawa S (1980) Freeze-fracture and etching studies on membrane damage on human erythrocytes caused by formation of intracellular ice. Cryobiology 17(4):351–362
24. Mazur P, Cole KW (1989) Roles of unfrozen fraction, salt concentration, and changes in cell volume in the survival of frozen human erythrocytes. Cryobiology 26(1):1–29
25. Miyazaki T, Nakatsuji N, Suemori H (2014) Optimization of slow cooling cryopreservation for human pluripotent stem cells. Genesis 52(1):49–55
26. Miyazaki T, Futaki S, Suemori H, Taniguchi Y, Yamada M, Kawasaki M, Hayashi M, Kumagai H, Nakatsuji N, Sekiguchi K, Kawase E (2012) Laminin E8 fragments support efficient adhesion and expansion of dissociated human pluripotent stem cells. Nat Commun 3:1236
27. Ji L, de Pablo JJ, Palecek SP (2004) Cryopreservation of adherent human embryonic stem cells. Biotechnol Bioeng 88(3):299–312
28. Ware CB, Nelson AM, Blau CA (2005) Controlled-rate freezing of human ES cells. Biotechniques 38(6):879–880, 882–873
29. Claassen DA, Desler MM, Rizzino A (2009) ROCK inhibition enhances the recovery and growth of cryopreserved human embryonic stem cells and human induced pluripotent stem cells. Mol Reprod Dev 76(8):722–732
30. Li X, Krawetz R, Liu S, Meng G, Rancourt DE (2009) ROCK inhibitor improves survival of cryopreserved serum/feeder-free single human embryonic stem cells. Hum Reprod 24(3):580–589
31. Martin-Ibanez R, Unger C, Stromberg A, Baker D, Canals JM, Hovatta O (2008) Novel cryopreservation method for dissociated human embryonic stem cells in the presence of a ROCK inhibitor. Hum Reprod 23(12):2744–2754
32. Mollamohammadi S, Taei A, Pakzad M, Totonchi M, Seifinejad A, Masoudi N, Baharvand H (2009) A simple and efficient cryopreservation method for feeder-free dissociated human induced pluripotent stem cells and human embryonic stem cells. Hum Reprod 24(10):2468–2476
33. Lee JY, Lee JE, Kim DK, Yoon TK, Chung HM, Lee DR (2010) High concentration of synthetic serum, stepwise equilibration and slow cooling as an efficient technique for large-scale cryopreservation of human embryonic stem cells. Fertil Steril 93(3):976–985

Cryopreservation in Closed Bag Systems as an Alternative to Clean Rooms for Preparations of Peripheral Blood Stem Cells

6

Silvia Spoerl, Robert Peter, and Angela M. Krackhardt

Abstract

Autologous and allogeneic stem cell transplantation (SCT) represents a therapeutic option widely used for hematopoietic malignancies. One important milestone in the development of this treatment strategy was the development of effective cryopreservation technologies resulting in a high quality with respect to cell viability as well as lack of contamination of the graft.

Stem cell preparations have been initially performed within standard laboratories as it is routinely still the case in many countries. With the emergence of cleanrooms, manufacturing of stem cell preparations within these facilities has become a new standard mandatory in Europe. However, due to high costs and laborious procedures, novel developments recently emerged using closed bag systems as reliable alternatives to conventional cleanrooms. Several hurdles needed to be overcome including the addition of the cryoprotectant dimethylsulfoxide (DMSO) as a relevant manipulation. As a result of the development, closed bag systems proved to be comparable in terms of product quality and patient outcome to cleanroom products. They also comply with the strict regulations of good manufacturing practice.

With closed systems being available, costs and efforts of a cleanroom facility may be substantially reduced in the future. The process can be easily extended for other cell preparations requiring minor modifications as donor lymphocyte preparations. Moreover, novel developments may provide solutions for the production of advanced-therapy medicinal products in closed systems.

S. Spoerl • R. Peter • A.M. Krackhardt (✉)
III. Medical Department, Hematology and Oncology, Klinikum rechts der Isar, Technische Universität München, Ismaningerstr. 22, 81675 München, Germany
e-mail: angela.krackhardt@tum.de

F. Karimi-Busheri, M. Weinfeld (eds.), *Biobanking and Cryopreservation of Stem Cells*, Advances in Experimental Medicine and Biology 951, DOI 10.1007/978-3-319-45457-3_6

Keywords

Cryopreservation • Closed bag systems • Clean room facility • Peripheral blood stem cells

Abbreviations

ATMP	Advanced therapy medicinal products
CCPS	Closed cryo prep set
CCFPS Closed	cryo frozen prep set
CD	Cluster of differentiation
DLI	Donor lymphocyte infusions
DMSO	Dimethylsulfoxide
EMA	European Medicines Agency
EU	European Union
FACT	Foundation for the accreditation of cellular therapy
FDA	Food and Drug Administration
GMP	Good manufacturing process
HES	Hydroxyethyl Starch
HEPA	High-efficiency particulate air
LC	Langerhans cells
MNC	Mononuclear cells
PBSC	Peripheral blood stem cells
PVA	Polyvinyl alcohol
SCT	Stem cell transplantation (SCT)
SOP	Standard operating procedures
ULPA	Ultra-low particulate air

6.1 Introduction

Cryopreservation of peripheral blood stem cells (PBSC) collected by apheresis has become an essential procedure for many institutions worldwide as the number of autologous and allogeneic stem cell transplantations is constantly rising. Although cryopreservation of PBSC is in principle a simple procedure, it needs to ensure a high quality with respect to PBSC viability and lack of contamination being highly essential for successful transplantation of the immune-compromised patient.

Depending on the transplantation setting, PBSC are collected by apheresis from a patient or from an allogeneic donor following prior stem cell mobilization. Eventually, multiple apheresis procedures may be necessary to obtain the adequate PBSC numbers needed for one or multiple PBSC transplantations.

Cryopreservation of PBSC gives high flexibility with respect to repeated transplantations as well as patient specific schedules. In case of patients with multiple myeloma, the number of PBSC to be collected may be sufficient for up to four transplantations. Cleanrooms have been the standard for cryopreservation for many years – especially within Europe. Their construction and maintenance are strictly regulated by SOP (standard operating procedures) guidelines according to GMP (good manufacturing practice) regulating and controlling all procedures conducted in the cleanroom environment [1].

Although the advantages of a cleanroom facility ensuring the high quality of collected PBSC are predominant, cleanrooms are quite expensive. The initial fixed costs for the establishment of a class B clean room are approximately 1 million euro. The variable yearly costs are about 35,000–100,000 euros depending upon space and instrumentation. Thus, novel developments reducing the costs but maintaining high product quality are urgently needed.

6.2 Collection and Cryopreservation of PBSC

Since the first allogeneic transplantations being performed in the US in the 1950s and 1960s [2, 3] cell based therapies have become substantial in the treatment of hematopoietic malignant diseases. Subsequently, novel therapeutic strategies using autologous stem cell transplantations developed [4]. To date, clinical indications for stem cell transplantation cover a broad range of hematopoietic malignancies but also solid tumors and non-malignant diseases [5–7]. Patients or healthy donors who donate PBSC undergo leukapheresis after stem cell mobilization. Apheresis is performed if the blood cell count contains a sufficient level of PBSC. The patients/donors blood is removed through a venous catheter and is transferred to a blood cell separator machine, where PBSC are separated by centrifugation. As anticoagulant, a citrate solution is commonly used. Once the required amount of PBSC is collected, DMSO is added to the cells at concentrations of 5–10 % and controlled rate freezing is started. The PBSC can be stored at temperatures below −130 °C for years. Before and after cell processing, cell viability and potential contamination is assessed as major aspects of quality control. When the cell product is required for an autologous or allogeneic transplantation, controlled thawing is performed prior to the transplantation and the cells are transfused to the patient.

6.3 Cryoprotectants

Cryoprotectant agents, controlled rate freezing and the usage of liquid nitrogen are the most important breakthrough developments enabling the storage of viable cells over a longer period of time. Cryoprotectants penetrate the cells and preserve their solute concentration, without being toxic to the cellular product. In 1948, Polge and colleagues accidently found that glycerol would enable fowl spermatozoa to survive freezing to −70 °C [8]. In 1959, Lovelock and Bishop [9] described that freezing damage could be prevented from cells by adding DMSO. Some years later, in 1965, Bouroncle et al. published the preservation of living cells in the presence of DMSO [10].

To date, DMSO is the most commonly used cryoprotectant. However, other substances are used for specific cell types or experimental purposes. Hydroxyethyl Starch (HES), a polymer with a high molecular weight, is used for cryopreservation of red blood cells [11]. Polyvinyl alcohol (PVA) [12] has been described as a new cryoprotectant and has also been applied for freezing of red blood cells. This agent might become an alternative to other cryoprotectants as DMSO and Glycerol in future. However, so far PVA was used only in experimental systems, thus, the potential clinical application needs to be further evaluated.

The addition of DMSO is, in fact, the only open manipulation during cryopreservation of PBSC. This has historically been performed without a cleanroom facility, as it is still performed outside of Europe. The DMSO concentration varies between different centers according to their internal standards and usually stays within a certain range of 5–10 % [13, 14]. The optimal concentration is still under debate although lower concentrations may result in reduced cell toxicity and side effects during and after transplantation [15]. It has been also published that the usage of 7.5 % DMSO was favorable compared to 10 % regarding engraftment of leukocytes, although platelet engraftment and transplantation-associated side effects did not seem to differ in both groups [16]. However, others report lack of statistical difference in therapeutic complications and engraftment for 5 % versus 10 % DMSO [17].

6.4 Controlled Freezing

With controlled-rate or slow freezing, biological materials are frozen down using programmable sequences. Slow freezing is protective for cellular products that have been pretreated with cryoprotectants such as glycerol or DMSO. A typical freezing program for PBSC would start at a temperature of about 2 °C. The temperature then decreases stepwise until a final temperature of −150 °C. The total freezing process takes about 1 h - in some protocols up to 2 h - and is supervised at all times and secured by a special alarm system. After the freezing process, the cellular products are transferred to Liquid Nitrogen (N_2) for further storage [18]. There are controversies about the need of a precooling step in the controlled freezing protocol. According to Dijkstra-Tiekstra et al., precooling was not necessary for the viability of cryopreserved cells, however the freezing rate had an influence on the recovery of white blood cells: slow-rate freezing revealed a higher cell recovery than fast-rate freezing [19]. The optimal cooling rate may additionally depend on the cell type [20].

6.5 Differences in the Processing of PBSC Between the US and Europe According to GMP

GMP guidelines regulate the manufacturing process of hematopoietic stem cells for clinical use. GMP guidelines aim to secure a high product quality for routine manufacturing processes by implementation of high standards for quality control. They were first implemented by the FDA (Food and Drug Administration) in the US and later in Europe concluded by the EU directive 2003/94 [21]. In Europe, the European Medicines Agency (EMA) is responsible for the development of guidelines and standards for medicinal products. In addition, every medicinal product for human use has to be granted a marketing authorization delivered by a competent regional authority. In the United States, the Foundation for the accreditation of cellular therapy (FACT) and the Food and Drug Administration (FDA) promote standards for cellular therapies according to evidence-based requirements.

6.6 Legal Developments in the EU

In 1989, the first edition of the EU guideline to good manufacturing practice of medicinal products for human and veterinarian use was published including an annex of the manufacture of sterile medicinal products. In January 1992, a second edition was released implementing Commission Directives 91/356 and 91/412. Twelve additional annexes were included. The rules governing medicinal products in the European Union contain guidance for the interpretation of the principles and guidelines of good manufacturing practices. They were laid down in Commission Directives 91/356/EEC and were amended by Commission Directive 2003/94/EC, of 8 October 2003, combining the guidelines of good manufacturing practice in respect of medicinal products for human use.

In this directive, high standards for quality control securing the safety of medicinal products are described. The manufacturing and processing of cellular products in clean room facilities is regulated by EudraLex ("The rules govering medicinal products in the European Union"), Volume 4 (EU Guidelines to good manufacturing practice), Annex 1 2008 (Manufacture of Sterile Medicinal Products). The principle in Annex I defines the manufacture of sterile products as subject to special requirements in order to minimize risks of microbiological contamination and of particulate and pyrogen contamination. It therefore requires the manufacturing of sterile products in clean areas entry to which should be through airlocks for personnel and/or for equipment and materials. Clean areas should be maintained to an appropriate cleanliness standard and supplied with air which has passed through filters of an appropriate efficiency.

The goal of production in a cleanroom environment is to ensure a low and controlled level of contamination (e.g. airborne microbes, aerosol particles) as regulated in the EU GMP guidelines Annex I. Cleanrooms are classified in grades A-D, defined by the maximum permitted number of particles per m^3 at different sizes (0,5 μm or 5 μm) at rest or in operation and limits for microbial contamination (in cfu/m^3), either the air sample, the settle plate, the contact plate and the glove print. According to common standards, the air entering a cleanroom facility is filtered and the air inside is recirculated through high-efficiency particulate air (HEPA) and/or ultra-low particulate air (ULPA) filters.

There are different standards for clean rooms: the European GMP classification (A–D), the British BS 5295 (class 2–4) and the ISO 14644-1 cleanroom standards (ISO 1-9). The ISO 14644-1 designation 1-9 is classified by the number of particles (>0.1 until 1 particles/m^3) and the microbial active air action levels (1–100 cfu/m^3), with ISO 1 being the "cleanest environment and ISO 9 referring to common room air. ISO designations provide uniform particle concentration values for cleanrooms in multiple industries.

A class A cleanroom is defined by a maximum of 3520 particles of a size of 0.5 μm per m^3 at rest and in operation and 20 particles at 5 μm m^3 at rest and in operation. A class B cleanroom allows 3520 particles of 0.5 μm at rest (29 particles at 5 μm) and 352,000 particles at 0.5 μm at rest and 2900 at 5 μm in operation. A class C cleanroom is defined by 352,000 particles of 0.5 μm (2900 at 5 μm) at rest and 3,520,000 of 0.5 μm (29,000 at of 5 μm) in operation. A class D clean room is also defined at rest with 3,520,000 particles of 0.5 μm and 29,000 particles of 5 μm, no limiting values are defined for a cleanroom class D in operation.

Cleanrooms are routinely monitored in operation, in terms of particles and microbial counts (based on the temperature and humidity), and they are further monitored on formal risk analysis. The personnel has to be trained in regards to hygiene and is equipped with special protective clothing such as hoods, face masks, gloves, boots, and coveralls. Desinfectants/detergants have to be available and sterilization has to be performed using common principles as moist heat, ethylene oxide, UV radiation. Aseptical filling is performed using a filter with 0,22 μm. Integrity testing of the filter needs to take place before and after using the filter.

Regulations according to different cleanroom standards are essential to maintain a certain quality level of the cellular products leaving a cleanroom facility in terms of sterility and integrity. They help to provide international standards, especially for products being transferred from the EU to the US or vice versa.

6.7 Cryopreservation in the Closed Bag System

With the high costs associated with the establishment and maintenance of a cleanroom facility, alternatives to conventional clean rooms such as the closed bag systems for cryopreservation are of high interest. Closed bag systems have been proposed for the preparation of diverse cell products. They consist of tightly connected sterile bags for collection and cryopreservation of PBSC, without the need of reopening the system within the cryopreservation procedures. All steps from stem cell collection until the freezing procedure are performed in a "closed system".

Closed bag systems for collection and storage of previously collected cells have been primarily described in an experimental animal model in 1979 by Fliedner et al. [22]. In 1980, the closed bag system was proposed for collection and freezing of human blood-derived mononuclear cells (MNC) as well as thawing the cryopreserved cells and washing them free of DMSO before transfusion to a patient [23]. A closed bag system has also been described for umbilical cord blood, as it was described in 1997 by Ademokun et al. [24]. This publication is supported by the work of Armitage et al. in 1999, presenting also a triple bag system for cord blood collection [25]. Both publications show favorable recoveries of nucleated cell counts, total progenitors and CD34-positives, which is especially important for cord blood where cell counts are usually low and good recovery rates are essential. In 2006, closed bag systems have been applied for cryopreservation of human mesenchymal stem cells [26]. Closed bag systems have been also proposed for cell therapeutics with potential use in cancer immunotherapy [27, 28]. Genetically modified, biologically functional CD34+ hematopoietic progenitor cell-derived Langerhans cells (LC) were grown in hydrophobic, closed culture bags. Those LC were comparable in terms of their viability and functionality compared to other LC conventionally grown in flasks or plates. However, in this system, the LC were used only experimentally and their clinical use needs to be further investigated [28].

Closed bag systems for cryopreservation of peripheral blood progenitor cells have been primarily described in 2007 in a preclinical study testing buffy coats and leukapheresis products of healthy donors [29]. In their study, Humpe et al. reported a closed system using a sterile filter to be equivalent to cleanroom-based methods in terms of quality of the cellular products and sterility [29]. The closed system described in this analysis consists of three cryopreservation bags, different tubing, a DMSO-resistant sterile filter, a tubing line for sterile connecting and a syringe for the removal of air.

The first functionally closed system involving routinely cryopreserved PBSC outside of a cleanroom facility for clinical use was published in 2015 [14]. This system has been accepted by legal authorities in the EU and is in process since 2010, so long-term results on a large patient cohort were available. Two systems for cryopreservation of peripheral blood stem cells (PBSC) within a functionally closed bag system were described: the "closed cryo prep set (CCPS) followed by a further development, the closed cryo frozen prep set (CCFPS) [14]. The CCPS consists of three components: two cryobags being connected to a tube system with a DMSO-resistant sterile filter, a syringe and a disposal bag. The system can be also used for manufacturing of donor lymphocyte infusions (DLI). In this context, the cell number can be individually adjusted by the treating clinician [14].

6.8 Comparison of Product Quality

Lack of contamination represents a crucial specification for product release. With respect to cleanroom facilities, divergent contamination rates are described. Donmez et al. found that in their setting, large-volume leukapheresis and high numbers of stem cell culture sampling were defined as risk factors for bacterial contamination [30]. Samples of the leukapheresis products were sent for bacterial culture from two time points: after processing and after thawing. Those samples were compared and microbial contamination was found in 5.7 % of the products after processing, and 3.66 % after thawing [30]. Kamble et al. found 4.5 % of their bone marrow and 3,9 % of their peripheral blood hematopoietic progenitor cell harvests to be contaminated. Both studies were conducted in cleanroom areas [31].

Few studies, in fact, compared contamination rates for manufacturing of PBSC in or outside of cleanroom facilities [32, 33]. In the study of Humpe et al., processing was performed according to a standard protocol under cleanroom class A conditions with surrounding class B conditions and appropriate monitoring (particulate and microbial). Outcome was compared to the processing in a normal laboratory without classified air class but with cryoprotective solution prepared under cleanroom conditions. Cellular integrity (viability) and cellular functionality did not differ substantially among both groups and no contaminations were reported. However, no clinical data are available, so no conclusion on clinical impact could be drawn in this context [29].

With respect to clinical application only two retrospective analyses compared PBSC cryopreservation in and outside the cleanroom. Cassens et al. reported in 2002 that there was no significant difference in contamination rates of clean benches (1.03 %) in comparison to clean rooms (1.32 %) [32]. This is in contrast to data published by Ritter et al. [33] demonstrating a contamination rate of 5.2 % in products from clean benches in contrast to 0.8 % in cleanroom derived compounds. Our group found only 1 out of 283 autologous and allogeneic cellular products being produced in the presented closed bag system outside of a cleanroom facility to be contaminated [14]. There are a number of factors potentially influencing the high contamination rate observed by Ritter et al. [33] including a high fluid volumes used for testing. In a study by

Weinstein et al., 5 ml versus 10 ml sample that was injected into blood culture flasks was compared and an increased rate of blood pathogens was detected when using a higher volume [34].

Of note, it seems that mostly dermal germs which usually do not cause bacteremia are the source of contamination of compounds as reported by diverse groups [31, 35]. Multiple factors can be associated with bacterial contamination including the presence of hair at the phlebotomy site [36] or desinfection of the involved skin [37] according to studies conducted by Perez et al. and Lee et al. demonstrating that bacterial contamination often occurs prior to the processing of the cellular product.

Thus, although it is undoubted that the risk should be minimized as best as possible, the role of infectious complications due to contaminated stem cell compounds seem to be overall of minor clinical significance independent of the usage of cleanroom facilities [38, 39]. Protective antibiotic treatment can be provided either by routine prophylaxis (sequential prophylaxis [40]) or according to previous sensitivity testing.

Apart from sterility, cell viability after cryopreservation is an important issue for cellular products which may depend on the defined processing conditions. Reduced viability of reinfused cells was shown to be associated with longer time to engraftment [41, 42] and viability post-thaw is important to assess in order to be able to predict potential delayed cell engraftment. In their study, Reich-Sloty and colleagues report that mostly granulocytes did not survive cryopreservation [43], other cell types as the CD34+ cells have clearly higher viability rates. We found in our previous study that cryopreservation of PBSC using a closed bag system was safe and effective with a high level of cellular viability and low contamination rates. When comparing the production in cleanroom areas and using closed bag systems, analysis of the viability of leukocytes and CD34+ stem cells and the recovery rate after manufacturing was found not to be compromised [14, 29].

6.9 Comparison of Transplantation Outcome

Cell products being manufactured in cleanroom facilities or with a closed bag system have been also analyzed in regard to their clinical outcome after transplantation. Mortality until neutrophil engraftment and duration until neutrophil/platelet engraftment after transplantation seem to be comparable for products being manufactured in a closed bag system compared to what is reported in the current literature [44–46]. However, there is few corresponding literature available and different factors may influence the outcome after transplantation which need to be taken into account as patient collective (older, heavily pretreated patients compared to younger ones without any significant comorbidities), antibiotic prophylaxis, the amount of transfused hematopoietic stem cells, remission status or conditioning treatment. Of note, mortality rates before engraftment seem to be comparable in the autologous as well as allogeneic setting using closed bag systems compared to centers using stem cell products having been prepared in cleanroom facilities [41, 42]. Future assessment will be necessary to further confirm these results.

6.10 Potential Use of Automatic Cell Manufacturing

GMP-compliant automatic cell manufacturing with the automat CliniMACS Prodigy [47] can be additionally used for a cleanroom-free processing of PBMC providing the advantage for additional major modifications. The automated system is able to save time of qualified staff, and more complex procedures as cell enrichment and depletion or final product formulation can be performed.

Hümmer et al. and Spohn et al. found that for this fully automatic system, recovery of target cells (e.g. CD34+ stem cells) after additional sorting was comparable to what was reported for the semiautomatic method [47, 48]. However, another study conducted by Stroncek et al. found CD34+ recovery and CD3+ T cell depletion to be lower than with the semiautomated CliniMACS Plus method [49].

Both, closed bag and automatic cell manufacturing systems require highly trained professional staff to supervise and validate the procedure. The closed bag system has to undergo several manual tests, especially tightness tests, before the manufacturing process can be started. Automatic cell manufacturing systems require high technical support and close monitoring. However, the financial efforts are still inferior in comparison to the costs of establishing and maintenance of a cleanroom facility.

6.11 Conclusions/Outcome

Closed bag systems have been investigated for cryopreservation of PBSC and they were shown to be safe in terms of product quality and further clinical outcome after transplantation. One major advantage is that they are clearly cost-saving for the manufacturing of PBSC compared to the high costs necessary for establishment and maintenance of a cleanroom facility. These processes can be extended to other minimally manipulated processes such as donor lymphocyte infusions (DLI). Currently, closed manufacturing processes within automatic cell manufacturing systems are developed and validated to be used for more complex cellular products as advanced medicinal-therapy medicinal products (ATMP). With respect to limited resources within the health care systems, these developments may become an important alternative to the usage of cleanroom facilities and may facilitate the clinical application of cellular therapies including ATMPs.

References

1. Smith L, Lowdell M (2003) Quality issues in stem cell and immunotherapy laboratories. Transfus Med 13:417–423
2. Thomas ED, Lochte H, Lu W, Ferrebee J (1957) Intravenous infusion of bone marrow in patients receiving radiation and chemotherapy. N Engl J Med 257:491–496
3. Bach F, Albertini R, Joo P, Anderson J, Bortin M (1968) Bone marrow transplantation in a patient with the Wiskott-Aldrich Syndrome. Lancet 2:1364–1966
4. Beaujean F, Hartmann O, Forestier CL, Bayet C, Duédari N, Parmentier C (1984) Successful infusion of 40 cryopreserved autologous bone-marrows. In vitro studies of the freezing procedure. Biomed Pharmacother 38:348–352
5. Mandanas R, Saez R, Epstein R, Confer D, Selby G (1998) Long-term results of autologous marrow transplantation for relapsed or refractory male or female germ cell tumors. Bone Marrow Transplant 21:569–576
6. Peinemann F, Smith L, Bartel C (2013) Autologous hematopoietic stem cell transplantation following high dose chemotherapy for non-rhabdomyosarcoma soft tissue sarcomas. Cochrane Database Syst Rev. doi:10.1002/14651858
7. Gupta N, Choudhary D, Sharma S, Khandelwal V, Dhamija M (2015) Haploidentical hematopoietic SCT for acquired severe aplastic anemia using post-transplant high-dose CY. Bone Marrow Transplant 50:155–156
8. Polge C, Smith A, Parkes A (1949) Revival of spermatozoa after vitrification and dehydration at low temperatures. Nature 164:666
9. Lovelock J, Bishop M (1959) Prevention of freezing damage to living cells by dimethyl sulphoxide. Nature 183:1394–1395
10. Bouroncle B (1965) Preservation of living cells at −79 degrees C with dimethyl sulfoxide. Proc Soc Exp Biol Med 119:958–961
11. Rittmeyer I, Nydegger U (1992) Influence of the cryoprotective agents glycerol and hydroxyethyl starch on red blood cell ATP and 2,3-diphosphoglyceric acid levels. Vox Sang 62:141–145
12. Deller R, Vatish M, Mitchell D, Gibson M (2014) Synthetic polymers enable non-vitreous cellular cryopreservation by reducing ice crystal growth during thawing. Nat Commun 5:3244
13. Smagur A, Mitrus I, Ciomber A et al (2015) Comparison of the cryoprotective solutions based on human albumin vs. autologous plasma: its effect on cell recovery, clonogenic potential of peripheral blood hematopoietic progenitor cells and engraftment after autologous transplantation. Vox Sang 108:417–424
14. Spoerl S, Peter R, Wäscher D et al (2015) Long-term experiences in cryopreservation of mobilized peripheral blood stem cells using a closed-bag system: a technology with potential for broader application. Transfusion 55:2702–2708
15. Morris C, Wreede L, Scholten M (2014) Should the standard dimethyl sulfoxide concentration be reduced? Results of a European Group for Blood and Marrow Transplantation prospective noninterventional study on usage and side effects of dimethyl sulfoxide. Transfusion 54:2514–2522

16. Mitrus I, Smagur A, Giebel S (2013) A faster reconstitution of hematopoiesis after autologous transplantation of hematopoietic cells cryopreserved in 7.5% dimethyl sulfoxide if compared to 10% dimethyl sulfoxide containing medium. Cryobiology 67:327–331
17. Akkok C, Liseth K, Nesthus I (2008) Autologous peripheral blood progenitor cells cryopreserved with 5 and 10 percent dimethyl sulfoxide alone give comparable hematopoietic reconstitution after transplantation. Transfusion 48:877–883
18. Abbruzzese L, Michieli M, Rupolo M et al (2010) A new freezing and storage procedure improves safety and viability of haematopoietic stem cells and neutrophil engraftment: a single institution experience. Vox Sang 98:172–180
19. Dijkstra-Tiekstra M, Setroikromo A, Kraan M, Gkoumassi E, Wildt-Eggen J (2014) Optimization of the freezing process for hematopoietic progenitor cells: effect of precooling, initial dimethyl sulfoxide concentration, freezing program, and storage in vapor-phase or liquid nitrogen on in vitro white blood cell quality. Transfusion 12:3155–3163
20. Gao D, Critser JK (2000) Mechanisms of cryoinjury in living cells. ILAR J 41:187–196
21. EU (2003) Annex I to the European Union (EU)-GMP-guidelines to production of sterile drugs. http://ec.europa.eu/enterprise/pharmaceuticals/eudralex/homev4.htm
22. Fliedner T, Calvo W, Körbling M, Nothdurft W, Pflieger H, Ross W (1979) Collection, storage and transfusion of blood stem cells for the treatment of hemopoietic failure. Blood Cells 5:313–328
23. Körbling M, Fliedner T, Rüber E, Pflieger H (1980) Description of a closed plastic bag system for the collection and cryopreservation of leukapheresis-derived blood mononuclear leukocytes and CFUc from human donors. Transfusion 20:293–300
24. Ademokun J, Chapman C, Dunn J et al (1997) Umbilical cord blood collection and separation for haematopoietic progenitor cell banking. Bone Marrow Transplant 19:1023–1028
25. Armitage S, Fehily D, Dickinson A, Chapman C, Navarrete C, Contreras M (1999) Cord blood banking: volume reduction of cord blood units using a semi-automated closed system. Bone Marrow Transplant 23:505–509
26. Mizuno N, Shiba H, Ozeki Y et al (2006) Human autologous serum obtained using a completely closed bag system as a substitute for foetal calf serum in human mesenchymal stem cell cultures. Cell Biol Int 6:521–524
27. Pincha M, Sundarasetty B, Salguero G et al (2012) Identity, potency, in vivo viability, and scaling up production of lentiviral vector-induced dendritic cells for melanoma immunotherapy. Hum Gene Ther Methods 1:38–55
28. Yuan J, Kendle R, Ireland J et al (2007) Scalable expansion of potent genetically modified human langerhans cells in a closed system for clinical applications. J Immunother 6:634–643
29. Humpe A, Jansen P, Tischer B et al (2007) Cryopreservation of cellular products in a closed-bag system with an incorporated dimethyl sulfoxide-resistant sterile filter outside of cleanroom facilities. Transfusion 47:1033–1041
30. Donmez A, Aydemir S, Arik B et al (2012) Risk factors for microbial contamination of peripheral blood stem cell products. Transfusion 52:777–781
31. Kamble R, Pant S, Selby G et al (2005) Microbial contamination of hematopoietic progenitor cell grafts-incidence, clinical outcome, and cost-effectiveness: an analysis of 735 grafts. Transfusion 45:874–878
32. Cassens U, Ahlke C, Garritsen H et al (2002) Processing of peripheral blood progenitor cell components in improved clean areas does not reduce the rate of microbial contamination. Transfusion 42:10–17
33. Ritter M, Schwedler J, Beyer J et al (2003) Bacterial contamination of ex vivo processed PBPC products under clean room conditions. Transfusion 43:1587–1595
34. Weinstein M, Mirrett S, Wilson M, Reimer L, Reller L (1994) Controlled evaluation of 5 versus 10 milliliters of blood cultured in aerobic BacT/Alert blood culture bottles. J Clin Microbiol 32:2103–2106
35. Vanneaux V, Foïs E, Robin M et al (2007) Microbial contamination of BM products before and after processing: a report of incidence and immediate adverse events in 257 grafts. Cytotherapy 9:508–513
36. Perez P, Bruneau C, Chassaigne M et al (2002) Multivariate analysis of determinants of bacterial contamination of whole-blood donations. Vox Sang 82:55–60
37. Lee C, Ho P, Chan N, Mak A, Hong J, Lin C (2002) Impact of donor arm skin disinfection on the bacterial contamination rate of platelet concentrates. Vox Sang 83:204–208
38. Nasser R, Hajjar I, Sandhaus L et al (1998) Routine cultures of bone marrow and peripheral stem cell harvests: clinical impact, cost analysis, and review. Clin Infect Dis 27:886–888
39. Padley D, Koontz F, Trigg M, Gingrich R, Strauss R (1996) Bacterial contamination rates following processing of bone marrow and peripheral blood progenitor cell preparations. Transfusion 36:53–56
40. Kim J, Goulston C, Zangari M, Tricot G, Boyer M, Hanson K (2014) Impact of a change in antibacterial prophylaxis on bacteremia and hospitalization rates following outpatient autologous peripheral blood stem cell transplantation for multiple myeloma. Transpl Infect Dis 16:421–429
41. Allan D, Keeney M, Howson-Jan K et al (2002) Number of viable CD34(+) cells reinfused predicts engraftment in autologous hematopoietic stem cell transplantation. Bone Marrow Transplant 29:967–972
42. Feugier P, Bensoussan D, Girard F et al (2003) Hematologic recovery after autologous PBPC trans-

plantation: importance of the number of postthaw CD34+ cells. Transfusion 43:878–884
43. Reich-Slotky R, Colovai A, Semidei-Pomales M et al (2008) Determining post-thaw CD34+ cell dose of cryopreserved haematopoietic progenitor cells demonstrates high recovery and confirms their integrity. Vox Sang 94:351–357
44. Gerds A, Fox-Geiman M, Dawravoo K et al (2010) Randomized phase III trial of pegfilgrastim versus filgrastim after autologous peripheral blood stem cell transplantation. Biol Blood Marrow Transplant 16:678–685
45. Wuchter P, Ran D, Bruckner T et al (2010) Poor mobilization of hematopoietic stem cells-definitions, incidence, risk factors, and impact on outcome of autologous transplantation. Biol Blood Marrow Transplant 16:490–499
46. Brown R, Adkins D, Goodnough L et al (1997) Factors that influence the collection and engraftment of allogeneic peripheral blood stem cells in patients with hematologic malignancies. J Clin Oncol 15:3067–3074
47. Hümmer C, Poppe C, Bunos M et al (2016) Automation of cellular therapy product manufacturing: results of a split validation comparing CD34 selection of peripheral blood stem cell apheresis product with a semi-manual vs an automatic procedure. J Trans Med 14:76
48. Spohn G, Wiercinska E, Karpova D et al (2015) Automated CD34+ cell isolation of peripheral blood stem cells apheresis product. Cytotherapy 17:1465–1471
49. Stroncek D, Tran M, Frodigh S et al (2016) Preliminary evaluation of a highly automated instrument for the selection of CD34+ cells from mobilized peripheral blood stem cell concentrates. Transfusion 56:511–517

7 Cryopreserved or Fresh Mesenchymal Stromal Cells: Only a Matter of Taste or Key to Unleash the Full Clinical Potential of MSC Therapy?

Guido Moll, Sven Geißler, Rusan Catar, Lech Ignatowicz, Martin J. Hoogduijn, Dirk Strunk, Karen Bieback, and Olle Ringdén

Abstract

Mesenchymal stromal cells (MSCs) harbor great therapeutic potential for numerous diseases. From early clinical trials, success and failure analysis, bench-to-bedside and back-to-bench approaches, there has been a great gain in knowledge, still leaving a number of questions to be answered regarding optimal manufacturing and quality of MSCs for clinical application. For treatment of many acute indications, cryobanking may remain a prerequisite, but great uncertainty exists considering the therapeutic value of freshly thawed (thawed) and continuously cultured (fresh) MSCs. The field has seen an explosion of new literature lately, outlining the relevance of the topic. MSCs appear to have compromised immunomodulatory activity directly after thawing for clinical application. This may provide a possible explanation for failure of early clinical trials. It is not clear if and how quickly MSCs recover their full therapeutic activity, and if the "cryo

G. Moll (✉)
Berlin-Brandenburg Center/School for Regenerative Therapies (BCRT/BSRT), Charité Universitätsmedizin, Berlin, Germany

Therapeutic Immunology (TIM), Department of Laboratory Medicine, Karolinska Institutet, Stockholm, Sweden
e-mail: guido.moll@charite.de

S. Geißler
Berlin-Brandenburg Center/School for Regenerative Therapies (BCRT/BSRT), Charité Universitätsmedizin, Berlin, Germany

Julius Wolff Institute (JWI), Berlin, Germany
e-mail: sven.geissler@charite.de

R. Catar
Department of Nephrology and Intensive Care Medicine, Charité Universitätsmedizin, Berlin, Germany
e-mail: rusan.catar@charite.de

L. Ignatowicz
Division of Dermatology and Venereology, Department of Clinical Sciences, Lund University, Lund, Sweden
e-mail: lech.ignatowicz@med.lu.se

M.J. Hoogduijn
Nephrology and Transplantation, Department of Internal Medicine, Erasmus Medical Center, Rotterdam, The Netherlands
e-mail: m.hoogduijn@erasmus.nl

F. Karimi-Busheri, M. Weinfeld (eds.), *Biobanking and Cryopreservation of Stem Cells*, Advances in Experimental Medicine and Biology 951, DOI 10.1007/978-3-319-45457-3_7

stun effect" is relevant for clinical success. Here, we will share our latest insights into the relevance of these observations for clinical practice that will be discussed in the context of the published literature. We argue that the differences of fresh and thawed MSCs are limited but significant. A key issue in evaluating potency differences is the time point of analysis after thawing. To date, prospective double-blinded randomized clinical studies to evaluate potency of both products are lacking, although recent progress was made with preclinical assessment. We suggest refocusing therapeutic MSC development on potency and safety assays with close resemblance of the clinical reality.

Keywords

Mesenchymal stromal cell • Cell therapy • Cryopreservation • Freeze-thawing • Freeze injury • Cell engraftment • Cryopreservation-induced cell death • T cells • Complement • Coagulation

Abbreviations

AT	Adipose tissue
ATMP	Advanced therapy medicinal product
BM	Bone marrow
CD3	Cluster of differentiation 3
CPA	Cryoprotective agent
DMSO	Dimethyl sulfoxide
FCS	Fetal calf serum
GvHD	Graft-versus-host disease
IBMIR	Instant blood-mediated inflammatory reaction
IFNg	Interferon gamma
IL1a	Interleukin 1 alpha
ISCT	International Society for Cellular Therapy
LN2	Liquid nitrogen
MOA	Mechanism of action
MSC	Mesenchymal stromal cell
NK cell	Natural killer cell
PBMC	Peripheral blood mononuclear cell
RNA	Ribonucleic acid
T1D	Type 1 diabetes
TNFa	Tumor necrosis factor alpha
UC	Umbilical cord

D. Strunk
Experimental and Clinical Cell Therapy Institute, Paracelsus Medical University, Salzburg, Austria
e-mail: dirk.strunk@pmu.ac.at

K. Bieback
Institute of Transfusion Medicine and Immunology, Medical Faculty Mannheim, Heidelberg University, Heidelberg, Germany

German Red Cross Blood Service Baden-Württemberg – Hessen, Mannheim, Germany
e-mail: Karen.bieback@medma.uni-heidelberg.de

O. Ringdén
Therapeutic Immunology (TIM), Department of Laboratory Medicine, Karolinska Institutet, Stockholm, Sweden

Division of Therapeutic Immunology (TIM, F79), Karolinska University Hospital, SE-14186 Stockholm, Sweden
e-mail: olle.ringden@ki.se

7.1 Introduction

Cryopreservation, often viewed as an "old school" discipline, is currently evolving at an accelerated rate [1], and has also attracted renewed attention in stem cell research [2]. Recently, Galipeau et al. initiated a critical discussion on clinical MSC cryopreservation [3], in order to understand the failure of a phase III clinical trial with MSCs [4]. This has now lead to broader discussion of this complicated matter [5–11].

Although the cells appear to be effective in many preclinical immunomodulatory and regenerative approaches, MSCs demonstrated suboptimal efficacy in a number of clinical trials [5]. Galipeau and others hypothesized [4–7], that this discrepant outcome is due to simple but substantial differences in the preparation and application of therapeutic MSCs in preclinical models compared to patient trials.

Parameters under scrutiny are: (1) the common use of fresh cells in preclinical models versus the predominant clinical use of thawed cells readily retrieved from cryostorage [3, 4], (2) the passage number of ex vivo-expanded cells, with minimal expanded cells (passage 1–4) used in European studies versus highly expanded cells (above passage 5) used in the failed company-sponsored phase three trial [4, 8, 12], and (3) the mode of application in preclinical and clinical settings (e.g. cell reconstitution, infusion buffers, and mode of cell injection) [6].

Consequently, means to optimize the conventional MSC product would be to use fresh minimal expanded cells reconstituted with a buffer optimized for systemic infusion [6]. This refined mode of cell graft preparation and application may be a first step to minimize the instant blood-mediated inflammatory reaction (IBMIR) and cell-mediated immunity [13–19], which appear to compromise therapeutic cell engraftment and bioactivity in vivo.

Galipeau et al. also highlighted that MSCs are transfused shortly after withdrawal from cryostorage [3], and that it appears to be more meaningful to assess their functional properties within this time frame [4], in order to predict their performance as pharmaceutical. During the following discourse, we will address the question if MSCs are really like sushi "fresh is best", and we will highlight novel approaches envisioned for development of more efficient MSC therapeutics.

7.2 MSC Therapy

MSC therapies are tested in more than 500 registered clinical studies [20, 21] and on their best way to revolutionize modern medicine. MSC-based therapeutics address a so far unmet clinical need for new regenerative and immunomodulatory therapies, based on their ability to regenerate defective tissues and balance detrimental immune responses [6, 22–24].

7.3 State of the 'Art'

Experimental MSC therapies have already been studied in a very broad range of treatment indications (Fig. 7.1a) (www.clinicaltrials.gov) [20], such as hematological and (auto)-immune disease (20 %), e.g. graft-versus-host disease (GvHD) and inflammatory bowel disease, bone and cartilage disease (19 %), neurological disease (18 %), cardiovascular disease (15 %), lung, liver, and kidney diseases (13 %), and many other indications (15 %).

MSCs have been mainly isolated from bone marrow (BM) (Fig. 7.1b) [25], but also other tissue sources, such as adipose tissue (AT) and perinatal tissues, have rapidly gained popularity since 2008, now accounting for 50 % of new drug submissions [26]. Particularly, extra-embryonic gestational tissues, such as placenta-derived fetal membranes (e.g. decidua) and umbilical cord (UC), are easily accessible without adverse donor site morbidity [27–30].

Studies on biodistribution and time course of MSC engraftment suggest that most cells are trapped in the lung microvasculature and destroyed shortly after infusion due to number of antagonizing mechanisms (Fig. 7.1c) [6, 31–33], e.g. triggering of innate immune cascades, natural

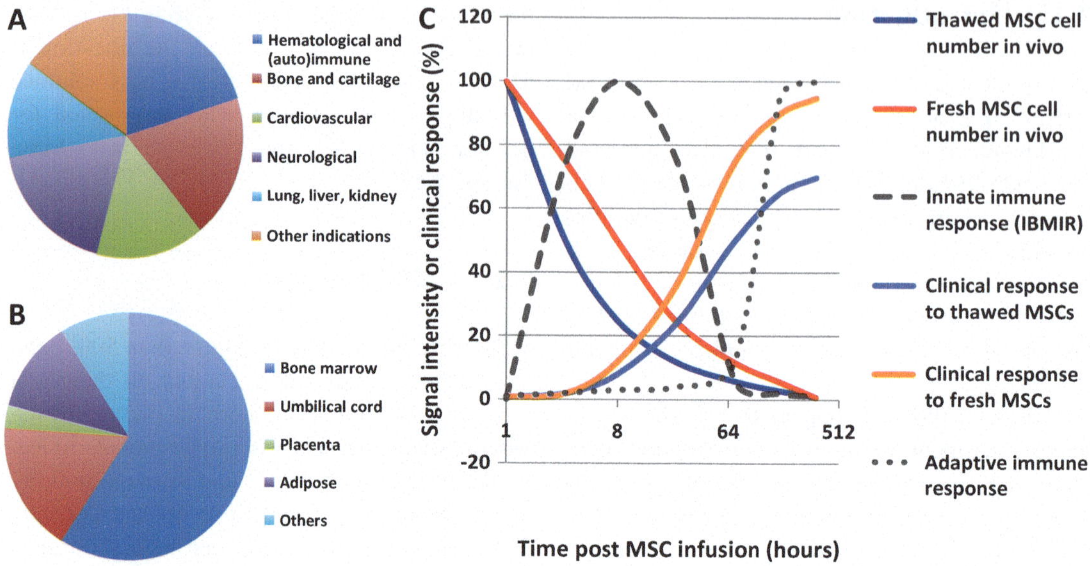

Fig. 7.1 Recent developments in MSC therapy. (**a**, **b**) Different treatment indications and tissue sources used for isolation of MSCs, and (**c**) time course of MSC engraftment, antagonizing mechanisms, and differential priming of therapeutic effects in response the fresh and thawed MSCs in vivo (extrapolated from e.g. [6, 13–16, 18, 81]). *Abbreviations*: *MSC* mesenchymal stromal cell, *IBMIR* instant blood-mediated inflammatory reaction

killer (NK) cells, and T- and B-cell-mediated adaptive immune responses [32]. Thus, it appears that MSC's therapeutic activity is not the result of long-term engraftment in vivo, but rather due to a modulatory action exerted within the first hours of infusion [18, 19].

At its latest follow-up in our and collaborating centers, 71 % of patients with steroid-resistant GvHD responded to MSC treatment [6, 34], with a trend for better response with lower passage (P1-2) fresh cells (Fig. 7.1c) [14]. The 1-year survival was better for patients receiving early compared to late passage cells (75 % vs. 21 %) [35], suggesting that fresh early passage MSCs have higher therapeutic value [13, 14, 35] and higher potency [6].

Initially, mostly BM-derived MSCs expanded in fetal calf serum (FCS) were used for treatment of acute GvHD [36]. Although FCS has been used as expansion medium for MSCs in the vast majority of studies, platelet lysate medium and human AB-serum have recently been used more frequently to substitute animal products [21, 37–45]. There seems to be a trend for more complete responders among patients treated with FCS-expanded MSCs [46]. However, recent meta-analysis on MSC therapy in GvHD found no relation between MSC age, expansion medium, MSC dose and the efficacy of MSC therapy [47].

Because BM aspiration is an invasive procedure with obvious discomfort for the donor, alternate MSC sources have been considered in several trials. AT-derived MSCs are increasingly used [48], and in recent years also UC-derived MSCs [49] and placenta-derived decidua stromal cells (DSCs) have been used to treat GvHD [29]. Promisingly, when the DSCs were dissolved and suspended in 5 % human albumin instead of human AB-plasma [15], survival for severe acute GvHD was improved in a pilot study [30].

A problem with evaluating these pilot studies is the lack of prospective randomized controls [47]. In a large study from the Osiris using BM-MSCs in a phase 3 acute GvHD trial complete response at 28 days was the same as in the placebo arm [50, 51]. However, among patients with liver GvHD, response to BM-MSCs was 76 % compared to 47 % in the placebo arm ($p=0.026$). Among patients with gastrointestinal GvHD, complete response was 78 % to

BM-MSCs compared to 64 % in the placebo arm ($p = 0.018$). Further, in many small clinical studies only one dose of MSCs has been used, and some studies indicate that repeated (e.g. at least weekly) doses may be superior to single-dose treatment [52]. With regard to dosing, it seems that 1×10^6/kg of MSCs seem to be sufficient according to available data [47].

With regard to comparison between fresh and thawed MSCs and role of MSC passage there are only few studies evaluating survival and response in patients with GvHD [14, 35]. To circumvent donor-to-donor heterogeneity and eliminate the risk that patients may respond to one batch of MSCs, but not to another, pooled cells from several donors have been used [53]. Recently, BM-MSCs from eight third-party donors were pooled and minimal expanded (P2) for treatment of steroid-refractory acute GVHD in children with excellent results [21].

7.4 Potency Assessment

In contrast to the more defined pharmacokinetics and dynamics of pharmaceuticals, novel MSC-based cellular therapeutics demonstrate a highly complex, multifactorial, conditional, and synergistic biological activity in vitro and in vivo [6, 8, 24, 54, 55]. Although being an endeavor for industrial product development, this could indeed be the essence of their superior beneficial activity in the above-mentioned treatment indications [24].

MSC potency testing may thus require a screening strategy for a panel of factors implicated in mediating their therapeutic activity (Fig. 7.2) [5, 24, 56]. For economical reasons, each individual panel to be tested could be adjusted from case to case, according to the specific treatment indication targeted, and the postulated mechanism of action (MOA) identified for this application. The figure summarizes a few of the many possible mediators implied to mediate MSC's beneficial therapeutic effects and potential readout methods.

The factors are categorized here into conventional cell-bound or secreted factors, and short-lived immunomodulatory metabolites. The first category maintains its activity on the cell surface or when secreted and deposited into the surrounding tissue matrix. The second category has to be continuously replenished through action of their parental enzymes.

When attributing MSC's paracrine effect primarily to short-lived metabolites, one may have to consider that their action would be most effective in a closed tissue compartment (e.g. solid tumor tissue compartment). Here, the enzymes can effectively "prepare/condition" the local environment, through depletion or enrichment of specific substrates and metabolites, respectively, temporarily rendering the neighboring or infiltrating immune cells quiescent.

It should also be considered that any stronger diffusion or mass exchange process (e.g. through blood circulation) might render the metabolite mechanism less effective. This could be counteracted by infusing/injecting sufficiently large amounts of cells with the potential to drive systemic immunomodulation via respective paracrine mechanisms.

Both resting and inflammation-activated MSCs should be studied [5, 8, 24, 55, 56], to obtain better representation of their bioactivity within a complex inflammatory environment. A recent International Society for Cellular Therapy (ISCT) position paper recommends a defined stimulus for immune activation [5], e.g. human interferon gamma (IFNg), or IFNg used together with tumor necrosis factor alpha (TNFa) or interleukin 1 alpha/beta (IL1a/b) [55]. The combination of IFNg with one of the other factors enhances their inhibitory effect [57], which will allow to readout their conditional response on a molecular level.

In a more pleiotropic approach, data obtained with these defined stimuli have been supplemented with data from MSCs exposed to complex stimuli [24, 55, 56, 58]. This can be achieved by coculture with mitogen- or alloantigen-stimulated peripheral blood mononuclear cells (PBMCs) or by coculture with cellular subsets (e.g. anti-CD3/CD28 stimulated T-cells) [59]. The advantage of these complex stimuli lies in the fact that the resulting MSC response is not

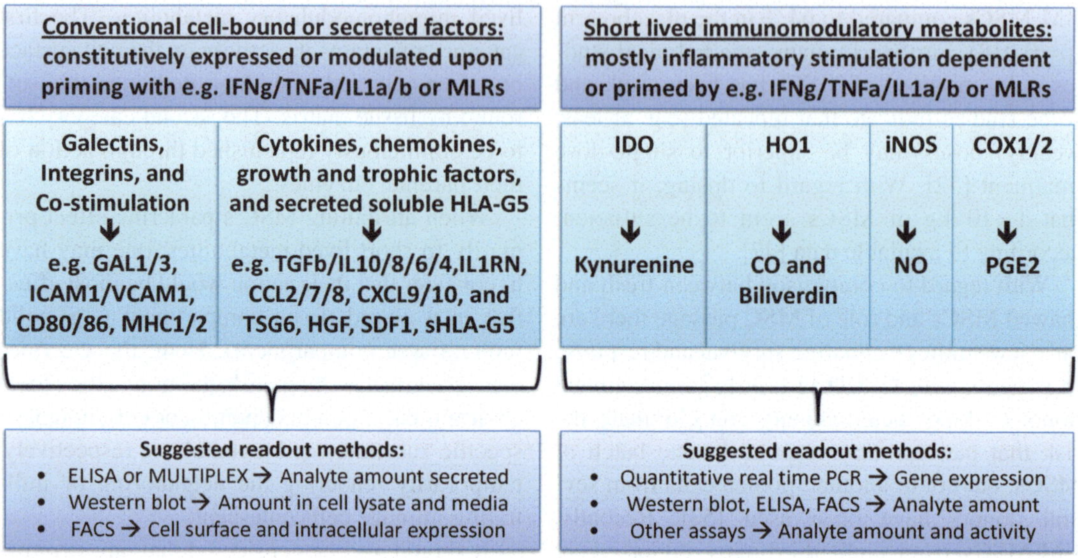

Fig. 7.2 Conventional readout strategy for MSC-derived therapeutic factors. The list of factors given in this figure does not aim to be a complete representation of all the possible paracrine mediators associated with MSC's therapeutic function. Both, resting and inflammation primed MSCs should be studied. We here also suggest investigating both, defined and complex inflammatory stimuli. *Abbreviations*: *CD80/86* cluster of differentiation 80/86, *GAL1/3* galectin 1 and 3, *MHC1/2* major histocompatibility complex 1 and 2, *ICAM1* intracellular adhesion molecule 1, *VCAM1* vascular cell adhesion protein 1, *IFNg* interferon gamma, *TNFa* tumor necrosis factor alpha, *IL1a/b* interleukin 1 alpha and beta, *MLR* mixed lymphocyte reaction, *TGFb* transforming growth factor beta, *IL1RN* IL1 receptor antagonist, *CCL2/7/8* chemokine C-C motif ligand 2, 7, and 8, *CXCL9/10* chemokine C-X-C motif 9 and 10, *TSG6* TNF-stimulated gene 6 protein, *HGF* hepatocyte growth factor, *SDF1* stroma cell-derived factor 1, *sHLA-G5* soluble human leukocyte antigen G5, *IDO* indoleamine 2,3-desoxygenase, *HO1* hemeoxygenase 1, *CO* carbon monoxide, *iNOS* inducible nitric oxide synthase, *NO* nitric oxide, *COX1/2* cyclic oxide synthase 1 and 2, *PGE2* prostaglandin E2, *ELISA* enzyme linked immunosorbent assay, *FACS* fluorescence-activated cell sorting, *PCR* polymerase chain reaction

limited to one specific pathway, but represents an aggregate multifactorial MSC response, more accurately depicting the highly complex molecular crosstalk in vivo.

The classical view that MSCs exert their function primarily through the paracrine secretion of small molecules has recently been contested by a newer hypothesis (Fig. 7.3). Upon systemic infusion or whole blood exposure MSCs activate the IBMIR (Fig. 7.3a), normally associated with adverse effects but here leading to beneficial reactions (Fig. 7.3b). Innate immunity promotes disintegration and release of bioactive microparticles [14], e.g. exosomes (70–150 nm), microvesicles (100 nm to 1 um), and apoptotic bodies (1–5 um) [14, 60].

These microparticles contain numerous immunomodulatory molecules, micro-RNAs, and transfer-RNAs [60–65], with the capacity to induce regenerative processes in vivo, e.g. angiogenesis and tissue self-renewal, and to modulate host-resident immune effector cells, e.g. MSC-derived microparticles can be engulfed by phagocytic cells, leading to a type-2 reprogramming into immunosuppressive and regenerative phenotypes (Fig. 7.3b).

For example, MSC's were shown to shed micro-RNA-containing exosomes that inhibit macrophage activation by suppressing Toll-like receptor signaling [65]. These exosomes were also shown to transfer mitochondria by means of fusion with macrophage, resulting in enhanced macrophage bioenergetics [65]. This study mechanistically links mitophagy and MSC survival with macrophage function, thereby providing a physiologically relevant context for the innate immunomodulatory activity of MSCs.

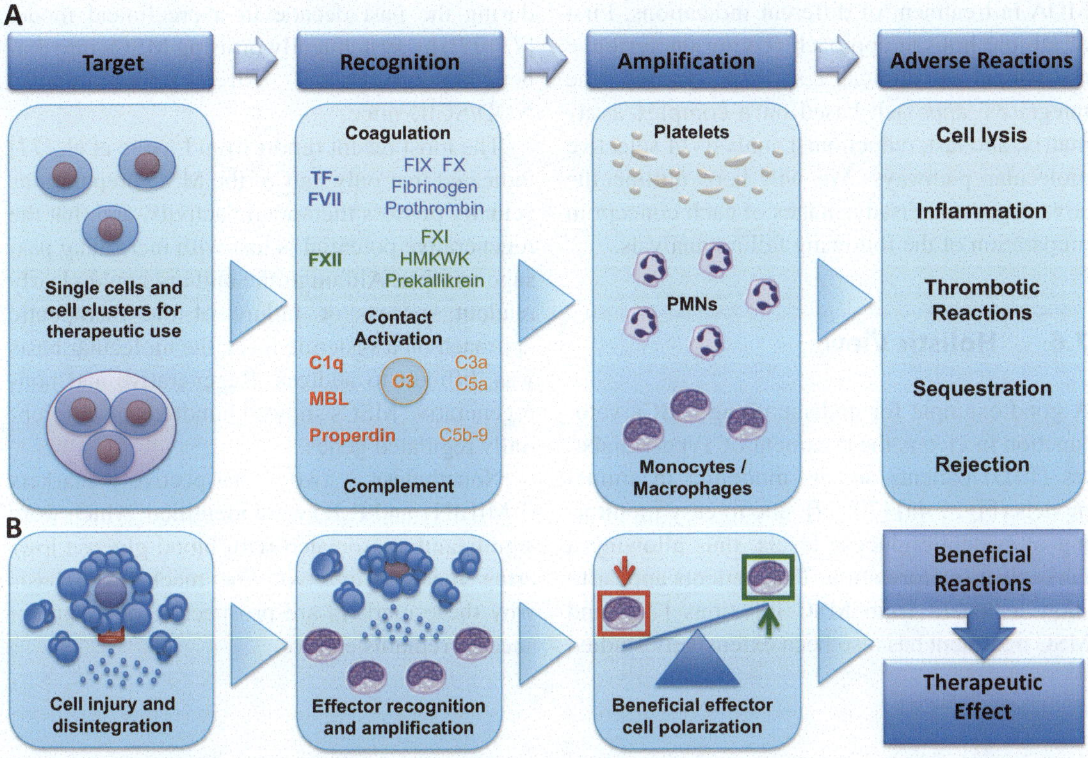

Fig. 7.3 Revised hypothesis for MSC therapeutic function in vivo. (**a**) Instant blood-mediated inflammatory reaction (IBMIR, this figure is adapted from incompatibility reactions triggered by innate immune responses to therapeutic cells [6]). Upon exposure to blood, recognition molecules belonging to different innate cascade systems target altered-self and non-self structures on cells. Factor VII (FVII), fibrinogen, and tissue factor (TF) are "recognition or trigger" molecules of the coagulation system, FXII and high molecular weight kininogen (HMWK) of the contact system, and C1q, mannose-binding lectin (MBL), and properdin of the complement system. The activation of each cascade system triggers multiple amplification reactions: activation of the coagulation cascade leads to generation of prothrombin and thrombin; activation of the contact system elicits generation of the potent vasoactive peptide bradykinin from HMWK; in the complement cascade, there is a powerful amplification of C3 that initiates generation of the anaphylatoxins C3a and C5a, as well as the lytic C5b-9 complex. These activation products trigger activation of platelets, polymorphonuclear leukocytes (PMNs), and monocytes/macrophages, which normally results in thrombotic and inflammatory reactions. These adverse events, together with complement-mediated cell lysis and coagulation-mediated sequestration may lead to rejection or serious damage to the cells. (**b**) The IBMIR response to MSCs, conventionally understood as adverse reaction, may actually promote beneficial reactions and MSC therapeutic effects. Cell injury and disintegration due to innate immune cascade recognition promotes the release of different types of bioactive molecules and microparticles from dying MSCs, thus greatly amplifying the initial signal, leading to signal recognition/amplification by multiple effector cell types, here represented by monocytes/macrophages, and eventually effector cell maturation/polarization into beneficial phenotypes, thus halting adverse immune reaction and promoting beneficial endogenous repair processes by reprogrammed host cells

7.5 Mechanism of Action

Conventional wisdom acquired during the development of pharmaceuticals may not always be suitable to assess or understand the complex in vivo MOA of MSCs and other cell-based advanced therapy medicinal products (ATMPs). The cellular and molecular mechanisms employed by MSCs to exert their therapeutic effects are manifold and their complicated cross-talk with host cells is highly complex [8, 22, 24, 55, 66–69].

The prior section on potency assessment introduces three major ideas to understand MSCs

MOA in treatment of different indications. First of all, the holistic approach relying on systemic assessment of clinical responses, second, the integrative approach based on a complex assay matrix, and last, reductionist analysis of selective molecular pathways. We will here outline the advantages and disadvantages of each concept in preparation of the following failure analysis.

7.6 Holistic View

A good example for understanding MSCs veto-function in vivo is the treatment of Type 1 diabetes (T1D) patients and its modeling in animal models (Fig. 7.4a) [70–77], due to easy monitoring of systemic glucose levels, thus allowing a fairly clear interpretation. T1D patients appear to show a benefit from MSC infusions [72], and MSC treatment has also been extensively studied during the past decade in a preclinical model [73–77], by systemically infusing MSCs into diabetic/hyperglycemic streptozotocin-induced NOD/SCID mice.

The most recent report from Lavoie et al. [77] indicated that only half of the MSC preparations actually possess therapeutic activity, and that the regenerative potential is lost with increasing passage number. Although the authors could clearly readout success or failure of the therapeutic approach on a systemic level, the molecular basis was difficult to address. Regenerative and non-regenerative MSCs showed hundreds of differentially regulated genes.

Nonetheless, two prospective markers (EMILIN1 and ILK) were identified, which were significantly associated with blood glucose lowering function in vivo. The mechanistic basis why these markers are prospective of treatment success remains elusive.

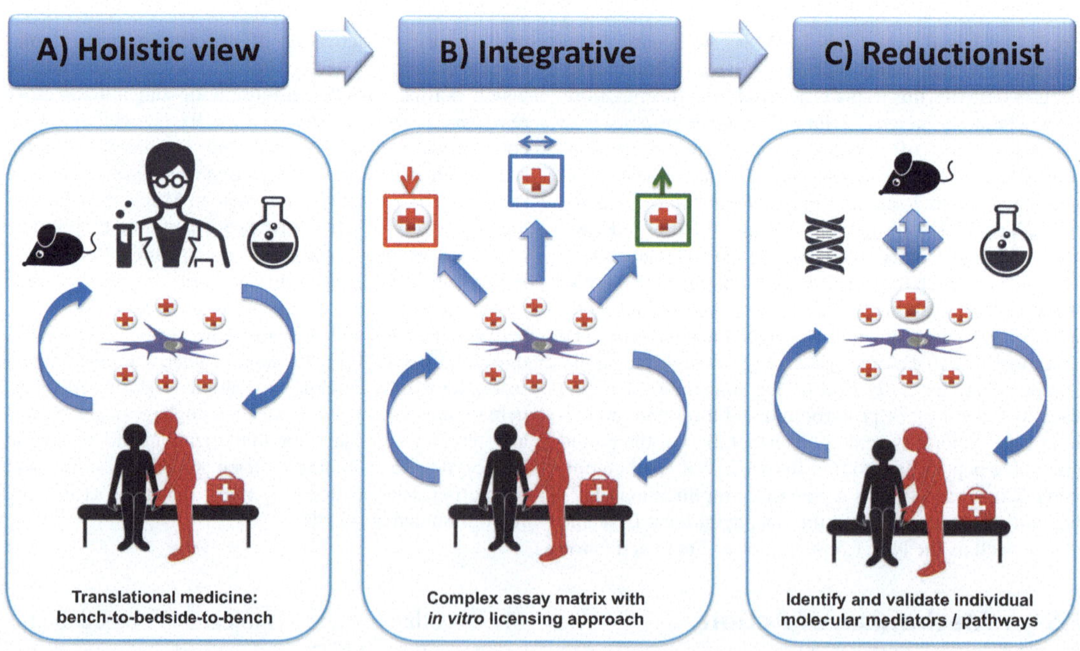

Fig. 7.4 Approaches to decipher MSC's mechanism of action. (**a**) Holistic view, from bench to bedside and back again, based on patient trials supported by modeling in preclinical animal models to decipher the therapeutic value of MSCs in patient oriented clinical studies, (**b**) integrative analysis, in vitro assessment of MSCs therapeutic properties through a complex assay matrix, supported by a robust in vitro licensing approach for studying resting and activated MSCs in order to replicate the clinical situation, and (**c**) reductionist approach, for identifying and validating individual molecular mediators and signaling pathways associated with MSCs therapeutic function

7.7 Integrative Analysis

MSC-like cell products share fundamental MOAs mediating their anti-inflammatory and repair properties [5, 8]. It is widely accepted, that the pharmaceutical effect of MSCs is predominantly mediated by both paracrine and contact-dependent factors, arising from intrinsic MSC physiological processes maintained after culture expansion. It is also accepted, that following in vivo delivery, MSCs respond to environmental cues, leading to additional cellular functionalities [18]. Thus, the ISCT suggested using an assay matrix, based on assessment of selected MSC markers robustly deployed by in vitro licensing (Fig. 7.4b) [5].

The ISCT workshop participants identified three preferred analytic methods that could inform a matrix assay approach: quantitative RNA analysis of selected gene products, flow cytometry analysis of functionally relevant surface markers and protein based assays of the secretome. The authors highlighted that the assay matrix approach could satisfy both mechanistic research, as well as development of release potency assays, to meet regulatory authority requirements for conduct of advanced clinical studies and their eventual registration.

7.8 Reductionist Approach

This approach aims to directly connect the disease status, interventional treatment, and consequent healing outcome with a molecular mediator/signaling pathway (Fig. 7.4c). Based on animal models and detailed mechanistic in vitro experiments one can elucidate cellular and molecular details of specific underlying mechanisms. The identification of these MSC-derived mediators could lead to markers for therapeutic potency and open new avenues for early supportive intervention. One of the greatest challenges is to find valid animal models appropriately mimicking human disease [78]. For example, MSCs were used in patients suffering from severe GvHD [34, 79], a remarkable success now validated by others [21]. Notably, subsequent studies in mouse models reported just after the first clinical study, did not demonstrate a therapeutic effect [78]. Since then, the scientific foundation of MSC suppression of GvHD in both humans and mice is increasingly understood, but if the mouse models had been tested first, the patients may have never been treated with MSCs [78].

7.9 Failure Analysis

The excitement about MSCs unique properties has led to a rapid clinical translation from initial preclinical proof-of-concept, to safety and efficacy analysis of therapeutic MSCs. Although MSC safety was documented in most cases, the rapid translation was overshadowed by a lack in efficacy of many first generation products [4]. Failure analysis, however, significantly contributed to improving our knowledge and to advance clinical trials. Major challenges remain [80], mainly related to issues in MSC production, e.g. cell product characterization, loss of potency with time in culture, and isolation methods that maximize per-cell/unit therapeutic efficacy. On the patient side, attention focused on the mechanistic basis for rapid cell elimination after systemic infusion, anti-donor immune responses to allogeneic MSCs and the primary mechanism of action in different clinical situations [80].

7.10 Engraftment and Bioactivity

The therapeutic efficacy of MSCs needs to be improved [6], and investigators wondered if increased cell persistence in vivo could promote a more sustainable therapeutic effect and supported the idea that a transient increase in cell survival would be beneficial in eliciting more sustainable responses in patients. Long-term engraftment of infused MSCs in patients is low to undetectable [81, 82]. Kinetics of MSC persistence in vivo indicate that most infused cells are rapidly cleared (Fig. 7.1c), due to embolisation and damage in the microvasculature [32, 83, 84]. This may result from the increase in MSC size and phenotypic changes accumulating after expansion [16, 85–87], triggering of innate immune cascades [13–17], and triggering of innate/adaptive cellular immune responses, e.g. NK-, T-, and B-cells [88]. Improved microvascular passage can be achieved by using smaller cells [89], by vasodilation [90, 91], and by using anticoagulants [17, 92]. Anticoagulation has also been shown to reduce anaphylactic reactions and microvascular toxicity in response to therapeutic stromal cells [16, 17, 93]. Experience with clinical islet cell and MSC infusion has shown that IBMIR damages many therapeutic cells [94–96], compromising engraftment and bioactivity, thus raising the need for multiple cell graft infusions [97].

7.11 Adverse Immune Reactions

The fast recognition, embolisation, and clearance of infused MSCs suggests the involvement of the IBMIR (Figs. 7.1c and 7.3a) [6, 13–18, 95], a sequence of incompatibility reactions of the innate immune system, observed after systemic introduction of cells and biomaterials into the blood circulation [95]. The IBMIR elicits its deleterious effects through a cascade of events being initiated by triggering of innate immune cascades and followed by subsequent effector cell infiltration and graft destruction [6].

Both MSCs pro-coagulant [13, 98, 99] and complement activating properties [100–102] have been implied to hamper graft performance [88, 103–105]. Known risk factors attributed to recognition of MSCs by IBMIR are induction of tissue factor expression [13, 16, 98, 99], long-term culture expansion [13, 16, 99], use of thawed cells from cryostorage [14, 15, 103], and potentially allogeneic mismatch [88, 101, 106]. Overcoming IBMIR-related graft damage could prove to be a key for maximizing the efficacy of cellular therapies [95, 96].

Apart from IBMIR, therapeutic MSCs may also become the target of innate and adaptive cellular immune responses such as NK-, T-, and B-cells [88]. Two major genetic barriers are embedded in every human, providing the genetic basis for transplant incompatibility and graft rejection: (1) the allogeneic-barrier between different human individuals, and (2) the xenogeneic-barrier between species. Both aspects are frequently reviewed in the context of MSC transplantation, e.g. when discussing the donor choice for clinical use or the choice of supplements for cell culture and processing [4, 24, 80, 107–109].

Although MSCs inhibit activation of NK cells in vitro, the cells can be targeted by activated NK cells in vivo [88]. Allogeneic MSCs show attenuated immunogenicity and may thus be rejected over time due to induction of adaptive immunity [107, 110, 111]. This limits their use in tissue replacements, but they may provide sufficient in vivo persistence for eliciting immunomodulation, with the safety advantage of being only transiently present. Furthermore, clinical MSCs do not express xeno or ABO antigens [15, 112, 113], but they could potentially be contaminated with these highly immunogenic carbohydrate antigens from culture supplements and washing buffers used for cell preparation [15, 114, 115].

7.12 Timing Is Everything

Timing is critical in MSC therapy, from the initial MSC isolation from adult-donors, to the best time points for patient treatment, and eventually choosing meaningful time points for readout of treatment response and conducting correlations to MSC product properties [8]. Thus, we will outline in the following paragraphs some of the most important aspects of timing in MSC therapy and strongly excuse any shortcomings due to brevity.

First of all, the time point of cell infusion in relation to the progress of the treated indication in patients may affect their therapeutic efficacy [8], e.g. in treatment of GvHD earlier diagnosis of disease-progression and fast intervention may yield a better response. Also younger patients may respond better than older patients [52]. As outlined above, patient clinical responses to MSC batches may not be robust and they may be transient in nature, both complicating patient response readout and MSC potency assessment.

Second, a number of time-dependent variables complicate MSCs production [6, 8].

Particularly intrinsic (cell donor age) and extrinsic cell ageing (ex vivo expansion process) may affect MSCs therapeutic properties, as reviewed elsewhere [116–120]. It is now well established that bone marrow stem cell content decreases with donor age [121], and that both, donor age and comorbidities, may affect cell efficacy [122]. If allogeneic cells are applicable, MSCs from younger donors may thus optionally be preferred for clinical use.

The most evident factor effecting MSCs potency appears to be the time in culture. Potency reduces after expansion and after repeated passaging [4, 8, 13, 35, 82, 83, 123–127], resulting in a gradual loss of progenitor properties and tissue forming capacity [123, 128], reduced long-term engraftment [82], lower clinical response and survival benefit [35, 125], and increased triggering of IBMIR [13], thus compromising engraftment and function.

Eventually, a major point of criticism raised is [3], that MSCs are mostly infused as thawed cells shortly after retrieval from cryostorage. Thus, it would appear more meaningful to also assess their functional properties within this time frame [4]. If thawed cells actually have different therapeutic qualities than fresh cells harvested directly from culture will be addressed in the following sections.

7.13 Fresh Versus Thawed Cells

Differences in clinical outcome may result from variables in MSC production and consequent therapeutic properties, as outlined above and reviewed elsewhere [6]. The product potency can be affected by MSC tissue origin, donor variation, culture time, supplements, and various other aspects of cell delivery, e.g. the use of fresh or thawed cells [4, 6, 9, 10].

Recent data indicate that it could be crucial for the outcome of MSC therapy, if the functional potency assessment of MSC products is conducted before or after cryobanking and consequent local or systemic therapeutic use (Fig. 7.5) [3]. Past experience has shown that hematopoietic stem cells can be cryopreserved with sufficient functional recovery [3, 129], but it is not clear yet if same also applies to MSCs [3]. Thus, we have established a working model for addressing issues of potential relevance (Fig. 7.6), to compare fresh and thawed MSCs produced with conventional protocols.

7.14 Clinical Cell Graft Preparation

Both, freshly harvested cells and freeze-thawed cells derived from cryostorage are usually detached from adherent 2D (e.g. single-layer or multi-stack tissue culture flasks) or 3D culture surfaces (e.g. advanced hollow-fiber bioreactors) by employing proteolytic enzymes such as trypsin (Fig. 7.6a). After reconstituting the cells as "single cell suspension" with an appropriate cryomedium, MSCs usually undergo the cryobanking procedure by controlled-rate freezing and cryobanking in liquid nitrogen or gas phase devices [9, 10].

While adherent cells in log-phase growth display optimal cell membrane physiology and metabolism, both detached fresh or freeze-thawed cells display substantial differences in physiology (Fig. 7.6b) [14, 15, 18, 59, 130–137]. Thawed cells readily recovered from cryostorage can show a higher degree of membrane asymmetry, phosphatidylserine exposure, heat shock protein expression, and actin-cytoskeletal disruption. Altogether, this may result in a relevant difference in susceptibility to innate and adaptive host immune recognition [14], and biodistribution following systemic delivery [135, 136], and may furthermore compromise MSC's intrinsic production of immunomodulatory molecules [130].

This may lead to cell injury and disintegration post infusion (Fig. 7.6c) [14], as characterized by cell shrinkage, cytoplasmic leakage, debris and apoptotic body formation, all induced by a panel of damage mechanisms (Fig. 7.6d), such as complement opsonization and cell membrane pore formation, T-cell cytolysis, and phagocytic engulfment [14, 136], which will attack particularly the readily thawed MSCs shortly after infusion [18, 136].

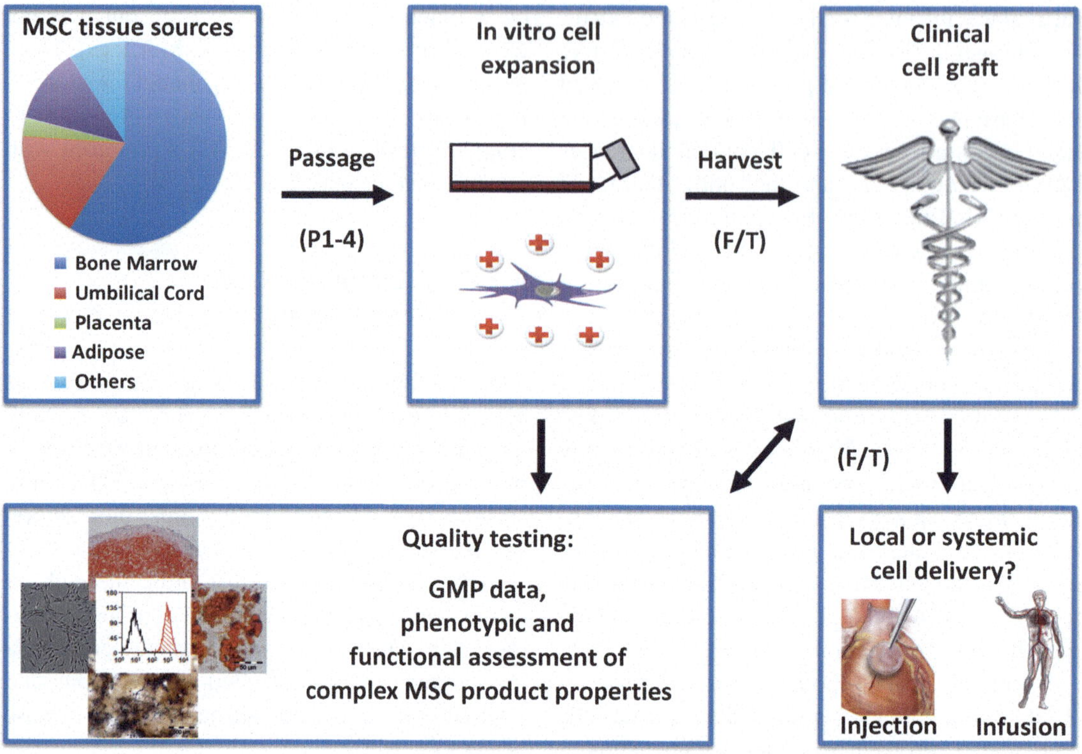

Fig. 7.5 Basic overview for the MSC production process. MSCs can be derived from different adult and prenatal tissue sources, upon primary isolation through selective adherence to tissue culture plastic, the cells are expanded in vitro for several passages (often P1-4), and then usually harvested via enzymatic detachment with proteolytic enzymes such as trypsin, and subjected to cryobanking. The clinical cell graft is subjected to extensive quality testing, such as phenotypic and functional assessment, before and optionally also after cryobanking. The cell graft can be delivered to patients via local injection or systemic infusion. *Abbreviations*: *GMP* good manufacturing process, *MSC* mesenchymal stromal cell, *F* fresh, *T* thawed, *P* passage

This complex mixture of viable and dying MSCs and cell-derived constituents (e.g. free or membrane engulfed MSC-derived bioactive molecules and membrane patterns) [14], can induce effector cell polarization into beneficial phenotypes (Fig. 7.6e) [100]. The efficacy of this process may or may not differ for fresh and thawed MSCs [18], depending on the exact disintegration rates and metabolic responsiveness of both cell types.

Interestingly, a series of very recent reviews ascribed similar immunomodulatory activity as described for MSCs also to the silent clearance of apoptotic cells [138, 139], with the potential to reduce GvHD, favor engraftment and reduce transplant rejection. Importantly, the outcome of apoptotic cell-based therapies may be highly dependent on the specific early apoptotic signals to be mediated and the appropriate apoptotic cell type to be used.

Other clinically relevant differences between fresh and thawed cells my lie in their noticeably different aggregation properties, antigen adsorption, and microvascular toxicity upon reconstitution as "single cell suspension" with various infusion buffers [15, 93, 103, 137], and after subsequent exposure to human whole blood following systemic delivery [13–16], resulting from their suboptimal hemocompatibility, particularly after thawing [14].

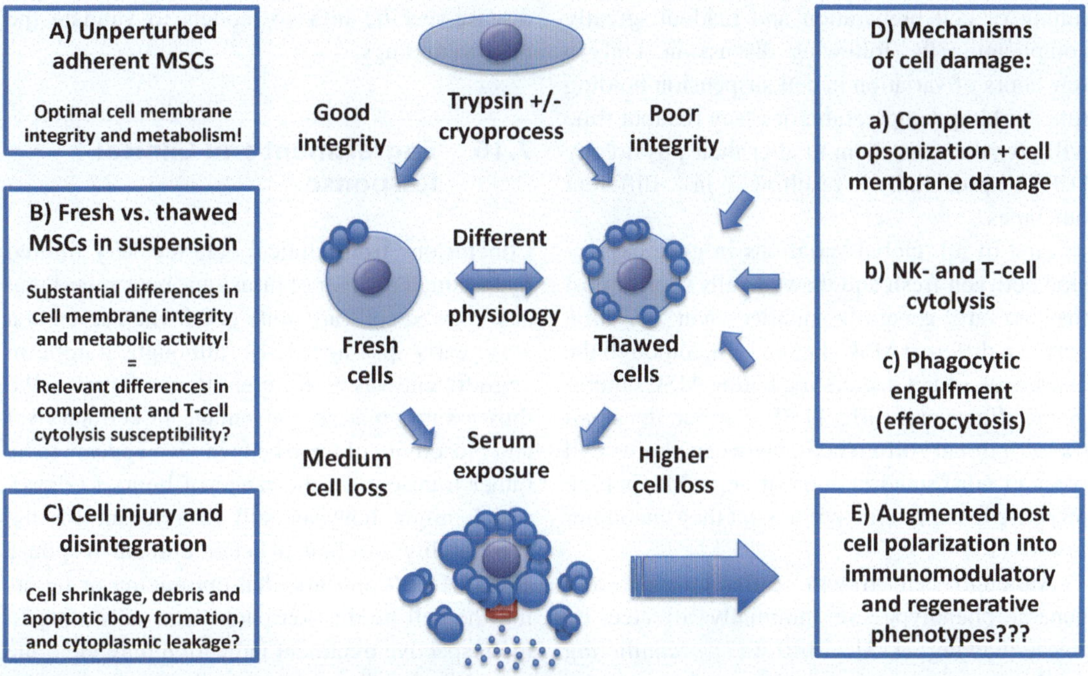

Fig. 7.6 Immunological analysis of fresh versus thawed MSCs

Table 7.1 Comparison of fresh and thawed cells in vitro and in vivo

Cellular properties in vitro and in vivo	No relevant difference	Relevant difference	Restored after cryorecovery?
Differentiation in vitro	[140–147]	[131]	
Cell growth in vitro	[140–146]	[134]	
Phenotype in vitro	[14, 131, 140–147]	[18, 130, 133–137]	[130]
Viability in vitro	[14, 134, 142–147]	[18, 130, 131, 133, 135, 137]	[130]
Bioactivity in vitro	[135, 147]	[14, 59, 130, 136]	[130, 136]
Bioactivity in vivo	[148]	[14, 18, 103, 132]	[132]
Engraftment in vivo	[14]	[135, 136]	[135]
Biodistribution in vivo	[14]	[135, 136]	[135]
Adverse events in vivo	[14]	[18, 103]	[103]

7.15 Cell Phenotype and Bioactivity

As a matter of fact, MSCs are subject to substantial physiological changes during the freeze-thawing procedure, which potentially affects a number of cellular properties (Table 7.1), such as in vitro cell growth, phenotype, differentiation, viability, bioactivity, and most importantly their in vivo bioactivity, biodistribution and safety profile [10]. Some researchers have observed altered properties of thawed MSCs [14, 15, 18, 59, 130–137], while others have documented little to no differences between both cell types (Table 7.1) [140–148].

Either way, many researchers documented that compromised functionality of thawed cells can be restored following a short subculture period post thawing [130, 132, 135, 136]. Thus, the answer may lie in the exact assay design and timing of cell preparation and readout, greatly complicating the following discussion. Only a few hours of variation in cell suspension holding time, holding temperature, or assay readout time will give the cells room to alter their physiology [131], potentially resulting in different outcomes.

First of all, global variations in gene expression between fresh and thawed cells shortly post thawing are generally smaller than variation between different MSC donors [18], although the expression of key factors mediating MSCs therapeutic effects may differ [130]. Further, the small transcriptional differences between both cell types in vitro stand in no relation to MSCs high response to the microenvironment they encounter in vivo [18].

Although cell growth, differentiation, and general phenotype are minimally affected by freeze-thawing [131, 140–147], conflicting results are reported for viability and bioactivity [14, 18, 130, 131, 133, 135–137]. In vitro differences in bioactivity between fresh and thawed cells can be found with earlier readout post thawing [14, 130], which replicates clinical reality of rapid cell infusion post cryoretrieval. MSCs recovering from the freeze-thawing process are impaired by the cellular heat shock response [130], but these differences are diluted at later time points [130, 147], indicating functional recovery during extended in vitro culture.

In contrast, to the in vitro situation, the in vivo readout strongly depends on the initial MSC holding temperature and time of cells in suspension before infusion [131], and the exact time point of analysis post infusion [18]. While MSC's transcriptional and translational readout shows little functional differences shortly (2 h) after cell infusion [18], readout at 8 h post infusion detected significant differences in the IBMIR response. Importantly, recent work by Cruz et al. highlights that fresh and thawed MSCs can be equally effective in vivo and that more disease-specific in vivo models should be tested [148].

To conclude, similarities and differences exist between fresh and thawed cells and timing is crucial for the observed outcome. Potency assessment for clinical use should more strongly engage disease-specific in vivo models to validate the in vitro findings.

7.16 Engraftment and Clinical Response

Conclusions from clinical data are very limited by the small number of infusions, but suggest that also thawed cells are quite good when given at a very early passage [14], although long-term engraftment was no different to our surprise. We thus assume that any advantage in cell survival and bioactivity of fresh cells in vivo appears to be rather transient (in the range of hours to days), which might however still be relevant for the cells ability to elicit a better clinical response (Fig. 7.1c). It appears, that optimizing or tailoring the cell production parameters according to the respective treatment indication may yield the most beneficial responses in the long run [18]. Some indications might allow for the use of fresh cells and it should then be considered (e.g. if yielding more robust responses), while other ones don't and we might have to find other ways to optimize the clinical efficacy [18].

7.17 Improvements in Process Design

Any improvements in process design need to anticipate the potential MOA of MSCs in different treatment indications. For many indications it has not been demonstrated to date if MSC's in vitro activity correlates with a therapeutic effect in vivo [6, 18]. Do the cells need to be active to elicit good clinical responses? If so, we may have to adjust the clinical application strategy accordingly, to give the cells optimal fitness for clinical use [9, 10].

If any detrimental effects can be overcome by optimizing the freeze-thawing protocol alone is difficult to say, since there might be certain limits to MSC biology [10]. Current data indicate that MSCs need to "reboot" after thawing and need time to restart their metabolism and other biochemical processes, thus one could consider to "jump-start" the cells by giving a revitalization cocktail, or use a cryorecovery approach as suggested by Galipeau [136].

Considering IBMIR and complement activation, it is worth testing different protocols focusing on DMSO, composition of cryomedia, and cell membrane physiology after thawing. Freezing of mammalian cells normally requires cell membrane permeabilization with DMSO forming pores [10], which is the likely reason for thawed MSC's altered surface topology and increased susceptibility to complement-mediated pattern recognition.

7.18 Optimized Freeze-Thawing Protocol

During the past decades, major developmental activities have focused on optimizing the freeze-thawing process of therapeutic cells [1, 9, 10], e.g. by improving freezing media, by refining vitrification, controlled rate freezing, and storage devices used for cell banking, and by developing new approaches for more controlled thawing conditions. Major efforts are currently invested into optimization of cryoprotective agents (CPAs) such as nontoxic nanoscale bio-inspired CPAs or freezing cells encapsulated in nanoliter droplets [149].

7.19 Pre-/Post-thaw Cellular Conditioning

Galipeau et al. studied pre-/post-thaw cellular conditioning and cryo-recovery of therapeutic cells before clinical use [135, 136]. For cryorecovery, the clinical MSCs could be produced and banked by conventional means, and to restart their metabolism before infusion, recovered by a 24–72-h subculture in GMP-compliant devices. The problem with thawed MSCs seems to be unresponsiveness directly after thawing, but if the MSCs are allowed to recover in culture they should be just as good as fresh cells. Further, the evaluation of various conditioning strategies did not yield any added value of alternative cryopreservation methods and formulations or autophagy and caspase inhibitor pretreatment, but IFNg-prelicensing prior to MSC freezing partially restored functionality in vitro and in vivo.

7.20 Continuous Production and Application

Currently, industrial models are mainly focusing on producing cryobanked cell batches, which are then released from the cell bank for clinical use upon demand. Alternatively, future developments may focus on the establishment of high-throughput online approaches, preferentially using fresh cells from bioreactors for direct use. This mode is applicable, if a therapeutic center has the capacity for online production and use of several hundred to thousand cell doses per year at specialized production and application sites, serving a larger area with several million inhabitants. Daily or weekly clinical demand could be estimated, and cell-harvesting windows established, for cells produced with advanced bioreactors. The cells could be harvested and prepared for infusion in the morning and used in the afternoon.

7.21 Conclusion

A hot debate is currently ongoing about the issue whether fresh therapeutic MSCs are more effective than cryopreserved MSCs. Thus, we here reviewed the current status of the field with regards to clinical use of MSCs, their MOA, potency assessment, failure analysis, and new improvements in process design. Although certain indications exist that readily thawed MSCs can be functionally impaired, too few data exist to date to make a final statement on the value of fresh and thawed MSCs. More preclinical and clinical assessment needs to be undertaken to evaluate the use of fresh and frozen cells with focus on individual clinical indications, as undertaken by Cruz et al. [148]. Recent clinical data also indicate that frozen cells are quite efficient when given at very early passage [14, 21], but clinical value may be increased by giving early passage fresh cells [14]. Use of cryobanked cells remains a prerequisite for many acute indications, but pre-/post-thaw cellular conditioning, culture recovery, and continuous cell production and application, are valuable options to be studied in the future. Other approaches to improve therapeutic efficacy may be to give pooled cells from several donors [21, 53], and to employ cell sources that may be more robust to freeze-thawing, such as placenta-derived DSCs. More research in vitro and especially in vivo, in well designed animal models and prospective clinical studies, is necessary to establish when it will be most applicable to give fresh or frozen stromal cells.

Acknowledgements Guido Moll's contributions were made possible by German Research Foundation (DFG) funding through the Berlin-Brandenburg School for Regenerative Therapies (BSRT, GSC203); Sven Geißler was supported by grants from the Berlin-Brandenburg Center for Regenerative Therapies (BCRT), the German Federal Ministry of Education and Research (BMBF, Fkz: 01EC1402B), and the DFG (GE2512/1-2); Lech Ignatowicz was supported by the Swedish Research Council (project 2012-1883); Olle Ringdén was supported by grants from the Swedish Research Council (K2014-64X-05971-34-4), the Swedish Cancer Society (CAN2013/671), the Children's Cancer Foundation (PR2013-0045), the Cancer Society in Stockholm (111293), and Karolinska Institutet.

References

1. Baust JM, Corwin WL, VanBuskirk R, Baust JG (2015) Biobanking: the future of cell preservation strategies. Adv Exp Med Biol 864:37–53
2. Stacey GN, Day JG (2014) Putting cells to sleep for future science. Nat Biotechnol 32:320–322
3. Galipeau J (2013) Concerns arising from MSC retrieval from cryostorage and effect on immune suppressive function and pharmaceutical usage in clinical trials. ISBT Sci Ser 8:100–101
4. Galipeau J (2013) The mesenchymal stromal cells dilemma-does a negative phase III trial of random donor mesenchymal stromal cells in steroid-resistant graft-versus-host disease represent a death knell or a bump in the road? Cytotherapy 15:2–8
5. Galipeau J, Krampera M, Barrett J, Dazzi F, Deans RJ, DeBruijn J, Dominici M, Fibbe WE, Gee AP, Gimble JM, Hematti P, Koh MB, LeBlanc K, Martin I, McNiece IK, Mendicino M, Oh S, Ortiz L, Phinney DG, Planat V, Shi Y, Stroncek DF, Viswanathan S, Weiss DJ, Sensebe L (2016) International Society for Cellular Therapy perspective on immune functional assays for mesenchymal stromal cells as potency release criterion for advanced phase clinical trials. Cytotherapy 18:151–159
6. Moll G, Le Blanc K (2015) Engineering more efficient multipotent mesenchymal stromal (stem) cells for systemic delivery as cellular therapy. ISBT Sci Ser 10:357–365
7. Luk F, de Witte SF, Bramer WM, Baan CC, Hoogduijn MJ (2015) Efficacy of immunotherapy with mesenchymal stem cells in man: a systematic review. Expert Rev Clin Immunol 11:617–636
8. Schepers K, Fibbe WE (2016) Unraveling mechanisms of mesenchymal stromal cell-mediated immunomodulation through patient monitoring and product characterization. Ann N Y Acad Sci 1370:15–23
9. Marquez-Curtis LA, Janowska-Wieczorek A, McGann LE, Elliott JA (2015) Mesenchymal stromal cells derived from various tissues: biological, clinical and cryopreservation aspects. Cryobiology 71:181–197
10. Yong KW, Wan Safwani WK, Xu F, Wan Abas WA, Choi JR, Pingguan-Murphy B (2015) Cryopreservation of human mesenchymal stem cells for clinical applications: current methods and challenges. Biopreserv Biobank 13:231–239
11. Kaipe H, Erkers T, Sadeghi B, Ringden O (2014) Stromal cells-are they really useful for GVHD? Bone Marrow Transplant 49:737–743
12. Martin PJ, Inamoto Y, Flowers ME, Carpenter PA (2012) Secondary treatment of acute graft-versus-host disease: a critical review. Biol Blood Marrow Transplant 18:982–988
13. Moll G, Rasmusson-Duprez I, von Bahr L, Connolly-Andersen AM, Elgue G, Funke L, Hamad OA, Lonnies H, Magnusson PU, Sanchez J, Teramura Y, Nilsson-Ekdahl K, Ringden O, Korsgren O, Nilsson B, Le Blanc K (2012) Are therapeutic human mesenchymal stromal cells compatible with human blood? Stem Cells 30:1565–1574
14. Moll G, Alm JJ, Davies LC, von Bahr L, Heldring N, Stenbeck-Funke L, Hamad OA, Hinsch R, Ignatowicz L, Locke M, Lonnies H, Lambris JD, Teramura Y, Nilsson-Ekdahl K, Nilsson B, Le Blanc K (2014) Do cryopreserved mesenchymal stromal cells display impaired immunomodulatory and therapeutic properties? Stem Cells 32:2430–2442
15. Moll G, Hult A, von Bahr L, Alm JJ, Heldring N, Hamad OA, Stenbeck-Funke L, Larsson S, Teramura Y, Roelofs H, Nilsson B, Fibbe WE, Olsson ML, Le Blanc K (2014) Do ABO blood group antigens hamper the therapeutic efficacy of mesenchymal stromal cells? PLoS ONE 9, e85040
16. Moll G, Ignatowicz L, Catar R, Luecht C, Sadeghi B, Hamad O, Jungebluth P, Dragun D, Schmidtchen A, Ringden O (2015) Different procoagulant activity

of therapeutic mesenchymal stromal cells derived from bone marrow and placental decidua. Stem Cells Dev 24:2269–2279
17. Gleeson BM, Martin K, Ali MT, Kumar AH, Pillai MG, Kumar SP, O'Sullivan JF, Whelan D, Stocca A, Khider W, Barry FP, O'Brien T, Caplice NM (2015) Bone marrow-derived mesenchymal stem cells have innate procoagulant activity and cause microvascular obstruction following intracoronary delivery: amelioration by anti-thrombin therapy. Stem Cells 33:2726–2737
18. Hoogduijn MJ, de Witte SF, Luk F, van den Hout-van Vroonhoven MC, Ignatowicz L, Catar R, Strini T, Korevaar SS, van IJcken WF, Betjes MG, Franquesa M, Moll G, Baan CC (2016) Effects of freeze-thawing and intravenous infusion on mesenchymal stromal cell gene expression. Stem Cells Dev 25:586–597
19. Hoogduijn MJ, Roemeling-van Rhijn M, Engela AU, Korevaar SS, Mensah FK, Franquesa M, de Bruin RW, Betjes MG, Weimar W, Baan CC (2013) Mesenchymal stem cells induce an inflammatory response after intravenous infusion. Stem Cells Dev 22:2825–2835
20. Squillaro T, Peluso G, Galderisi U (2016) Clinical trials with mesenchymal stem cells: an update. Cell Transplant 25:829–848
21. Kuci Z, Bonig H, Kreyenberg H, Bunos M, Jauch A, Janssen JW, Skific M, Michel K, Eising B, Lucchini G, Bakhtiar S, Greil J, Lang P, Basu O, von Luettichau I, Schultz A, Sykora KW, Jarisch A, Soerensen J, Salzmann-Manrique E, Seifried E, Klingebiel T, Bader P, Kuci S (2016) Mesenchymal stromal cells generated from pooled mononuclear cells of multiple bone marrow donors as a rescue therapy for children with severe steroid-refractory graft versus host disease: a multicenter survey. Haematologica 101(8):985–994
22. Uccelli A, Moretta L, Pistoia V (2008) Mesenchymal stem cells in health and disease. Nat Rev Immunol 8:726–736
23. Le Blanc K, Mougiakakos D (2012) Multipotent mesenchymal stromal cells and the innate immune system. Nat Rev Immunol 12:383–396
24. Doorn J, Moll G, Le Blanc K, van Blitterswijk C, de Boer J (2012) Therapeutic applications of mesenchymal stromal cells: paracrine effects and potential improvements. Tissue Eng B Rev 18:101–115
25. Mendicino M, Bailey AM, Wonnacott K, Puri RK, Bauer SR (2014) MSC-based product characterization for clinical trials: an FDA perspective. Cell Stem Cell 14:141–145
26. Kern S, Eichler H, Stoeve J, Kluter H, Bieback K (2006) Comparative analysis of mesenchymal stem cells from bone marrow, umbilical cord blood, or adipose tissue. Stem Cells 24:1294–1301
27. Stubbendorff M, Deuse T, Hua X, Phan TT, Bieback K, Atkinson K, Eiermann TH, Velden J, Schroder C, Reichenspurner H, Robbins RC, Volk HD, Schrepfer S (2013) Immunological properties of extraembryonic human mesenchymal stromal cells derived from gestational tissue. Stem Cells Dev 22:2619–2629
28. Parolini O, Alviano F, Bagnara GP, Bilic G, Buhring HJ, Evangelista M, Hennerbichler S, Liu B, Magatti M, Mao N, Miki T, Marongiu F, Nakajima H, Nikaido T, Portmann-Lanz CB, Sankar V, Soncini M, Stadler G, Surbek D, Takahashi TA, Redl H, Sakuragawa N, Wolbank S, Zeisberger S, Zisch A, Strom SC (2008) Concise review: isolation and characterization of cells from human term placenta: outcome of the first international Workshop on Placenta Derived Stem Cells. Stem Cells 26:300–311
29. Ringden O, Erkers T, Nava S, Uzunel M, Iwarsson E, Conrad R, Westgren M, Mattsson J, Kaipe H (2013) Fetal membrane cells for treatment of steroid-refractory acute graft-versus-host disease. Stem Cells 31:592–601
30. Ringdén O (2014) Placenta-derived decidual stromal cells – a novel therapy for graft-versus-host disease, haemorrhages and toxicity after allogeneic haematopoietic stem cell transplantation
31. Wang H, Liang X, Xu ZP, Crawford DH, Liu X, Roberts MS (2016) A physiologically based kinetic model for elucidating the in vivo distribution of administered mesenchymal stem cells. Sci Rep 6:22293
32. Leibacher J, Henschler R (2016) Biodistribution, migration and homing of systemically applied mesenchymal stem/stromal cells. Stem Cell Res Ther 7:7
33. Eggenhofer E, Benseler V, Kroemer A, Popp FC, Geissler EK, Schlitt HJ, Baan CC, Dahlke MH, Hoogduijn MJ (2012) Mesenchymal stem cells are short-lived and do not migrate beyond the lungs after intravenous infusion. Front Immunol 3:297
34. Le Blanc K, Frassoni F, Ball L, Locatelli F, Roelofs H, Lewis I, Lanino E, Sundberg B, Bernardo ME, Remberger M, Dini G, Egeler RM, Bacigalupo A, Fibbe W, Ringden O (2008) Mesenchymal stem cells for treatment of steroid-resistant, severe, acute graft-versus-host disease: a phase II study. Lancet 371:1579–1586
35. von Bahr L, Sundberg B, Lonnies L, Sander B, Karbach H, Hagglund H, Ljungman P, Gustafsson B, Karlsson H, Le Blanc K, Ringden O (2012) Long-term complications, immunologic effects, and role of passage for outcome in mesenchymal stromal cell therapy. Biol Blood Marrow Transplant J Am Soc Blood Marrow Transplant 18:557–564
36. Ringden O, Uzunel M, Rasmusson I, Remberger M, Sundberg B, Lonnies H, Marschall HU, Dlugosz A, Szakos A, Hassan Z, Omazic B, Aschan J, Barkholt L, Le Blanc K (2006) Mesenchymal stem cells for treatment of therapy-resistant graft-versus-host disease. Transplantation 81:1390–1397
37. von Dalowski F, Kramer M, Wermke M, Wehner R, Rollig C, Alakel N, Stolzel F, Parmentier S, Sockel K, Krech M, Schmitz M, Platzbecker U, Schetelig J,

Bornhauser M, von Bonin M (2016) Mesenchymal stromal cells for treatment of acute steroid-refractory graft versus host disease: clinical responses and long-term outcome. Stem Cells 34:357–366
38. Perez-Simon JA, Lopez-Villar O, Andreu EJ, Rifon J, Muntion S, Diez Campelo M, Sanchez-Guijo FM, Martinez C, Valcarcel D, Canizo CD (2011) Mesenchymal stem cells expanded in vitro with human serum for the treatment of acute and chronic graft-versus-host disease: results of a phase I/II clinical trial. Haematologica 96:1072–1076
39. Arima N, Nakamura F, Fukunaga A, Hirata H, Machida H, Kouno S, Ohgushi H (2010) Single intra-arterial injection of mesenchymal stromal cells for treatment of steroid-refractory acute graft-versus-host disease: a pilot study. Cytotherapy 12:265–268
40. Kinzebach S, Dietz L, Kluter H, Thierse HJ, Bieback K (2013) Functional and differential proteomic analyses to identify platelet derived factors affecting ex vivo expansion of mesenchymal stromal cells. BMC Cell Biol 14:48
41. Dreher L, Elvers-Hornung S, Brinkmann I, Huck V, Henschler R, Gloe T, Kluter H, Bieback K (2013) Cultivation in human serum reduces adipose tissue-derived mesenchymal stromal cell adhesion to laminin and endothelium and reduces capillary entrapment. Stem Cells Dev 22:791–803
42. Bieback K, Hecker A, Schlechter T, Hofmann I, Brousos N, Redmer T, Besser D, Kluter H, Muller AM, Becker M (2012) Replicative aging and differentiation potential of human adipose tissue-derived mesenchymal stromal cells expanded in pooled human or fetal bovine serum. Cytotherapy 14:570–583
43. Bieback K, Ha VA, Hecker A, Grassl M, Kinzebach S, Solz H, Sticht C, Kluter H, Bugert P (2010) Altered gene expression in human adipose stem cells cultured with fetal bovine serum compared to human supplements. Tissue Eng A 16:3467–3484
44. Bieback K, Hecker A, Kocaomer A, Lannert H, Schallmoser K, Strunk D, Kluter H (2009) Human alternatives to fetal bovine serum for the expansion of mesenchymal stromal cells from bone marrow. Stem Cells 27:2331–2341
45. Kocaoemer A, Kern S, Kluter H, Bieback K (2007) Human AB serum and thrombin-activated platelet-rich plasma are suitable alternatives to fetal calf serum for the expansion of mesenchymal stem cells from adipose tissue. Stem Cells 25:1270–1278
46. Gross G, Häupl T (2013) Stem cell-dependent therapies, mesenchymal stem cells in chronic inflammatory disorders
47. Hashmi S, Ahmed M, Murad MH, Litzow MR, Adams RH, Ball LM, Prasad VK, Kebriaei P, Ringden O (2016) Survival after mesenchymal stromal cell therapy in steroid-refractory acute graft-versus-host disease: systematic review and meta-analysis. Lancet Haematol 3:e45–e52
48. Fang B, Song Y, Zhao RC, Han Q, Lin Q (2007) Using human adipose tissue-derived mesenchymal stem cells as salvage therapy for hepatic graft-versus-host disease resembling acute hepatitis. Transplant Proc 39:1710–1713
49. Wu KH, Chan CK, Tsai C, Chang YH, Sieber M, Chiu TH, Ho M, Peng CT, Wu HP, Huang JL (2011) Effective treatment of severe steroid-resistant acute graft-versus-host disease with umbilical cord-derived mesenchymal stem cells. Transplantation 91:1412–1416
50. Martin PJ, Uberti JP, Soiffer RJ, Klingemann H, Waller EK, Daly AS, Herrmann RP, Kebriaei P (2010) Prochymal improves response rates in patients with Steroid-Refractory Acute Graft Versus Host Disease (SR-GVHD) involving the liver and gut: results of a randomized, placebo-controlled, multicenter phase III trial in GVHD. Biol Blood Marrow Transplant 16:S169–S170
51. Kurtzberg J, Prockop S, Teira P, Bittencourt H, Lewis V, Chan KW, Horn B, Yu L, Talano JA, Nemecek E, Mills CR, Chaudhury S (2014) Allogeneic human mesenchymal stem cell therapy (remestemcel-L, prochymal) as a rescue agent for severe refractory acute graft-versus-host disease in pediatric patients. Biol Blood Marrow Transplant J Am Soc Blood Marrow Transplant 20:229–235
52. Ball LM, Bernardo ME, Roelofs H, van Tol MJ, Contoli B, Zwaginga JJ, Avanzini MA, Conforti A, Bertaina A, Giorgiani G, der Zijde CM J-v, Zecca M, Le Blanc K, Frassoni F, Egeler RM, Fibbe WE, Lankester AC, Locatelli F (2013) Multiple infusions of mesenchymal stromal cells induce sustained remission in children with steroid-refractory, grade III–IV acute graft-versus-host disease. Br J Haematol 163:501–509
53. Ringden O, Leblanc K (2011) Pooled MSCs for treatment of severe hemorrhage. Bone Marrow Transplant 46:1158–1160
54. Jitschin R, Mougiakakos D, Von Bahr L, Volkl S, Moll G, Ringden O, Kiessling R, Linder S, Le Blanc K (2013) Alterations in the cellular immune compartment of patients treated with third-party mesenchymal stromal cells following allogeneic hematopoietic stem cell transplantation. Stem Cells 31:1715–1725
55. Krampera M (2011) Mesenchymal stromal cell 'licensing': a multistep process. Leukemia 25:1408–1414
56. Galipeau J, Krampera M (2015) The challenge of defining mesenchymal stromal cell potency assays and their potential use as release criteria. Cytotherapy 17:125–127
57. Ren G, Zhang L, Zhao X, Xu G, Zhang Y, Roberts AI, Zhao RC, Shi Y (2008) Mesenchymal stem cell-mediated immunosuppression occurs via concerted action of chemokines and nitric oxide. Cell Stem Cell 2:141–150
58. Reinke S, Geissler S, Taylor WR, Schmidt-Bleek K, Juelke K, Schwachmeyer V, Dahne M, Hartwig T, Akyuz L, Meisel C, Unterwalder N, Singh NB, Reinke P, Haas NP, Volk HD, Duda GN (2013) Terminally differentiated CD8(+) T cells negatively

affect bone regeneration in humans. Sci Transl Med 5:177ra136

59. Ketterl N, Brachtl G, Schuh C, Bieback K, Schallmoser K, Reinisch A, Strunk D (2015) A robust potency assay highlights significant donor variation of human mesenchymal stem/progenitor cell immune modulatory capacity and extended radio-resistance. Stem Cell Res Ther 6:236
60. Börger V, Bremer M, Görgens A, Giebel B (2016) Mesenchymal stem/stromal cell-derived extracellular vesicles as a new approach in stem cell therapy. ISBT Sci Ser 11:228–234
61. Rani S, Ryan AE, Griffin MD, Ritter T (2015) Mesenchymal stem cell-derived extracellular vesicles: toward cell-free therapeutic applications. Mol Ther 23:812–823
62. Anderson JD, Johansson HJ, Graham CS, Vesterlund M, Pham MT, Bramlett CS, Montgomery EN, Mellema MS, Bardini RL, Contreras Z, Hoon M, Bauer G, Fink KD, Fury B, Hendrix KJ, Chedin F, El-Andaloussi S, Hwang B, Mulligan MS, Lehtio J, Nolta JA (2016) Comprehensive proteomic analysis of mesenchymal stem cell exosomes reveals modulation of angiogenesis via nuclear factor-KappaB signaling. Stem Cells 34:601–613
63. Koniusz S, Andrzejewska A, Muraca M, Srivastava AK, Janowski M, Lukomska B (2016) Extracellular vesicles in physiology, pathology and therapy of the immune and central nervous system, with focus on extracellular vesicles derived from mesenchymal stem cells as therapeutic tools. Front Cell Neurosci 10:109
64. Merino-Gonzalez C, Zuniga FA, Escudero C, Ormazabal V, Reyes C, Nova-Lamperti E, Salomon C, Aguayo C (2016) Mesenchymal stem cell-derived extracellular vesicles promote angiogenesis: potencial clinical application. Front Physiol 7:24
65. Phinney DG, Di Giuseppe M, Njah J, Sala E, Shiva S, St Croix CM, Stolz DB, Watkins SC, Di YP, Leikauf GD, Kolls J, Riches DW, Deiuliis G, Kaminski N, Boregowda SV, McKenna DH, Ortiz LA (2015) Mesenchymal stem cells use extracellular vesicles to outsource mitophagy and shuttle microRNAs. Nat Commun 6:8472
66. Bernardo ME, Locatelli F, Fibbe WE (2009) Mesenchymal stromal cells. Ann N Y Acad Sci 1176:101–117
67. Singer NG, Caplan AI (2011) Mesenchymal stem cells: mechanisms of inflammation. Annu Rev Pathol 6:457–478
68. Bernardo ME, Fibbe WE (2013) Mesenchymal stromal cells: sensors and switchers of inflammation. Cell Stem Cell 13:392–402
69. Ma S, Xie N, Li W, Yuan B, Shi Y, Wang Y (2014) Immunobiology of mesenchymal stem cells. Cell Death Differ 21:216–225
70. Chhabra P, Brayman KL (2013) Stem cell therapy to cure type 1 diabetes: from hype to hope. Stem Cells Transl Med 2:328–336
71. Hematti P, Kim J, Stein AP, Kaufman D (2013) Potential role of mesenchymal stromal cells in pancreatic islet transplantation. Transplant Rev (Orlando) 27:21–29
72. Carlsson PO, Schwarcz E, Korsgren O, Le Blanc K (2015) Preserved beta-cell function in type 1 diabetes by mesenchymal stromal cells. Diabetes 64:587–592
73. Lee RH, Seo MJ, Reger RL, Spees JL, Pulin AA, Olson SD, Prockop DJ (2006) Multipotent stromal cells from human marrow home to and promote repair of pancreatic islets and renal glomeruli in diabetic NOD/scid mice. Proc Natl Acad Sci U S A 103:17438–17443
74. Urban VS, Kiss J, Kovacs J, Gocza E, Vas V, Monostori E, Uher F (2008) Mesenchymal stem cells cooperate with bone marrow cells in therapy of diabetes. Stem Cells 26:244–253
75. Ezquer FE, Ezquer ME, Parrau DB, Carpio D, Yanez AJ, Conget PA (2008) Systemic administration of multipotent mesenchymal stromal cells reverts hyperglycemia and prevents nephropathy in type 1 diabetic mice. Biol Blood Marrow Transplant 14:631–640
76. Ho JH, Tseng TC, Ma WH, Ong WK, Chen YF, Chen MH, Lin MW, Hong CY, Lee OK (2012) Multiple intravenous transplantations of mesenchymal stem cells effectively restore long-term blood glucose homeostasis by hepatic engraftment and beta-cell differentiation in streptozocin-induced diabetic mice. Cell Transplant 21:997–1009
77. Lavoie JR, Creskey MC, Muradia G, Bell GI, Sherman SE, Gao J, Stewart DJ, Cyr TD, Hess DA, Rosu-Myles M (2016) EMILIN-1 and ILK are novel markers of islet regenerative function in human multipotent mesenchymal stromal cells. Stem Cells 34:2249–2255
78. Horwitz EM (2013) Advancing regenerative medicine the translational way. Sci Transl Med 5:177fs179
79. Le Blanc K, Rasmusson I, Sundberg B, Gotherstrom C, Hassan M, Uzunel M, Ringden O (2004) Treatment of severe acute graft-versus-host disease with third party haploidentical mesenchymal stem cells. Lancet 363:1439–1441
80. Griffin MD, Elliman SJ, Cahill E, English K, Ceredig R, Ritter T (2013) Concise review: adult mesenchymal stromal cell therapy for inflammatory diseases: how well are we joining the dots? Stem Cells 31:2033–2041
81. von Bahr L, Batsis I, Moll G, Hagg M, Szakos A, Sundberg B, Uzunel M, Ringden O, Le Blanc K (2012) Analysis of tissues following mesenchymal stromal cell therapy in humans indicates limited long-term engraftment and no ectopic tissue formation. Stem Cells 30:1575–1578
82. Horwitz EM, Gordon PL, Koo WK, Marx JC, Neel MD, McNall RY, Muul L, Hofmann T (2002) Isolated allogeneic bone marrow-derived mesenchymal cells engraft and stimulate growth in children

with osteogenesis imperfecta: implications for cell therapy of bone. Proc Natl Acad Sci U S A 99:8932–8937
83. Wagner B, Henschler R (2013) Fate of intravenously injected mesenchymal stem cells and significance for clinical application. Adv Biochem Eng Biotechnol 130:19–37
84. Erkers T, Kaipe H, Nava S, Mollden P, Gustafsson B, Axelsson R, Ringden O (2015) Treatment of severe chronic graft-versus-host disease with decidual stromal cells and tracing with (111)indium radiolabeling. Stem Cells Dev 24:253–263
85. Karp JM, Leng Teo GS (2009) Mesenchymal stem cell homing: the devil is in the details. Cell Stem Cell 4:206–216
86. Braun J, Kurtz A, Barutcu N, Bodo J, Thiel A, Dong J (2013) Concerted regulation of CD34 and CD105 accompanies mesenchymal stromal cell derivation from human adventitial stromal cell. Stem Cells Dev 22:815–827
87. Hoogduijn MJ, van den Beukel JC, Wiersma LC, Ijzer J (2013) Morphology and size of stem cells from mouse and whale: observational study. BMJ 347:f6833
88. Ankrum JA, Ong JF, Karp JM (2014) Mesenchymal stem cells: immune evasive, not immune privileged. Nat Biotechnol 32:252–260
89. Fischer UM, Harting MT, Jimenez F, Monzon-Posadas WO, Xue H, Savitz SI, Laine GA, Cox CS Jr (2009) Pulmonary passage is a major obstacle for intravenous stem cell delivery: the pulmonary first-pass effect. Stem Cells Dev 18:683–692
90. Gao J, Dennis JE, Muzic RF, Lundberg M, Caplan AI (2001) The dynamic in vivo distribution of bone marrow-derived mesenchymal stem cells after infusion. Cells Tissues Organs 169:12–20
91. Schrepfer S, Deuse T, Reichenspurner H, Fischbein MP, Robbins RC, Pelletier MP (2007) Stem cell transplantation: the lung barrier. Transplant Proc 39:573–576
92. Yukawa H, Watanabe M, Kaji N, Okamoto Y, Tokeshi M, Miyamoto Y, Noguchi H, Baba Y, Hayashi S (2012) Monitoring transplanted adipose tissue-derived stem cells combined with heparin in the liver by fluorescence imaging using quantum dots. Biomaterials 33:2177–2186
93. Deak E, Ruster B, Keller L, Eckert K, Fichtner I, Seifried E, Henschler R (2010) Suspension medium influences interaction of mesenchymal stromal cells with endothelium and pulmonary toxicity after transplantation in mice. Cytotherapy 12:260–264
94. Bennet W, Sundberg B, Groth CG, Brendel MD, Brandhorst D, Brandhorst H, Bretzel RG, Elgue G, Larsson R, Nilsson B, Korsgren O (1999) Incompatibility between human blood and isolated islets of Langerhans: a finding with implications for clinical intraportal islet transplantation? Diabetes 48:1907–1914
95. Nilsson B, Korsgren O, Lambris JD, Ekdahl KN (2010) Can cells and biomaterials in therapeutic medicine be shielded from innate immune recognition? Trends Immunol 31:32–38
96. Nilsson B, Ekdahl KN, Korsgren O (2011) Control of instant blood-mediated inflammatory reaction to improve islets of Langerhans engraftment. Curr Opin Organ Transplant 16:620–626
97. Shapiro AM, Lakey JR, Ryan EA, Korbutt GS, Toth E, Warnock GL, Kneteman NM, Rajotte RV (2000) Islet transplantation in seven patients with type 1 diabetes mellitus using a glucocorticoid-free immunosuppressive regimen. N Engl J Med 343:230–238
98. Stephenne X, Nicastro E, Eeckhoudt S, Hermans C, Nyabi O, Lombard C, Najimi M, Sokal E (2012) Bivalirudin in combination with heparin to control mesenchymal cell procoagulant activity. PLoS ONE 7, e42819
99. Tatsumi K, Ohashi K, Matsubara Y, Kohori A, Ohno T, Kakidachi H, Horii A, Kanegae K, Utoh R, Iwata T, Okano T (2013) Tissue factor triggers procoagulation in transplanted mesenchymal stem cells leading to thromboembolism. Biochem Biophys Res Commun 431:203–209
100. Moll G, Jitschin R, von Bahr L, Rasmusson-Duprez I, Sundberg B, Lonnies L, Elgue G, Nilsson-Ekdahl K, Mougiakakos D, Lambris JD, Ringden O, Le Blanc K, Nilsson B (2011) Mesenchymal stromal cells engage complement and complement receptor bearing innate effector cells to modulate immune responses. PLoS ONE 6, e21703
101. Li Y, Lin F (2012) Mesenchymal stem cells are injured by complement after their contact with serum. Blood 120:3436–3443
102. Soland MA, Bego M, Colletti E, Zanjani ED, St Jeor S, Porada CD, Almeida-Porada G (2013) Mesenchymal stem cells engineered to inhibit complement-mediated damage. PLoS ONE 8, e60461
103. Quimby JM, Webb TL, Habenicht LM, Dow SW (2013) Safety and efficacy of intravenous infusion of allogeneic cryopreserved mesenchymal stem cells for treatment of chronic kidney disease in cats: results of three sequential pilot studies. Stem Cell Res Ther 4:48
104. Bianco P, Barker R, Brustle O, Cattaneo E, Clevers H, Daley GQ, De Luca M, Goldstein L, Lindvall O, Mummery C, Robey PG, Sattler de Sousa EBC, Smith A (2013) Regulation of stem cell therapies under attack in Europe: for whom the bell tolls. EMBO J 32:1489–1495
105. Bianco P, Cao X, Frenette PS, Mao JJ, Robey PG, Simmons PJ, Wang CY (2013) The meaning, the sense and the significance: translating the science of mesenchymal stem cells into medicine. Nat Med 19:35–42
106. Alagesan S, Griffin MD (2014) Autologous and allogeneic mesenchymal stem cells in organ transplanta-

tion: what do we know about their safety and efficacy? Curr Opin Organ Transplant 19:65–72
107. Millard SM, Fisk NM (2013) Mesenchymal stem cells for systemic therapy: shotgun approach or magic bullets? Bioessays 35:173–182
108. Tolar J, Le Blanc K, Keating A, Blazar BR (2010) Concise review: hitting the right spot with mesenchymal stromal cells. Stem Cells 28:1446–1455
109. Ankrum J, Karp JM (2010) Mesenchymal stem cell therapy: two steps forward, one step back. Trends Mol Med 16:203–209
110. Kaipe H, Carlson LM, Erkers T, Nava S, Mollden P, Gustafsson B, Qian H, Li X, Hashimoto T, Sadeghi B, Alheim M, Ringden O (2015) Immunogenicity of decidual stromal cells in an epidermolysis bullosa patient and in allogeneic hematopoietic stem cell transplantation patients. Stem Cells Dev 24:1471–1482
111. Crop MJ, Korevaar SS, de Kuiper R, IJzermans JN, van Besouw NM, Baan CC, Weimar W, Hoogduijn MJ (2011) Human mesenchymal stem cells are susceptible to lysis by CD8(+) T cells and NK cells. Cell Transplant 20:1547–1559
112. Sundin M, Ringden O, Sundberg B, Nava S, Gotherstrom C, Le Blanc K (2007) No alloantibodies against mesenchymal stromal cells, but presence of anti-fetal calf serum antibodies, after transplantation in allogeneic hematopoietic stem cell recipients. Haematologica 92:1208–1215
113. Schafer R, Schnaidt M, Klaffschenkel RA, Siegel G, Schule M, Radlein MA, Hermanutz-Klein U, Ayturan M, Buadze M, Gassner C, Danielyan L, Kluba T, Northoff H, Flegel WA (2011) Expression of blood group genes by mesenchymal stem cells. Br J Haematol 153:520–528
114. Spees JL, Gregory CA, Singh H, Tucker HA, Peister A, Lynch PJ, Hsu SC, Smith J, Prockop DJ (2004) Internalized antigens must be removed to prepare hypoimmunogenic mesenchymal stem cells for cell and gene therapy. Mol Ther 9:747–756
115. Heiskanen A, Satomaa T, Tiitinen S, Laitinen A, Mannelin S, Impola U, Mikkola M, Olsson C, Miller-Podraza H, Blomqvist M, Olonen A, Salo H, Lehenkari P, Tuuri T, Otonkoski T, Natunen J, Saarinen J, Laine J (2007) N-glycolylneuraminic acid xenoantigen contamination of human embryonic and mesenchymal stem cells is substantially reversible. Stem Cells 25:197–202
116. Geissler S, Textor M, Schmidt-Bleek K, Klein O, Thiele M, Ellinghaus A, Jacobi D, Ode A, Perka C, Dienelt A, Klose J, Kasper G, Duda GN, Strube P (2013) In serum veritas-in serum sanitas? Cell non-autonomous aging compromises differentiation and survival of mesenchymal stromal cells via the oxidative stress pathway. Cell Death Dis 4, e970
117. Dimmeler S, Leri A (2008) Aging and disease as modifiers of efficacy of cell therapy. Circ Res 102:1319–1330
118. Dimmeler S, Ding S, Rando TA, Trounson A (2014) Translational strategies and challenges in regenerative medicine. Nat Med 20:814–821
119. Efimenko AY, Kochegura TN, Akopyan ZA, Parfyonova YV (2015) Autologous stem cell therapy: how aging and chronic diseases affect stem and progenitor cells. Biores Open Access 4:26–38
120. Kundrotas G, Gasperskaja E, Slapsyte G, Gudleviciene Z, Krasko J, Stumbryte A, Liudkeviciene R (2016) Identity, proliferation capacity, genomic stability and novel senescence markers of mesenchymal stem cells isolated from low volume of human bone marrow. Oncotarget 7:10788–10802
121. Caplan AI (1994) The mesengenic process. Clin Plast Surg 21:429–435
122. Stolzing A, Jones E, McGonagle D, Scutt A (2008) Age-related changes in human bone marrow-derived mesenchymal stem cells: consequences for cell therapies. Mech Ageing Dev 129:163–173
123. Banfi A, Muraglia A, Dozin B, Mastrogiacomo M, Cancedda R, Quarto R (2000) Proliferation kinetics and differentiation potential of ex vivo expanded human bone marrow stromal cells: implications for their use in cell therapy. Exp Hematol 28:707–715
124. Javazon EH, Beggs KJ, Flake AW (2004) Mesenchymal stem cells: paradoxes of passaging. Exp Hematol 32:414–425
125. Crisostomo PR, Wang M, Wairiuko GM, Morrell ED, Terrell AM, Seshadri P, Nam UH, Meldrum DR (2006) High passage number of stem cells adversely affects stem cell activation and myocardial protection. Shock 26:575–580
126. Binato R, de Souza Fernandez T, Lazzarotto-Silva C, Du Rocher B, Mencalha A, Pizzatti L, Bouzas LF, Abdelhay E (2013) Stability of human mesenchymal stem cells during in vitro culture: considerations for cell therapy. Cell Prolif 46:10–22
127. Bertolo A, Mehr M, Janner-Jametti T, Graumann U, Aebli N, Baur M, Ferguson SJ, Stoyanov JV (2016) An in vitro expansion score for tissue-engineering applications with human bone marrow-derived mesenchymal stem cells. J Tissue Eng Regen Med 10:149–161
128. Wagner W, Bork S, Lepperdinger G, Joussen S, Ma N, Strunk D, Koch C (2010) How to track cellular aging of mesenchymal stromal cells? Aging (Albany NY) 2:224–230
129. Berz D, McCormack EM, Winer ES, Colvin GA, Quesenberry PJ (2007) Cryopreservation of hematopoietic stem cells. Am J Hematol 82:463–472
130. Francois M, Copland IB, Yuan S, Romieu-Mourez R, Waller EK, Galipeau J (2012) Cryopreserved mesenchymal stromal cells display impaired immunosuppressive properties as a result of heat-shock response and impaired interferon-gamma licensing. Cytotherapy 14:147–152

131. Pal R, Hanwate M, Totey SM (2008) Effect of holding time, temperature and different parenteral solutions on viability and functionality of adult bone marrow-derived mesenchymal stem cells before transplantation. J Tissue Eng Regen Med 2:436–444
132. Otsuru S, Hofmann TJ, Raman P, Olson TS, Guess AJ, Dominici M, Horwitz EM (2015) Genomic and functional comparison of mesenchymal stromal cells prepared using two isolation methods. Cytotherapy 17:262–270
133. Davies OG, Smith AJ, Cooper PR, Shelton RM, Scheven BA (2014) The effects of cryopreservation on cells isolated from adipose, bone marrow and dental pulp tissues. Cryobiology 69:342–347
134. Pollock K, Sumstad D, Kadidlo D, McKenna DH, Hubel A (2015) Clinical mesenchymal stromal cell products undergo functional changes in response to freezing. Cytotherapy 17:38–45
135. Chinnadurai R, Garcia MA, Sakurai Y, Lam WA, Kirk AD, Galipeau J, Copland IB (2014) Actin cytoskeletal disruption following cryopreservation alters the biodistribution of human mesenchymal stromal cells in vivo. Stem Cell Rep 3:60–72
136. Chinnadurai R, Copland IB, Garcia MA, Petersen C, Lewis C, Waller N, Kirk AD, Galipeau J (2016) Cryopreserved MSCs are susceptible to T-cell mediated apoptosis which is partly rescued by IFNγ licensing. Stem Cells
137. Cui LL, Kinnunen T, Boltze J, Nystedt J, Jolkkonen J (2016) Clumping and viability of bone marrow derived mesenchymal stromal cells under different preparation procedures: a flow cytometry-based in vitro study. Stem Cells Int 2016:1764938
138. Morelli AE, Larregina AT (2016) Concise review: mechanisms behind apoptotic cell-based therapies against transplant rejection and graft versus host disease. Stem Cells 34:1142–1150
139. Saas P, Daguindau E, Perruche S (2016) Concise review: apoptotic cell-based therapies -rationale, preclinical results and future clinical developments. Stem Cells 34:1464–1473
140. Bruder SP, Jaiswal N, Haynesworth SE (1997) Growth kinetics, self-renewal, and the osteogenic potential of purified human mesenchymal stem cells during extensive subcultivation and following cryopreservation. J Cell Biochem 64:278–294
141. Gonda K, Shigeura T, Sato T, Matsumoto D, Suga H, Inoue K, Aoi N, Kato H, Sato K, Murase S, Koshima I, Yoshimura K (2008) Preserved proliferative capacity and multipotency of human adipose-derived stem cells after long-term cryopreservation. Plast Reconstr Surg 121:401–410
142. Liu G, Zhou H, Li Y, Li G, Cui L, Liu W, Cao Y (2008) Evaluation of the viability and osteogenic differentiation of cryopreserved human adipose-derived stem cells. Cryobiology 57:18–24
143. Ginis I, Grinblat B, Shirvan MH (2012) Evaluation of bone marrow-derived mesenchymal stem cells after cryopreservation and hypothermic storage in clinically safe medium. Tissue Eng Part C Methods 18:453–463
144. Mamidi MK, Nathan KG, Singh G, Thrichelvam ST, Mohd Yusof NA, Fakharuzi NA, Zakaria Z, Bhonde R, Das AK, Majumdar AS (2012) Comparative cellular and molecular analyses of pooled bone marrow multipotent mesenchymal stromal cells during continuous passaging and after successive cryopreservation. J Cell Biochem 113:3153–3164
145. Minonzio G, Corazza M, Mariotta L, Gola M, Zanzi M, Gandolfi E, De Fazio D, Soldati G (2014) Frozen adipose-derived mesenchymal stem cells maintain high capability to grow and differentiate. Cryobiology 69:211–216
146. Yong KW, Pingguan-Murphy B, Xu F, Abas WA, Choi JR, Omar SZ, Azmi MA, Chua KH, Wan Safwani WK (2015) Phenotypic and functional characterization of long-term cryopreserved human adipose-derived stem cells. Sci Rep 5:9596
147. Luetzkendorf J, Nerger K, Hering J, Moegel A, Hoffmann K, Hoefers C, Mueller-Tidow C, Mueller LP (2015) Cryopreservation does not alter main characteristics of good manufacturing process-grade human multipotent mesenchymal stromal cells including immunomodulating potential and lack of malignant transformation. Cytotherapy 17:186–198
148. Cruz FF, Borg ZD, Goodwin M, Sokocevic D, Wagner D, McKenna DH, Rocco PR, Weiss DJ (2015) Freshly thawed and continuously cultured human bone marrow-derived mesenchymal stromal cells comparably ameliorate allergic airways inflammation in immunocompetent mice. Stem Cells Transl Med 4:615–624
149. Guven S, Demirci U (2012) Integrating nanoscale technologies with cryogenics: a step towards improved biopreservation. Nanomedicine (Lond) 7:1787–1789

8 Biobanking of Human Mesenchymal Stem Cells: Future Strategy to Facilitate Clinical Applications

Kar Wey Yong, Jane Ru Choi, and Wan Kamarul Zaman Wan Safwani

Abstract

Human mesenchymal stem cells (hMSCs), a type of adult stem cells that hold great potential in clinical applications (e.g., regenerative medicine and cell-based therapy) due to their ability to differentiate into multiple types of specialized cells and secrete soluble factors which can initiate tissue repair and regulate immune response. hMSCs need to be expanded *in vitro* or cryopreserved to obtain sufficient cell numbers required for clinical applications. However, long-term *in vitro* culture-expanded hMSCs may raise some biosafety concerns (e.g., chromosomal abnormality and malignant transformation) and compromised functional properties, limiting their use in clinical applications. To avoid those adverse effects, it is essential to cryopreserve hMSCs at early passage and pool them for off-the-shelf use in clinical applications. However, the existing cryopreservation methods for hMSCs have some notable limitations. To address these limitations, several approaches have to be taken in order to produce healthy and efficacious cryopreserved hMSCs for clinical trials, which remains challenging to date. Therefore, a noteworthy amount of resources has been utilized in research in optimization of the cryopreservation methods, development of freezing devices, and formulation of cryopreservation media to ensure that hMSCs maintain their therapeutic characteristics without raising biosafety concerns following cryopreservation. Biobanking of hMSCs would be a crucial strategy to facilitate clinical applications in the future.

K.W. Yong • J.R. Choi • W.K.Z. Wan Safwani (✉)
Department of Biomedical Engineering, Faculty of Engineering, University of Malaya,
Lembah Pantai, 50603 Kuala Lumpur, Malaysia
e-mail: ronald_yong88@yahoo.com; janeru129@hotmail.com; wansafwani@um.edu.my

F. Karimi-Busheri, M. Weinfeld (eds.), *Biobanking and Cryopreservation of Stem Cells*, Advances in Experimental Medicine and Biology 951, DOI 10.1007/978-3-319-45457-3_8

Keywords

Human mesenchymal stem cells • Cryopreservation • Clinical trials • Clinical applications • Slow freezing • Vitrification • Limitations • Challenges

Abbreviations

hMSCs	Human mesenchymal stem cells
CPA	Cryoprotective agent
DMSO	Dimethyl sulfoxide
FBS	Fetal bovine serum
GvHD	Graft-versus-host disease

8.1 Introduction

Regenerative medicine, a growing area of medicine, aims to treat diseases which are not easily treated, e.g., cardiovascular diseases and diabetes, by repairing the injured tissues and restoring their functions [1]. Human stem cells, particularly human mesenchymal stem cells (hMSCs), hold great promise in regenerative medicine due to their multilineage differentiation potential and paracrine function (ability to secrete soluble factors), which can regulate immune response and initiate tissue repair or regeneration [2–5]. Further, hMSCs are relatively easy to be isolated from various tissues in human body (e.g., adipose tissues, bone marrow and periosteum), and ethical concerns related to the use of embryonic stem cells can be avoided [6–8]. To date, hMSCs have been applied in clinical trials targeting clinical disorders such as graft-versus-host disease (GvHD), stroke and cardiovascular diseases [9]. Such clinical applications may require trillions of hMSCs, which can be produced through extensive *in vitro* cell expansion [10]. However, long-term *in vitro* cell expansion may raise biosafety concerns (e.g., chromosomal aberration and malignant transformation) and reduce therapeutic potency of hMSCs [11, 12]. It has been reported that patients treated with hMSCs at early passage have improved clinical outcome over those treated with late passage cells [13]. Therefore, an effective method of long-term preservation and storage of hMSCs at early passage is essential.

Cryopreservation is the only method used to preserve and store cells, including hMSCs, for a long-term period [14]. Cryopreservation adopts a principle which utilizes ultralow temperatures (approximately −196 °C, e.g., in liquid nitrogen) to halt the metabolic activity of cells while maintaining their life and cell functionality [15, 16]. Cryopreservation allows transportation and pooling of cells to reach cell numbers required for clinical applications while maintaining their functional properties [17]. It also allows the completion of quality control and safety testing of cells prior to clinical applications [18, 19]. A bank of efficacious and healthy cryopreserved cells would be available as a ready off-the-shelf supply, allowing better timing of therapy [15]. Further, cryopreservation produces a bank of cells at specific passages with intact functional properties and genetic characteristics. These validated cells can be used to initiate new experiments, maximize the long-term use of cells, and minimize experimental variation [20].

A successful cryopreservation requires the development of protocols that achieves high cell recovery and maintain cell functionality, including proliferation potential, multi-lineage differentiation potential, immunomodulatory property, migration ability and secretory profile, which are identical to that of the cells at pre-storage state [21]. As the utilization of cryopreserved cells continues to increase, researchers are demanding a standardized protocol which can achieve higher cell recovery and the cells recovered are biochemically and physiologically identical to that of pre-storage state at the structural, genomic, proteomic and functional levels [22]. Further, the cells should be xeno-free, nontoxic and immune privileged in human [14]. However, many challenges remain thus far, including sub-optimal cell recovery, addition of toxic or animal components in the cryopreservation medium and potential contamination [23]. Therefore, a remarkable amount of resources has been utilized in research in optimization of the cryopreservation methods, development of freezing devices, and formulation of cryopreservation media, in order to ensure that hMSCs maintain their therapeutic characteristics without raising biosafety concerns following cryopreservation [15].

8.2 Clinical Applications of hMSCs

The first clinical trial of *in vitro* culture-expanded hMSCs isolated from bone marrow was conducted in patients with hematologic malignancies in year 1995 [24]. Since then, the clinical application of hMSCs has been further explored. From year 2010 to 2014, the clinical trials database (http://www.clinicaltrials.gov) demonstrated 685 registered clinical trials of hMSCs in different clinical phases to evaluate their potential therapeutic applications worldwide. The number of registered clinical trial of hMSCs has been rising since year 2010 (Fig. 8.1a). Most of these trials are aimed to treat cardiovascular diseases (e.g., myocardial ischemia and stroke), musculoskeletal diseases (e.g., osteoarthritis and rheumatic diseases), neurological diseases (e.g., spinal cord injury and amyotrophic lateral sclerosis) and immune system diseases (e.g., GvHD and multiple sclerosis) (Fig. 8.1b). Majority of these trials are in Phase I (safety evaluation), Phase II (establishment of the efficacy of hMSCs against a placebo), or a mixture of Phases I/II studies. Only an inconsiderable number of clinical trials are in Phase III (establishment of the efficacy of hMSCs against many standard or well-known treatments) or a mixture of Phases II/III or Phase IV (description of the additional information such as treatment's risk and optimal dosage) (Fig. 8.1c). In general, most trials have reported that no sign of significant adverse effects was seen in the patients receiving hMSCs-based therapies [25]. Further, many completed clinical trials have showed the efficacy of hMSCs in treating clinical disorders, including GvHD, amyotrophic sclerosis, liver cirrhosis, stroke and acute myocardial infarction [26].

As of 31st December 2014, there are 23 registered clinical trials of cryopreserved hMSCs (derived from bone marrow) in patients with clinical disorders such as myocardial infarction, GvHD and Crohn' disease. The first clinical trial using infusion of cryopreserved hMSCs (named as Prochymal® or remestemcel-L) in patients with GvHD was conducted in year 2005. Prochymal is an off-the-shelf stem cell product, which is frozen and stored in Plasma-Lyte®A (an isotonic solution for intravenous infusion) containing 5 % human serum albumin and 10 % DMSO. For administration, Prochymal was thawed and diluted with Plasma-Lyte®A to achieve the cell concentration needed for infusion and to reduce final concentration of DMSO in the infused product to 3.75 %. The viability of cells after thawing was at least 70 %, as determined by Trypan blue assay [27–30]. Prochymal has been approved by Canada as the first adult stem cell-based therapy to treat GvHD. No identifiable adverse effect or acute toxicity was seen in the patients receiving this therapy [27–29]. Prochymal is currently undergoing Phase III trials for refractory Crohn's disease, and Phase II trials for the treatment of myocardial infarction, diabetes mellitus type 1 and chronic obstructive pulmonary disease. On the other hand, MultiStem®, another product of cryopreserved hMSCs (derived from bone marrow), which is also stored frozen in a mixture of PlasmaLyte, human serum albumin and DMSO, undergoes Phase II trials for ischemic stroke, acute myocardial infarction, acute respiratory distress syndrome, and inflammatory bowel diseases (e.g., ulcerative colitis) [31].

Taken together, hMSCs have only been approved for the treatment of GvHD, whereas significant efforts are required in clinical trials in order to prompt the use of hMSCs in the treat-

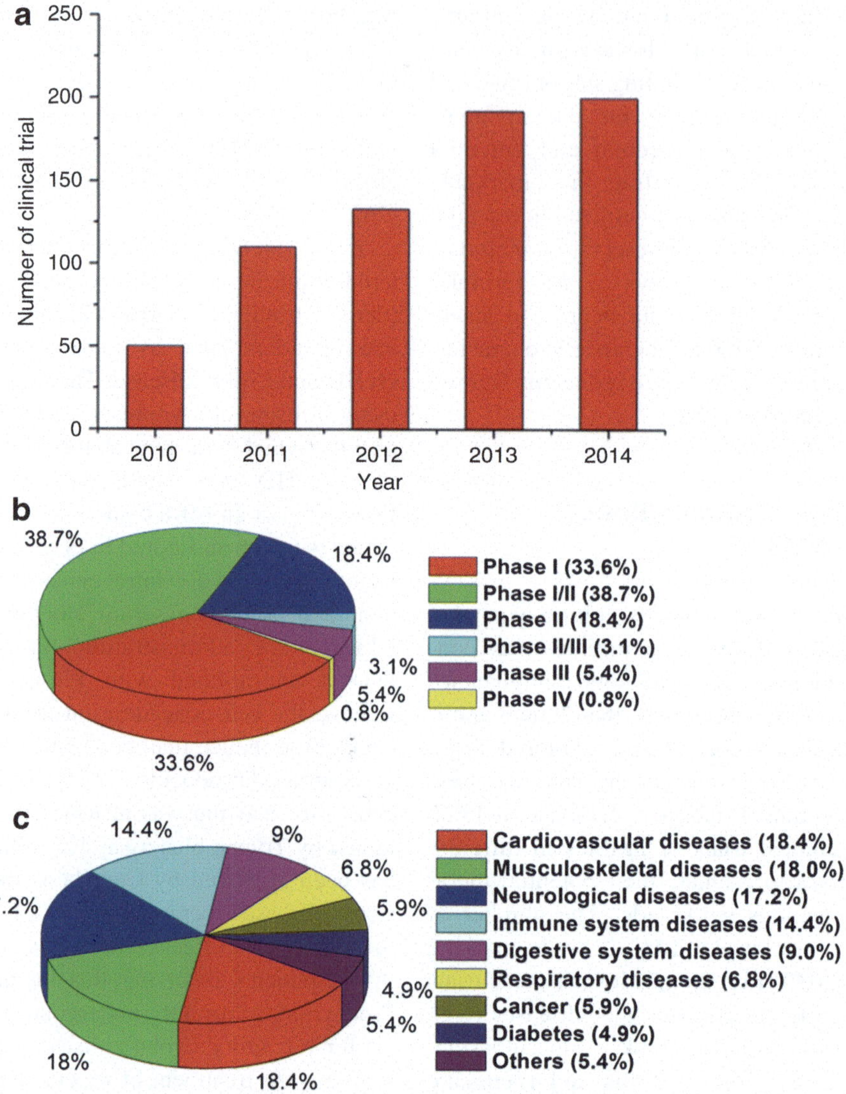

Fig. 8.1 (**a**) Number of registered clinical trials of hMSCs from year 2010 to 2014. (**b**) Clinical trials of hMSCs classified by phase. (**c**) Clinical trials of hMSCs classified by category of studied disease (Data from ClinicalTrials.gov)

ment of other clinical disorders in the future. Since cryopreserved hMSCs offer great potential for clinical application, most research has been focused on optimization of cryopreservation condition of hMSCs (including hMSCs isolated from umbilical cord, adipose tissues and amniotic fluid other than bone marrow) in order to produce more off-the-shelf hMSCs for clinical trial and future clinical applications.

8.3 Existing Methods of Cryopreservation of hMSCs

8.3.1 Slow Freezing

The slow freezing method is the preferred method of cryopreservation of high volume cells such as hMSCs and cell lines [32]. With a freezing rate of

1 °C/min (an optimal cooling rate for mammalian cells) [33], a large number of cells can be frozen in a cryovial at a low concentration (<1.5 M) of cryoproctective agents (CPAs) [34]. This freezing rate allows cellular dehydration and reduces the risk of intracellular ice formation. To achieve such a low freezing rate, direct contact of cells with non-sterile liquid nitrogen is not required during freezing, thus avoiding potential contamination with other microorganism or pathogens [14]. Therefore, a large amount of cryopreserved cells which are contamination-free can be pooled to provide sufficient cells for off-the-shelf clinical use. A uniform freezing rate can be generated in a passive or active manner. Passive freezing approach utilizes a commercially available freezing container filled with alcohol, e.g., "Mr. Frosty" (temperature is lowered at freezing rate of 1 °C/min due to the slow freezing property of isopropanol), whereas active freezing approach utilizes a programmable controlled rate freezer [22]. However, it has been reported that both freezing approaches, display similar potential to maintain phenotype, viability, and functional properties of hMSCs [35]. With the advance of technology a programmed freezer termed "Cell Alive System" (CAS) has been developed, which vibrates the water molecules and cells using alternating magnetic field and electric field during the process of freezing to prevent intra- and extra-cellular ice formation [36]. It has been reported that the risk of freeze injury to MSCs can be further reduced by using this system [37]. Among various types of slow freezing methods, freezing protocol which adopts passive freezing approach is favorable for cryopreservation of hMSCs due to low cost and high cryopreservation efficiency.

To preserve hMSCs efficiently using a slow freezing method, it is essential to optimize and determine the ideal CPAs used to preserve hMSCs. The roles of CPAs in cell cryopreservation are to stabilize the cell membrane, minimize osmotic stress to the cells, and protect cells against intracellular and extracellular ice crystal formation which are both harmful to cells [38, 39]. CPAs are divided into permeating CPAs (e.g., dimethyl sulfoxide (DMSO) and glycerol) and non-permeating CPAs (e.g., sucrose, trehalose and fetal bovine serum (FBS)). Permeating CPAs can penetrate the cell membrane due to their low molecular weight and remove the water from the cells to prevent the formation of intracellular ice [40–42]. Non-permeating CPAs is unable to pass through the cell membrane but it is able to protect the cell membrane by forming a viscous glassy shell around the outer surface of cells and regulate intracellular and extracellular osmotic pressure [40, 41] (Fig. 8.2).

Among the CPAs, DMSO with relatively high membrane permeability and FBS being rich in protein and growth factors is commonly utilized for cell cryopreservation [43]. DMSO with a concentration of 10 % (v/v) combined with FBS (20–90 %) (v/v) has been commonly used to preserve hMSCs [44–46], but both exhibit a considerable disadvantages. DMSO is cytotoxic at the temperatures beyond 4 °C [47]. The clinical uses of cells preserved in 10 % DMSO have caused many adverse effects (e.g., neurotoxicity and respiratory depression) in recipients [48, 49]. DMSO may induce undesired differentiation of stem cells to neuronal-like cells [50]. On the other hand, FBS contains a number of proteins and peptides which can initiate xenogeneic immune responses [51, 52]. In addition, FBS could possibly transfer pathogens [47]. Therefore, it is recommended to minimize or exclude the use of FBS as part of the cryopreservation medium [53]. Overall, an ideal CPA should maintain functional properties and high survival rate of cells after thawing, and allow for cell transplantation without raising biosafety issues (e.g., xenogeneic immune response, cytotoxicity and tumourigenesis).

8.3.2 Vitrification

Vitrification is a process which requires a very high freezing rate to convert a cell-laden CPA suspension directly from its aqueous phase to a glass state upon contact with liquid nitrogen [54]. Generally, two approaches have been used to preserve cells using vitrification – equilibrium vitrification and non-equilibrium vitrification. In

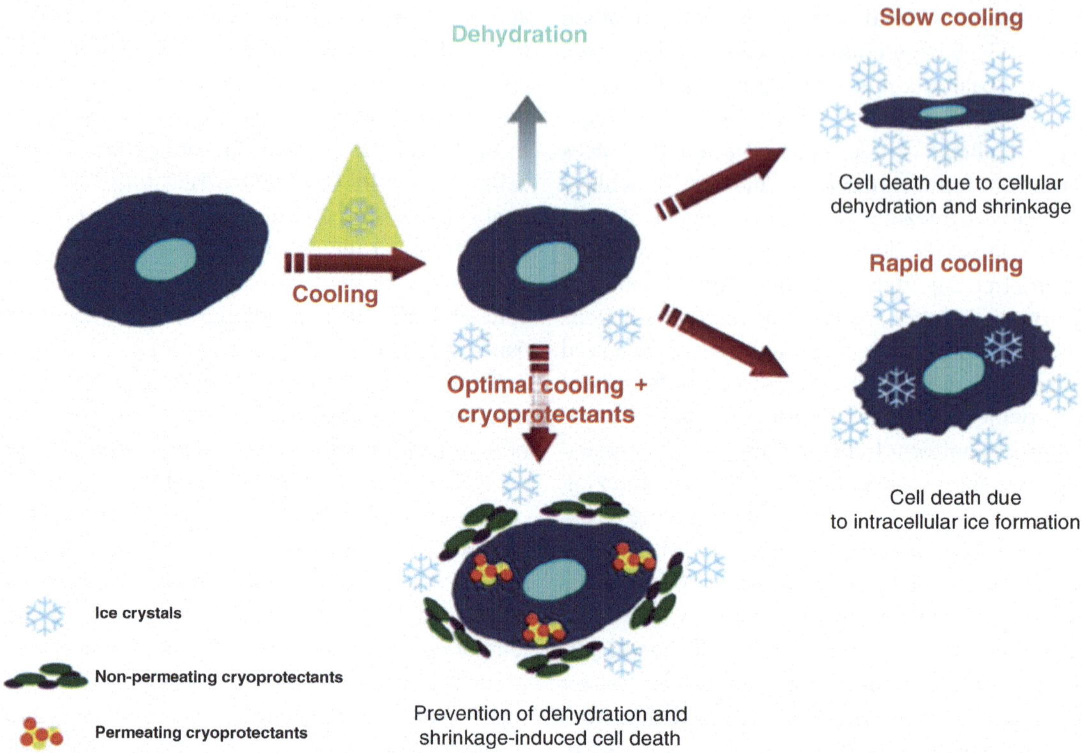

Fig. 8.2 The role of CPAs in cell cryopreservation. Cryopreservation requires CPAs (either permeating or non-permeating) to protect cells from freeze injury (e.g., cell death) by avoiding the formation of intra- and extracellular ice crystals which are both harmful to cells during freezing (Reproduced from Martin-Ibanez et al. [65])

equilibrium vitrification, cells are exposed to multimolar CPA mixtures in a stepwise manner, which helps to reduce chemical toxicity, followed by immersion into liquid nitrogen [55]. hMSCs have been vitrified using this approach, which could successfully maintain their functional properties and viability [56–58]. However, this approach has high potential to cause osmotic shock and chemical toxicity to cells [55].

In non-equilibrium vitrification, extremely high freezing rates has been performed with only one CPA at a low concentration for cell preservation [55]. In this approach, vitrification systems are divided into carrier-based systems and carrier-free systems. Cryoloops (700,000 °C/min), quartz microcapillaries (250,000 °C/min) and plastic straws (2500 °C/min) are among the carriers that have been developed for vitrification with each of them offering a different freezing rate [34]. The utilization of higher freezing rate allows vitrification using CPA at a lower concentration, thereby reducing the risk of osmotic damage and chemical toxicity to the cells. On the other hand, carrier-free systems which can generate micro-droplets of cell-laden CPA followed by ejection into liquid nitrogen, have been developed to further increase freezing rate of vitrification [34]. hMSCs have been vitrified using a carrier-free system without the need for directly contacting the cells with liquid nitrogen [59], and the results seem promising.

8.3.3 Slow Freezing Versus Vitrification

Post-thaw viability of cells was reported to be higher using vitrification compared to slow freezing method as the high freezing rate reduces the time of intra- and extracellular ice formation [60]. However, vitrification which adopts an approach of direct cell-to-liquid nitrogen contact, may lead to potential pathogenic contamination

[61]. The Hepatitis B virus and *Aspergillus sp.* are among the pathogens which have been reported to contaminate liquid nitrogen [61, 62]. The source of contamination may come from non-sterile liquid nitrogen itself or cross-contamination from infected samples in the liquid nitrogen storage tank [63]. Therefore, good operational skill is required for vitrification. Large scale cryopreservation of cells is impractical. Vitrification is usually applied to small volume cells such as embryonic stem cells, embryos and oocytes because it requires low volume of cell suspension to achieve high freezing rate [64]. Therefore, it is not recommended to cryopreservation of hMSCs in large volumes. Further, the recycling of cells is labor-intensive as it needs manual picking up of individual cell colonies [60], which may lead to the cell loss. The advantages and disadvantages of slow freezing and vitrification are described in brief in Table 8.1. Collectively, the slow freezing method is preferable because cryopreservation of hMSCs can be performed on a large scale, the post-thaw viability of cells is high, and it avoids the risk of pathogenic contamination, allowing cryopreserved hMSCs to be used for clinical applications with low risk of pathogen transmission.

Table 8.1 Comparison between slow freezing and vitrification method

No	Aspect	Slow freezing	Vitrification
1.	Concentration of CPAs required	Low	High
2.	Risk of freeze injury	High	Low
3.	Post-thaw viability	High	Higher
4.	Risk of toxicity of CPAs	Low	High
5.	Potential pathogenic contamination	Low	High
6.	Operational skill	Easy	Good operational skill is needed

Adapted from Yong et al. [14]
CPAs cryoprotective agents

8.4 Challenges in Cryopreservation of hMSCs for Clinical Applications

8.4.1 Slow Freezing

Today, the issue of adverse effects in recipients of cells cryopreserved in 10 % DMSO and FBS using a slow freezing method is challenging [60, 65]. Although the transplantation protocol has included the step of post-thaw washing of the cells to remove DMSO prior to transplantation [66, 67], but washing the cells by centrifugation prior to removal of DMSO can cause a significant cell loss. Further, total removal of DMSO from cryopreserved cells is time-consuming and complex [68]. Therefore, the development of an alternative cryopreservation medium or a cryopreservation medium consisting of a reduced concentration of DMSO is required [69]. To date, the introduction of non-toxic polymers (e.g., polyvinylpyrrolidone) and disaccharides (e.g., sucrose and trehalose) as a CPA [35, 70], have not achieved the aim of replacing DMSO for cryopreservation of hMSCs as they are less efficient than DMSO in maintaining the viability of hMSCs, possibly due to their inability to penetrate cell membrane. However, when trehalose encapsulated in genipin-cross-linked Pluronic F127-chitosan nanoparticles was delivered into hMSCs for cryopreservation, it was found that cells are capable to maintain their functional properties and high viability following cryopreservation [71]. This indicates that non-permeating CPAs, e.g., trehalose, may be more efficient when they are delivered intracellularly for cryopreservation of hMSCs. Further investigation is needed to assess the biodegradability and cytotoxicity of the genipin-cross-linked Pluronic F127-chitosan nanoparticles, and explore more efficient methods to deliver CPAs intracellularly. Further investigation is also required to explore alternative CPAs to replace DMSO completely.

Recently, it has been proposed that xeno-free cryopreservation medium can be used to preserve hMSCs [14]. For instance, 5 % human albumin solution has been proposed to replace FBS for cryopreservation of hMSCs [72]. Further, our recent study has demonstrated that hMSCs preserved in xeno-free cryopreservation medium consists of only 5 % DMSO were capable to maintain high viability and their functional properties following cryopreservation [73]. However, it has been reported that cells (e.g., ESCs and lymphocytes) cryopreserved in DMSO may display chromosomal abnormality and changes in telomere length, which might in turn lead to tumour formation [74, 75]. These concerns have established the need for assessment of tumourigenic potential and chromosomal abnormality in cryopreserved hMSCs prior to clinical applications. It has been demonstrated that cryopreservation does not alter chromosome structures and numbers in hMSCs, indicating no sign of chromosomal abnormality [45, 76–78]. Further, our recent study has demonstrated that cryopreserved hMSCs including those preserved in 5 % DMSO display low tumourigenic potential, indicating their low risk in tumourigenesis [79]. These results indicate that hMSCs preserved in 5 % DMSO which have low risk in inducing xenogeneic immune response and cytotoxicity, could be an ideal cell source for clinical applications in the future. Further investigation is needed to assess immunomodulatory property, migration ability and secretory profile of hMSCs preserved in 5 % DMSO following cryopreservation.

8.4.2 Vitrification

To vitrify cells for clinical applications, many issues need to be addressed, including small scale cryopreservation, high risk of contamination and cell loss due to the ineffective collection of frozen cells. With the advance of technology, the development of an ejection-based micro-droplet generation system permits efficient vitrification of micro-droplets in a continuous and high throughput manner, which has significantly improved the performance for cryopreservation of hMSCs. Further, a closed, sterile, and automated vitrification system (e.g., ejection-based micro-droplet generation system) which allows vitrification of cells via indirect contact with liquid nitrogen should be developed to avoid potential contamination and reduce cell loss [32, 59]. Xu's group has developed a device which can vitrify micro-droplets of cell-laden CPA printed on a silver film via boiling heat transfer without the need for direct contact with liquid nitrogen [59]. hMSCs subjected to vitrification using such approach were found to maintain cell phenotype, high viability and osteogenic differentiation potential. This device allows vitrification with low risk of contamination and cell loss, indicating its potential use in vitrification of hMSCs for clinical applications. Further investigation is needed to perform further assessment of cell functionality in hMSCs subjected to vitrification using this device. In addition, a sterile polyetrafluoroethylene cartridge with filteration and ultraviolet radiation has been suggested to sterilize liquid nitrogen prior to vitrification of cells since liquid nitrogen has been reported to be potentially contaminated with infectious agents [80].

8.4.3 Thawing

Besides freezing, it is essential to select the optimum method to thaw the frozen cells. The standard thawing method is to warm the cells at a rapid rate (>100 °C/min) at 37 °C in a water bath until all the ice melted [68], resulting in recovery of a high numbers of viable cells following thawing. Thawing frozen cells at such a rapid rate could minimize ice recrystallization and cellular exposure time to high concentrations of CPA [22]. A dry warming procedure has been proposed to thaw the frozen cells to avoid possible contamination of cells with microorganism presence in the water bath [81]. Therefore, a controlled-rate thawing chamber was developed to thaw frozen cells at a warming rate of 10 °C/min, resulting in a high viability of cells, which is comparable to those thawed with the standard method [82]. After thawing, the cells are washed by centrifugation to remove CPAs, which are particularly toxic (e.g., DMSO) prior to clinical applications. However, the process of washing may result in a significant cell loss which may in turn affect the clinical outcome. In the future, it is essential to develop a controlled-rate dry warming device which is low cost, reliable and portable for thawing frozen cells, and a method which can remove CPAs while minimizing cell loss for effective clinical applications.

8.5 Conclusion and Future Perspectives

Nowadays, the demand for organ transplantation has been rising rapidly due to the increasing incidence of chronic diseases (e.g., liver cirrhosis and myocardial ischemia), which lead to the end-stage failure of many vital organs (e.g., liver and heart). The supply of organs from deceased donors has remained low and insufficient to meet the increasing demand. So, shortage of organs for transplantation has become a major crisis worldwide. To solve the organ shortage problem, regenerative medicine which emphasizes on the use of human stem cells in the treatment, has evolved rapidly. Among various types of stem cells, hMSCs possess paracrine function and immunosuppressive property, which support their use in various clinical settings. Implantation of such multi-functional hMSCs may treat the fatal diseases while addressing the problems of organ shortage crisis. However, long-term culture-expanded hMSCs may raise biosafety concerns and demonstrate decreased therapeutic potency, limiting their widespread use. Cryopreservation could retain the therapeutic characteristics of hMSCs with low risk of biosafety issues, indicating their potential use in clinical applications. Therefore, the development of biobank of hMSCs is essential as healthy and efficacious cryopreserved cells can be pooled for off-the-shelf use when needed, allowing better timing of therapy. To date, many challenges remains in addressing the limitations of the existing cryopreservation protocols (from freezing process to thawing process and therapeutic administration), including sub-optimal cell recovery, potential contamination and addition of toxic or animal-based agents into the cryopreservation medium. A standardized and safe cryopreservation protocol of hMSCs which can achieve high recovery of cells and maintain cell functionality should be established. In the future, biobanking of hMSCs would be a crucial strategy to facilitate clinical applications.

Acknowledgements This report was supported by University of Malaya Research Grant (UMRG RP040B-15HTM) from the Ministry of Education Malaysia.

References

1. Heathman TR, Glyn VA, Picken A, Rafiq QA, Coopman K, Nienow AW, Kara B, Hewitt CJ (2015) Expansion, harvest and cryopreservation of human mesenchymal stem cells in a serum-free microcarrier process. Biotechnol Bioeng 112(8):1696–1707
2. Caplan AI, Dennis JE (2006) Mesenchymal stem cells as trophic mediators. J Cell Biochem 98(5):1076–1084
3. Doulatov S, Daley GQ (2013) Development. A stem cell perspective on cellular engineering. Science 342(6159):700–702
4. Choi JR, Pingguan-Murphy B, Wan Abas WA, Noor Azmi MA, Omar SZ, Chua KH, Wan Safwani WK (2014) Impact of low oxygen tension on stemness, proliferation and differentiation potential of human adipose-derived stem cells. Biochem Biophys Res Commun 448(2):218–224
5. Choi JR, Pingguan-Murphy B, Abas WABW, Azmi MAN, Omar SZ, Chua KH, Safwani WKZW (2014) Hypoxia promotes growth and viability of human adipose-derived stem cells with increased growth factors secretion. J Asian Sci Res 4(7):328–338
6. Lindroos B, Suuronen R, Miettinen S (2011) The potential of adipose stem cells in regenerative medicine. Stem Cell Rev 7(2):269–291
7. Callihan P, Mumaw J, Machacek DW, Stice SL, Hooks SB (2011) Regulation of stem cell pluripotency and differentiation by G protein coupled receptors. Pharmacol Ther 129(3):290–306
8. Choi JR, Pingguan-Murphy B, Wan Abas WA, Yong KW, Poon CT, Noor Azmi MA, Omar SZ, Chua KH, Xu F, Wan Safwani WK (2015) In situ normoxia enhances survival and proliferation rate of human adipose tissue-derived stromal cells without increasing the risk of tumourigenesis. PLoS One 10(1):e0115034
9. Heathman TR, Nienow AW, McCall MJ, Coopman K, Kara B, Hewitt CJ (2015) The translation of cell-based therapies: clinical landscape and manufacturing challenges. Regen Med 10(1):49–64

10. Rowley J, Abraham E, Campbell A, Brandwein H, Oh S (2012) Meeting lot-size challenges of manufacturing adherent cells for therapy. BioProcess Int 10(3):7
11. Roemeling-van Rhijn M, de Klein A, Douben H, Pan Q, van der Laan LJ, Ijzermans JN, Betjes MG, Baan CC, Weimar W, Hoogduijn MJ (2013) Culture expansion induces non-tumorigenic aneuploidy in adipose tissue-derived mesenchymal stromal cells. Cytotherapy 15(11):1352–1361
12. Rosland GV, Svendsen A, Torsvik A, Sobala E, McCormack E, Immervoll H, Mysliwietz J, Tonn JC, Goldbrunner R, Lonning PE, Bjerkvig R, Schichor C (2009) Long-term cultures of bone marrow-derived human mesenchymal stem cells frequently undergo spontaneous malignant transformation. Cancer Res 69(13):5331–5339
13. Moll G, Alm JJ, Davies LC, von Bahr L, Heldring N, Stenbeck-Funke L, Hamad OA, Hinsch R, Ignatowicz L, Locke M, Lonnies H, Lambris JD, Teramura Y, Nilsson-Ekdahl K, Nilsson B, Le Blanc K (2014) Do cryopreserved mesenchymal stromal cells display impaired immunomodulatory and therapeutic properties? Stem Cells 32(9):2430–2442
14. Yong KW, Wan Safwani WK, Xu F, Wan Abas WA, Choi JR, Pingguan-Murphy B (2015) Cryopreservation of human mesenchymal stem cells for clinical applications: current methods and challenges. Biopreserv Biobank 13(4):231–239
15. Marquez-Curtis LA, Janowska-Wieczorek A, McGann LE, Elliott JA (2015) Mesenchymal stromal cells derived from various tissues: biological, clinical and cryopreservation aspects. Cryobiology 71(2):181–197
16. Rowley SD (1992) Hematopoietic stem cell cryopreservation: a review of current techniques. J Hematother 1(3):233–250
17. Davies OG, Smith AJ, Cooper PR, Shelton RM, Scheven BA (2014) The effects of cryopreservation on cells isolated from adipose, bone marrow and dental pulp tissues. Cryobiology 69(2):342–347
18. Hubel A (1997) Parameters of cell freezing: implications for the cryopreservation of stem cells. Transfus Med Rev 11(3):224–233
19. Hubel A (2011) Advancing the preservation of cellular therapy products. Transfusion 51(4):1537–2995
20. Stubban C, Katkov I, Loring JF, Wesselschmidt RL (2007) Cryopreservation of human embryonic stem cells. In: Loring JF, Wesselschmidt RL, Schwartz PH (eds) Human stem cell manual. Academic, Oxford, pp 47–55
21. Baust JG (2006) Concepts in biopreservation. In: Baust JG, Baust JM (eds) Advances in biopreservation. CRC Press, Boca Raton, pp 1–14
22. Baust JM, Corwin WL, VanBuskirk R, Baust JG (2015) Biobanking: the future of cell preservation strategies. In: Feridoun KB (ed) Biobanking in the 21st century. Springer, Cham, pp 37–53
23. Baust JM (2006) Properties of cells and tissues influencing preservation outcome: molecular basis of preservation-induced cell death. In: Baust JG, Baust JM (eds) Advances in biopreservation. CRC Press, Boca Raton, pp 63–88
24. Lazarus HM, Haynesworth SE, Gerson SL, Rosenthal NS, Caplan AI (1995) Ex vivo expansion and subsequent infusion of human bone marrow-derived stromal progenitor cells (mesenchymal progenitor cells): implications for therapeutic use. Bone Marrow Transplant 16(4):557–564
25. Otto WR, Wright NA (2011) Mesenchymal stem cells: from experiment to clinic. Fibrogenesis Tissue Repair 4(20):1755–1536
26. Wang S, Qu X, Zhao RC (2012) Clinical applications of mesenchymal stem cells. J Hematol Oncol 5(19):1756–8722
27. Prasad VK, Lucas KG, Kleiner GI, Talano JA, Jacobsohn D, Broadwater G, Monroy R, Kurtzberg J (2011) Efficacy and safety of ex vivo cultured adult human mesenchymal stem cells (prochymal) in pediatric patients with severe refractory acute graft-versus-host disease in a compassionate use study. Biol Blood Marrow Transplant 17(4):534–541
28. Kurtzberg J, Prockop S, Teira P, Bittencourt H, Lewis V, Chan KW, Horn B, Yu L, Talano JA, Nemecek E, Mills CR, Chaudhury S (2014) Allogeneic human mesenchymal stem cell therapy (remestemcel-L, prochymal) as a rescue agent for severe refractory acute graft-versus-host disease in pediatric patients. Biol Blood Marrow Transplant 20(2):229–235
29. Kebriaei P, Isola L, Bahceci E, Holland K, Rowley S, McGuirk J, Devetten M, Jansen J, Herzig R, Schuster M, Monroy R, Uberti J (2009) Adult human mesenchymal stem cells added to corticosteroid therapy for the treatment of acute graft-versus-host disease. Biol Blood Marrow Transplant 15(7):804–811
30. Hare JM, Traverse JH, Henry TD, Dib N, Strumpf RK, Schulman SP, Gerstenblith G, DeMaria AN, Denktas AE, Gammon RS, Hermiller JB Jr, Reisman MA, Schaer GL, Sherman W (2009) A randomized, double-blind, placebo-controlled, dose-escalation study of intravenous adult human mesenchymal stem cells (prochymal) after acute myocardial infarction. J Am Coll Cardiol 54(24):2277–2286
31. Vaes B, Van't Hof W, Deans R, Pinxteren J (2012) Application of multiStem ((R)) allogeneic cells for immunomodulatory therapy: clinical progress and pre-clinical challenges in prophylaxis for graft versus host disease. Front Immunol 3(345):1–9
32. Guven S, Demirci U (2012) Integrating nanoscale technologies with cryogenics: a step towards improved biopreservation. Nanomedicine (London, England) 7(12):1787–1789
33. Hubel A, Spindler R, Skubitz AP (2014) Storage of human biospecimens: selection of the optimal storage temperature. Biopreserv Biobank 12(3):165–175
34. Zhang X, Catalano PN, Gurkan UA, Khimji I, Demirci U (2011) Emerging technologies in medical applications of minimum volume vitrification. Nanomedicine (London, England) 6(6):1115–1129

35. Janz Fde L, Debes Ade A, Cavaglieri Rde C, Duarte SA, Romao CM, Moron AF, Zugaib M, Bydlowski SP (2012) Evaluation of distinct freezing methods and cryoprotectants for human amniotic fluid stem cells cryopreservation. J Biomed Biotechnol 2012:649353
36. Wowk B (2012) Electric and magnetic fields in cryopreservation. Cryobiology 64(3):301–303; author reply 304–305
37. Lee SY, Huang GW, Shiung JN, Huang YH, Jeng JH, Kuo TF, Yang JC, Yang WC (2012) Magnetic cryopreservation for dental pulp stem cells. Cells Tissues Organs 196(1):23–33
38. McGann LE (1978) Differing actions of penetrating and nonpenetrating cryoprotective agents. Cryobiology 15(4):382–390
39. Rudolph AS, Crowe JH (1985) Membrane stabilization during freezing: the role of two natural cryoprotectants, trehalose and proline. Cryobiology 22(4):367–377
40. Karlsson JO (2002) Cryopreservation: freezing and vitrification. Science 296(5568):655–656
41. Karlsson JO, Toner M (1996) Long-term storage of tissues by cryopreservation: critical issues. Biomaterials 17(3):243–256
42. Berz D, McCormack EM, Winer ES, Colvin GA, Quesenberry PJ (2007) Cryopreservation of hematopoietic stem cells. Am J Hematol 82(6):463–472
43. Shivakumar SB, Bharti D, Jang SJ, Hwang SC, Park JK, Shin JK, Byun JH, Park BW, Rho GJ (2015) Cryopreservation of human Wharton's jelly-derived mesenchymal stem cells following controlled rate freezing protocol using different cryoprotectants; a comparative study. Int J Stem Cells 8(2):155–169
44. Liu G, Zhou H, Li Y, Li G, Cui L, Liu W, Cao Y (2008) Evaluation of the viability and osteogenic differentiation of cryopreserved human adipose-derived stem cells. Cryobiology 57(1):18–24
45. Miranda-Sayago JM, Fernandez-Arcas N, Benito C, Reyes-Engel A, Herrero JR, Alonso A (2012) Evaluation of a low cost cryopreservation system on the biology of human amniotic fluid-derived mesenchymal stromal cells. Cryobiology 64(3):160–166
46. Zhang HT, Chen H, Zhao H, Dai YW, Xu RX (2011) Neural stem cells differentiation ability of human umbilical cord mesenchymal stromal cells is not altered by cryopreservation. Neurosci Lett 487(1):118–122
47. Zambelli A, Poggi G, Da Prada G, Pedrazzoli P, Cuomo A, Miotti D, Perotti C, Preti P, Robustelli della Cuna G (1998) Clinical toxicity of cryopreserved circulating progenitor cells infusion. Anticancer Res 18(6B):4705–4708
48. Benekli M, Anderson B, Wentling D, Bernstein S, Czuczman M, McCarthy P (2000) Severe respiratory depression after dimethylsulphoxide-containing autologous stem cell infusion in a patient with AL amyloidosis. Bone Marrow Transplant 25(12):1299–1301
49. Windrum P, Morris TC (2003) Severe neurotoxicity because of dimethyl sulphoxide following peripheral blood stem cell transplantation. Bone Marrow Transplant 31(4):315
50. Woodbury D, Schwarz EJ, Prockop DJ, Black IB (2000) Adult rat and human bone marrow stromal cells differentiate into neurons. J Neurosci Res 61(4):364–370
51. Tuschong L, Soenen SL, Blaese RM, Candotti F, Muul LM (2002) Immune response to fetal calf serum by two adenosine deaminase-deficient patients after T cell gene therapy. Hum Gene Ther 13(13):1605–1610
52. Mackensen A, Drager R, Schlesier M, Mertelsmann R, Lindemann A (2000) Presence of IgE antibodies to bovine serum albumin in a patient developing anaphylaxis after vaccination with human peptide-pulsed dendritic cells. Cancer Immunol Immunother 49(3):152–156
53. Balci D, Can A (2013) The assessment of cryopreservation conditions for human umbilical cord stroma-derived mesenchymal stem cells towards a potential use for stem cell banking. Curr Stem Cell Res Ther 8(1):60–72
54. Rall WF, Fahy GM (1985) Ice-free cryopreservation of mouse embryos at −196 degrees C by vitrification. Nature 313(6003):573–575
55. Hunt CJ (2011) Cryopreservation of human stem cells for clinical application: a review. Transfus Med Hemother 38(2):107–123
56. Moon JH, Lee JR, Jee BC, Suh CS, Kim SH, Lim HJ, Kim HK (2008) Successful vitrification of human amnion-derived mesenchymal stem cells. Hum Reprod 23(8):1760–1770
57. Massood E, Maryam K, Parvin S, Mojgan M, Noureddin NM (2013) Vitrification of human umbilical cord Wharton's jelly-derived mesenchymal stem cells. Cryo Letters 34(5):471–480
58. Wu Y, Wen F, Gouk SS, Lee EH, Kuleshova L (2015) Cryopreservation strategy for tissue engineering constructs consisting of human mesenhymal stem cells and hydrogel biomaterials. Cryo Letters 36(5):325–335
59. Shi M, Ling K, Yong KW, Li Y, Feng S, Zhang X, Pingguan-Murphy B, Lu TJ, Xu F (2015) High-throughput non-contact vitrification of cell-laden droplets based on cell printing. Sci Rep 5(17928)
60. Li Y, Tan JC, Li LS (2010) Comparison of three methods for cryopreservation of human embryonic stem cells. Fertil Steril 93(3):999–1005
61. Tedder RS, Zuckerman MA, Goldstone AH, Hawkins AE, Fielding A, Briggs EM, Irwin D, Blair S, Gorman AM, Patterson KG et al (1995) Hepatitis B transmission from contaminated cryopreservation tank. Lancet 346(8968):137–140
62. Fountain D, Ralston M, Higgins N, Gorlin JB, Uhl L, Wheeler C, Antin JH, Churchill WH, Benjamin RJ (1997) Liquid nitrogen freezers: a potential source of microbial contamination of hematopoietic stem cell components. Transfusion 37(6):585–591
63. AbdelHafez F, Xu J, Goldberg J, Desai N (2011) Vitrification in open and closed carriers at different cell

stages: assessment of embryo survival, development, DNA integrity and stability during vapor phase storage for transport. BMC Biotechnol 11(29):1472–6750
64. Song YS, Adler D, Xu F, Kayaalp E, Nureddin A, Anchan RM, Maas RL, Demirci U (2010) Vitrification and levitation of a liquid droplet on liquid nitrogen. Proc Natl Acad Sci U S A 107(10):4596–4600
65. Martin-Ibanez R, Hovatta O, Canals JM (2012) Cryopreservation of human pluripotent stem cells: are we going in the right direction? In: Katkov II (ed) Current frontiers in cryobiology. InTech, Rijeka, pp 139–166
66. Fraser JK, Cairo MS, Wagner EL, McCurdy PR, Baxter-Lowe LA, Carter SL, Kernan NA, Lill MC, Slone V, Wagner JE, Wallas CH, Kurtzberg J (1998) Cord Blood Transplantation Study (COBLT): cord blood bank standard operating procedures. J Hematother 7(6):521–561
67. Rubinstein P, Dobrila L, Rosenfield RE, Adamson JW, Migliaccio G, Migliaccio AR, Taylor PE, Stevens CE (1995) Processing and cryopreservation of placental/umbilical cord blood for unrelated bone marrow reconstitution. Proc Natl Acad Sci U S A 92(22):10119–10122
68. Thirumala S, Goebel WS, Woods EJ (2009) Clinical grade adult stem cell banking. Organogenesis 5(3):143–154
69. Son JH, Heo YJ, Park MY, Kim HH, Lee KS (2010) Optimization of cryopreservation condition for hematopoietic stem cells from umbilical cord blood. Cryobiology 60(3):287–292
70. Thirumala S, Wu X, Gimble JM, Devireddy RV (2010) Evaluation of polyvinylpyrrolidone as a cryoprotectant for adipose tissue-derived adult stem cells. Tissue Eng Part C Methods 16(4):783–792
71. Rao W, Huang H, Wang H, Zhao S, Dumbleton J, Zhao G, He X (2015) Nanoparticle-mediated intracellular delivery enables cryopreservation of human adipose-derived stem cells using trehalose as the sole cryoprotectant. ACS Appl Mater Interfaces 7(8):5017–5028
72. Verdanova M, Pytlik R, Kalbacova MH (2014) Evaluation of sericin as a fetal bovine serum-replacing cryoprotectant during freezing of human mesenchymal stromal cells and human osteoblast-like cells. Biopreserv Biobank 12(2):99–105
73. Yong KW, Pingguan-Murphy B, Xu F, Abas WA, Choi JR, Omar SZ, Azmi MA, Chua KH, Wan Safwani WK (2015) Phenotypic and functional characterization of long-term cryopreserved human adipose-derived stem cells. Sci Rep 5:9596
74. Diaferia GR, Dessi SS, Deblasio P, Biunno I (2008) Is stem cell chromosomes stability affected by cryopreservation conditions? Cytotechnology 58(1):11–16
75. Jenkins EC, Ye L, Silverman WP (2012) Does the cryogenic freezing process cause shorter telomeres? Cryobiology 65(1):72–73
76. de Lima Prata K, de Santis GC, Orellana MD, Palma PV, Brassesco MS, Covas DT (2012) Cryopreservation of umbilical cord mesenchymal cells in xenofree conditions. Cytotherapy 14(6):694–700
77. Luetzkendorf J, Nerger K, Hering J, Moegel A, Hoffmann K, Hoefers C, Mueller-Tidow C, Mueller LP (2015) Cryopreservation does not alter main characteristics of Good Manufacturing Process-grade human multipotent mesenchymal stromal cells including immunomodulating potential and lack of malignant transformation. Cytotherapy 17(2):186–198
78. Angelo PC, Ferreira AC, Fonseca VD, Frade SP, Ferreira CS, Malta FS, Pereira AK, Leite HV, Brum AP, Pardini VC, Gomes KB, Cabral AC (2012) Cryopreservation does not alter karyotype, multipotency, or NANOG/SOX2 gene expression of amniotic fluid mesenchymal stem cells. Genet Mol Res GMR 11(2):1002–1012
79. Yong KW, Safwani WK, Xu F, Zhang X, Choi JR, Abas WA, Omar SZ, Azmi MA, Chua KH, Pingguan-Murphy B (2016) Assessment of tumourigenic potential in long-term cryopreserved human adipose-derived stem cells. J Tissue Eng Regen Med 12(10)
80. Parmegiani L, Cognigni GE, Filicori M (2009) Ultraviolet sterilization of liquid nitrogen prior to vitrification. Hum Reprod 24(11):2969
81. Rollig C, Babatz J, Wagner I, Maiwald A, Schwarze V, Ehninger G, Bornhauser M (2002) Thawing of cryopreserved mobilized peripheral blood – comparison between waterbath and dry warming device. Cytotherapy 4(6):551–555
82. Thirumala S, Zvonic S, Floyd E, Gimble JM, Devireddy RV (2005) Effect of various freezing parameters on the immediate post-thaw membrane integrity of adipose tissue derived adult stem cells. Biotechnol Prog 21(5):1511–1524

9 Menstrual Blood-Derived Stem Cells: *In Vitro* and *In Vivo* Characterization of Functional Effects

Maria Carolina Oliveira Rodrigues,
Trenton Lippert, Hung Nguyen,
Sussannah Kaelber, Paul R. Sanberg,
and Cesar V. Borlongan

Abstract

Accumulating evidence has demonstrated that menstrual blood stands as a viable source of stem cells. Menstrual blood-derived stem cells (MenSCs) are morphologically and functionally similar to cells directly extracted from the endometrium, and present dual expression of mesenchymal and embryonic cell markers, thus becoming interesting tools for regenerative medicine. Functional reports show higher proliferative and self-renewal capacities than bone marrow-derived stem cells, as well as successful differentiation into hepatocyte-like cells, glial-like cells, endometrial stroma-like cells, among others. Moreover, menstrual blood stem cells may be used with increased efficiency in reprogramming techniques for induced Pluripotent Stem cell (iPS) generation. Experimental studies have shown successful treatment of stroke, colitis, limb ischemia, coronary disease, Duchenne's muscular atrophy and streptozotocin-induced type 1 diabetes animal models with MenSCs. As we envision an off-the-shelf product for cell therapy, cryopreserved MenSCs appear as a feasible clinical product. Clinical applications, although still very limited, have great potential and ongoing studies should be disclosed in the near future.

M.C.O. Rodrigues (✉)
Division of Clinical Immunology, Ribeirão Preto Medical School, University of São Paulo, Ribeirão Preto, Brazil
e-mail: mcarolor@usp.br

T. Lippert • H. Nguyen • S. Kaelber • P.R. Sanberg
C.V. Borlongan (✉)
Department of Neurosurgery and Brain Repair, Morsani College of Medicine, University of South Florida, 12901 Bruce B. Downs Blvd, Tampa, FL 33612, USA
e-mail: cborlong@health.usf.edu

F. Karimi-Busheri, M. Weinfeld (eds.), *Biobanking and Cryopreservation of Stem Cells*, Advances in Experimental Medicine and Biology 951, DOI 10.1007/978-3-319-45457-3_9

Keywords
Endometrium • Regenerative medicine • Endometrial stem cells • Menstrual blood stem cells • Pluripotent stem cells • Immunomodulation • Stem cell therapy

Abbreviations

MenSCs	Menstrual blood-derived stem cells
iPS	Induced Pluripotent Stem cell
MHC	Major histocompatibility complex
ERCs	Endometrial regenerative cells
hTERT	Human telomerase reverse transcriptase
MMPs	Matrix metalloproteases
VEGF	Vascular endothelial growth factor
BDNF	Brain-derived neurotrophic factor
DOPAC	Dopamine and dihydroxyphenylacetic acid

9.1 Introduction

In the past decades, several cell types from different sources have been investigated as therapeutic tools in experimental, translational and clinical research. Stem cells have been especially explored for their cytoprotective, regenerative and immunomodulatory capacities [1]. Adult, rather than embryonic stem cells, or those obtained by iPS-generating techniques, have been currently preferred in studies that aim clinical translation, despite their more restricted differentiation potential. Indeed, teratogenicity and ethical concerns, high costs and low efficiency stand as obstacles still to be surpassed before the more immature stem cells are enrolled in clinical trials.

Cells derived from bone marrow and cord-blood have been the focus of the initial studies on cell therapy, due to the already established experience with bone marrow transplantation. Later, easier access sources began to be examined, mainly in disposable tissues such as the placenta, amniotic fluid and membrane, umbilical cord tissue, dental pulp, and adipose tissue, among others [2–5]. Of particular relevance are the mesenchymal cells, which have been more extensively studied and hold good *in vivo* repairing potential. Mesenchymal cells do not spontaneously express class II major histocompatibility complex (MHC) molecules, what confers them low immunogenicity and ability to be tolerated by the host immune system [1]. Therefore, they have been tested as tools for treatment of autoimmune, inflammatory and degenerative diseases.

Successive studies have reported differences in mesenchymal cells from different sources and how they may be more or less suitable for treatment of specific disorders according to their tissue of origin [6–8]. In this scenario, menstrual blood has recently been investigated as source of stem cells for therapeutic purposes, the *in vitro* and preclinical studies revealing great potential, perhaps overcoming that of cells from other sources. This review addresses current and already established knowledge about *in vitro* and *in vivo* research on menstrual blood-derived stem cells, with emphasis on morphological and functional characteristics. Also discussed are the future perspectives of such cells, especially concerning clinical applications.

9.2 Endometrium and Menstrual Blood-Derived Stem Cells

More than 30 years ago, stem cells were first described to be present in the endometrium [9]. The intense tissue remodeling observed during menses and pregnancy mirrors the endometrial potential to renew itself, effect that has been ascribed to local stem cell activity [10]. In addition, endometrial cells from patients with endometriosis are remarkably able to migrate and graft to distantly located tissues [11]. In the endometrium, epithelial cells can be found in the surface epithelium as well as within the tubular glands, which anatomically continue to the interface with the myometrium.

The remaining tissue consists of stromal cells, smooth muscle cells, endothelial cells and leukocytes [12]. *The functionalis*, which can be considered the upper layer of the endometrium, contains glands that are held together loosely by stromal tissue. The basalis contains dense stroma, basal regions of the glands, supportive vasculature and lymphoid aggregates [13]. The *functionalis* is eliminated monthly during menstruation, while the remaining, *basalis*, is able to renew the endometrium, under hormonal influence. At each cycle, progenitor cells migrate from the *basalis*, proliferate, and regenerate the *functionalis*. In post-menopause women, progenitor cells can be found in the *basalis*, as a possible explanation to menstrual cycle reactivation through hormonal replacement [14].

The endometrium contains three types of stem cells: the epithelial progenitor cells, mesenchymal cells and endothelial progenitor cells [11]. Although multipotent stem cells may be isolated and cultured from endometrial biopsies and menstrual blood, it has been proposed that epithelial progenitor cells cannot be obtained from menstrual blood and most likely reside in the *basalis* layer [15, 16]. In fact, epithelial progenitors can be found in the endometrial *basalis* in post-menopause women, indicating a possible source of stem cells when menstrual blood is not any more available [14].

In 2004, Chan et al. isolated epithelial progenitors and stromal cells derived from the endometrium [17]. In the laboratory, both the isolated epithelial progenitors and the stromal cells were cloned and amplified, but phenotypic markers within the epithelial cells were lost and a feeder layer was needed over cell passages. Later, similar cells were isolated from menstrual blood [18–20] and named endometrial regenerative cells (ERCs). These cells differentiated into tissues from the three germ layers, indicating their *in vitro* multipotentiality. In 2008, Patel et al. isolated stem cells from menstrual blood (MenSCs), showing *in vitro* clonogenic properties as well as the differentiation of tissues derived from mesoderm and ectoderm [21]. They also demonstrated markers of pluripotency, such as Oct-4, SSEA-4 and c-kit, which were expressed by the MenSCs and are usually found in immature cell types, such as embryonic stem cells. Furthermore, telomerase activity and human telomerase reverse transcriptase (hTERT) expression in those cells resembled that of embryonic stem cells [21, 22]. In a subpopulation of the MenSCs, unexpectedly high clonogenic and proliferative potential was confirmed, besides expression of pluripotency markers (e.g., SSEA-4) [20]. Also, MenSCs were efficiently and stably gene modified by retroviral transfection, indicating that these cells could become suitable tools for gene delivery. Finally, levels of matrix metalloproteases (MMPs) in MenSCs are several fold above those presented by mesenchymal cells derived from cord blood, indicating a specific ability for remodeling [18].

Studies have investigated the *in vitro* potential of the MenSC cell fate, in particular demonstrating the stem cells derived from menstrual blood differentiated into functional hepatocyte-like cells, expressing hepatic surface markers (e.g., albumin, cytokeratin-18 and alpha fetoprotein) and genes [23, 24]. Menstrual blood-derived cells were also differentiated into nucleus pulposus-like cells in a co-culture system, the final product confirmed by phenotypic and gene expression evaluations [25]. Similarly, menstrual blood progenitors were differentiated into glial-like cells [26]. Moreover, MenSCs have higher capability to be differentiated towards cardiomyocytes than bone marrow-derived cells and *in vivo* differentiation of menstrual blood cells into cardiomyocytes has been reported [27, 28]. Finally, chondrogenic differentiation of MenSCs seeded on a nanofibrous scaffold, strongly reactive to anti-collagen II antibodies, has been reported [29]. Menstrual blood-derived stem cells may also be used with increased efficiency in reprogramming techniques for iPS generation [30, 31].

Although numerous studies have been published on MenSC properties in the past years, some questions still remain unanswered. Since the initial investigations, it has been shown that more than one single cell population may be isolated from either endometrial tissue or menstrual blood, and the multiple available reports do not clearly indicate which is the most important cell type for regenerative purposes. Cervello et al. [32] isolated epithelial and stromal-cell enriched side populations through flow cytometry of Hoechst- stained endometrium cells [32]. *In vitro*, the cells were characterized and showed high proliferative potentials, especially when they were exposed to an environment containing hypoxic conditions, similar to that of the endometrial environment. However, when these cells were administered to immunodeficient mice, limited proliferation and differentiation were observed, demonstrating a divergence of *in vitro* and *in vivo* effects. Masuda et al. (2010) implanted similar cells in female mice under the kidney capsule, and observed human tissue development in some of the animals under estrogen stimulation [33]. Side-population cells successfully differentiated into glandular epithelial, stromal and, for the first time, endothelial cells, with presence of CD31 and human vimetin co-expressing in small and medium sized vessels. However, although they were detectable, the *in vivo* differentiation capacity was considered poor. For better proliferative results, cells were combined with the endometrial cells of the remaining population (i.e., main population.). Combined with reported data from literature, these findings suggest that the endometrium has multiple factors that can cooperate for therapeutic properties of this tissue, rather than from a single cell type. In fact, it has been previously proposed that even the endometrial niche itself may contribute to the quality of cell activity [11].

The angiogenic potential of the endometrium-derived cells is also pertinent for vascular growth and remodeling investigation. Disorders such as stroke, cardiac ischemia and chronic vascular insufficiency, among others, may stand as perfect targets for these cells. In fact, successful results have been achieved by preclinical studies on myocardial infarct, limb ischemia, and stroke [25, 34, 35]. Angiogenic effects are also proposed in the treatment of severe skin burns, perhaps associating cells to intelligent artificial films [34, 36].

Another question concerns the classification criteria for menstrual blood-derived stem cells. Murphy et al. [34] speculate that the ERCs they isolated share properties that overlap and may even be considered equivalent to cells reported by other studies, but may not be the same endometrial stromal cells described by Taylor, in 2004 [18, 21, 37, 38]. The ERCs show low concentrations of the STRO-1 mesenchymal cell marker, differentiate into a wider range of tissues and exhibit higher proliferative capacity than cord blood-derived cells [18]. When compared with bone marrow-derived mesenchymal cells, ERCs also presented higher lymphocyte proliferative inhibitory potential, and different gene expression profiles, which were mostly towards inflammatory and immune pathways in the ERCs, while stem cell and cancer signaling in the bone marrow-derived mesenchymal cells [39]. Taylor postulates that the stromal cells observed within the endometrium are derived from the bone marrow, as observed in female recipients of allogeneic bone marrow transplantation that have donor cells detected in tissues from endometrial biopsies. These findings were later duplicated in female rats, which presented bone marrow-derived GFP-positive cells in the endometrium long after transplantation [40]. In practical matters, however, although endometrial-derived cells are grouped through different phenotypical, differentiation and proliferative criteria, when applied *in vivo* they have similar effects and comparable therapeutic abilities to advance repair.

9.3 Experimental Applications of Stem Cells Derived from the Endometrium

Progenitor cells derived directly from the endometrium or indirectly, from menstrual blood, have been investigated for their *in vivo* immunomodulatory and restorative effects (see Table 9.1).

Table 9.1 Main preclinical studies reporting use of menstrual blood and endometrial-derived cells

Animal model	Main in vivo effect	Administration route	Reference
Endometrial tissue-derived cells			
Parkinson's disease	Dopamine and dihydroxyphenylacetic acid (DOPAC) production	Local (brain)	[37]
Experimental allergic encephalitis	Prevented development of neuroinflammation (prophylactic effect)	Intraperitoneally	[41]
Pelvic organ prolapse	Decreased chronic inflammation and fibrosis	Local (pelvis)	[42]
Menstrual blood-derived cells			
Duchenne muscular dystrophy	In vivo expression of dystrophin/possible fusion to myocytes	Intramuscular	[22]
Coronary artery ligation	Improved heart function, decreased infracted areas	Local (heart)	[28]
Chronic limb ischemia	Prevented limb amputation	Intramuscular	[34]
Glioma	Reduction in 50% of tumor size	iv and intratumoral	[43]
Stroke	Neuroprotection	Iv and local (striatum)	[35]
Liver cirrhosis	Restored liver function	Intrasplenic	[24]
Colitis	Decreased disease activity	iv	[46]
Streptozotocin-induced diabetes	Extended survival and improved glycemic control	iv	[47]

Abbreviation: *iv* intravenous

In an aggressive glioma mouse model, ERCs injected either by intravenous or intratumoral routes successfully reduced tumor size in almost 50% [43]. Among possible mechanisms, the authors considered secretion of inhibitory factors by the injected cells, selective angiogenesis in favor of benign and against malignant tissue, and cell-tumor interaction stimulating the differentiation of the tumoral tissue.

Borlongan et al. evaluated the effects of menstrual blood-derived cell transplantation in experimental stroke [35]. Stromal-like menstrual blood cells were selected for CD117 or C-kit receptors by isolating and expanding the stem cells. Those cells selected indicate a high probability of proliferation, migration and survival [44]. The embryonic-like stem cells expressed *in vitro* phenotypic markers, such as Oct4, SSeA and Nanog. Up to nine passages of cultures were maintained, which provided evidence of the safety and reliability of those cells, and part of the cells were induced to express neural markers (MAP2 and Nestin).

When menstrual blood cells were added to cultured rat neurons that had been exposed to a hypoxic insult, the menstrual blood cells exerted neuroprotective effects. Behavioral tests were administered and stroke animals that received menstrual blood cells displayed less neurological deficits, irrespective of the injection site, i.e. systemic or local administration into the striatum.

Analyses of the central nervous tissue in the transplanted stroke rats revealed that the grafted cells did not exhibit markers of cell lineage differentiation. Results showed that the human cells, although detected and some migrating to areas other than the injected, expressed their original stemness markers. Altogether, these results indicate that cell differentiation was not the main therapeutic operating mechanism and that other pathways may be involved in the observed functional recovery.

In addition, vascular endothelial growth factor (VEGF), brain-derived neurotrophic factor (BDNF), and neurotrophin-3 were detected as possible neuroprotective trophic factors released by ERCs. These results may be supported by a recent *in vitro* study reporting that menstrual blood-derived stem cells express the neuronal marker β-III-tubulin and endogenous BDNF in higher levels than bone marrow and adipose tissue-derived cells [45].

In a different CNS disorder context, emphasizing more restorative than neuroprotective effects, Wolff et al. reported the use of endometrial-derived cells in a Parkinson's disease immunocompetent mouse model [37]. *In vitro* differentiated dopamine-producing cells expressing neural markers nestin and tyrosine hydroxylase were injected into the brains of the animals. Labeled cells were detected in the brains up to 5 weeks after administered. Therapeutic potential was recognized *in vivo* when the migration, differentiation and production of dopamine were detected. These cells have the potential to functionally restore the damaged tissues, either through cell replacement or endogenous repair.

An experimental colitis mouse model improved after treatment with intravenous injections of human menstrual blood-derived ERCs [46]. Despite short follow-up of 14 days, treated animals presented decreased disease activity scores, reduced inflammatory cytokine levels and higher regulatory cell frequencies, indicating a systemic beneficial effect of cell infusions. In addition, histological evaluations evidenced significant modulation of the inflammatory infiltrate.

Streptozotocin-induced diabetes mice were treated with one single intravenous injection of human menstrual blood-derived stem cells [47]. Treated mice presented longer survival and better glycemic control when compared with untreated diabetic mice. In accordance with previous reports, MenSCs presented high proliferation, short doubling time, and increased expression of the immature cell marker Oct-4. Although migration of labeled MenSCs to the pancreata and islet beta-cell regeneration were detected, there were no evidence of cell transdifferentiation into pancreatic tissue, suggesting that paracrine mechanisms prevailed.

Menstrual blood-derived cells were also tested in a rat coronary ligation model [28]. In this study, menstrual blood-derived mesenchymal cells were co-cultured with fetal cardiomyocytes separated by an athelocollagen membrane (no cell-cell contact), and *in vitro* differentiated into beating cardiac precursor-like cells. Subsequently, non-differentiated menstrual blood-derived mesenchymal cells were injected into and marginal to the ischemic area of the heart, 2 weeks after coronary artery ligation. Clinical and histological effects were compared to bone marrow mesenchymal cell and culture medium or fibroblast injections. Animals treated with MenSCs presented higher heart function recovery rates, as well as smaller infarcted areas, than those from the other treatment groups. High in situ cardiomyogenic transdifferentiation was also observed. Also investigating the vessel-regenerating potential of the ERCs, Murphy et al. injected MenSCs into the hind-limb muscle just distal to the level of artery ligation, in a chronic limb ischemia rat model [34]. Multiple cell injections were administered shortly after the vascular obstruction – immediately, 2 and 4 days after ligation – and showed a protective effect, preventing limb necrosis and amputation. Finally, MenSC differentiated *in vitro* into hepatocyte-like cells were injected intrasplenically and restored liver function, normalizing albumin and transaminase levels [24].

9.4 The Advantages and Disadvantages of Cryopreservation of Menstrual Blood-Derived Stem Cells

Cryopreservation can be utilized to preserve previously expanded stem cells for future use, as shown in Fig. 9.1. MenSCs can be cryopreserved for extended period of time. There are both advantages and disadvantages of cryopreservation of stem cells. These pros and cons will eventually determine if the MenSCs should be cryopreserved. Table 9.2 summarizes the advantages and disadvantages of cryopreservation.

The first advantage of cryopreservation is that the stem cells can be stored for extended period

Fig. 9.1 Menstrual blood can be harvested and processed to obtain stem cells. The MenSCs are expanded and then cryopreserved until needed, when the cells are thawed and prepared for transplantation

Table 9.2 Advantages and disadvantages of cryopreservation of stem cells

Advantages	Disadvantages
Extended storage time	Cell viability might be low due to freezing and thawing
Ready to use for acute diseases	Functional properties of stem cells might change due to cryopreservation
Autologous transplant	High cost for storage
Quality control	Potential bioethics dilemma
Easy for distribution	

of time. Hence, it can be readily available for expansion and use with relatively short period of time. This is extremely important for acute time sensitive disorders such as stroke and traumatic brain injury. Many preclinical studies have demonstrated that the earlier the injection of stem cells, the better the recovery is. Therefore, cryopreserved stem cells can shorten the waiting period so that they can be used within the critical therapeutic period. In addition, cryopreservation of stem cells allows autologous transplant. In clinical settings, autologous transplant is always more preferable than heterologous transplant. Another advantage of cryopreserved stem cells is that the quality of the stem cells can be ensured. The stem cells can be phenotyped, karyotyped and scanned for quality and any abnormality prior to transplantation. These tests can ensure the integrity of the stem cells for transplantation. Finally, cryopreservation of stem cells allows them to become 'off the shelf' products. These cells can be distributed to many small hospitals and clinics nationwide and even worldwide instead of concentrated in small number of major facilities, which have the equipment and the experience to harvest and process the stem cells.

Despite many advantages of cryopreserved stem cells, there are also disadvantages that are worth taken into consideration. Cryopreservation involves freezing and thawing processes of the stem cells. This raises the concern of the cells' viability since many cells may not survive the processes. One of the solutions for this issue is to store a large numbers of cells, which in turn raises the cost associated with stem cell treatment. Another concern with cryopreservation of stem cells is that the function of the stem cells might change because of the freezing and thawing processes. Since freezing and thawing are forms of stress to the cells, this might change the properties of the stem cells. Further investigations might help to elucidate the effects of cryopreservation on the stem cells. Finally, the cost of processing and storing of the stem cells are high. This might create a bioethical issue that the majority of the population would not be able to afford the treatment and only a small niche of wealthy people could have access to this type of therapy. It is important for both scientists and clinicians to consider these issues so they can be addressed.

9.5 Clinical Applications of Menstrual Blood-Derived Stem Cells

Only in the last decade has the potential of menstrual blood-derived stem cells in regenerative medicine began to be explored. Therefore, very few clinical studies are available, most of them still ongoing (see Table 9.3). The only clinical study yet published evaluated safety aspects of endometrial-derived stromal cell administration in patients with multiple sclerosis [48]. Four patients were treated with intrathecal injections

Table 9.3 Clinical trials evaluating menstrual blood-derived cells

Treated condition	Outcomes	Number of patients	Status	Administration route	Reference
Multiple sclerosis	Safety and neurological function	4	Completed	Intrathecal and iv	[48]
Congestive heart failure	Safety and cardiac function	17 out of 60 planned	Ongoing/recruiting	Local – coronary sinus delivery	[49]
Liver cirrhosis	Safety and liver function	50 planned	Ongoing/recruiting	iv	[51]
H7N9 bird flu lung injury	Safety and lung injury	20 planned	Ongoing/recruiting	iv	[52]
Critical limb ischemia	Safety, vascularization and frequency of amputation	15 planned	Ongoing/unknown	Intramuscular	[53]
Type 1 diabetes	Safety, glucose metabolism, beta cell function	50 planned	Ongoing/unknown	Pancreatic artery or iv	[54]
Embryo implantation	Endometrial receptivity	60 planned	Ongoing/unknown	Intrauterine	[55]

Abbreviation: *iv* intravenous

of 16–30 million cells and one of the patients also received an additional intravenous injection of the stem cells. No adverse events were registered, in accordance with the pre-clinical studies, and functional stabilization was reported. However, due to the nature of multiple sclerosis and the worsening progression of the illness, any conclusions about the effectiveness of the treatment could be considered premature due to the longest follow-up reaching only 12 months.

Ongoing trials aim to evaluate the effect of MenSC injections in several clinical conditions. Menstrual blood-derived stem cells have been tested in congestive heart failure patients, as a follow-up of the preclinical study published by Hida et al. [28, 49]. This double-blind, randomized phase II study intends to include 60 patients, divided into placebo and treatment groups. Allogeneic MenSCs (Medistem Inc, California, USA) are injected through a minimally invasive and previously described retrograde delivery technique [50]. The last communication, in 2013, reported that 17 patients had been enrolled, with safety being so far analyzed and confirmed [49].

In another trial, female liver cirrhosis patients receive intravenous injections of MenSCs twice a week, for 2 weeks [51]. Patients will be followed for 2 years and evaluated for safety and efficacy of the cell therapy upon liver function and progression of the cirrhosis. Another trial evaluates efficacy of MenSC intravenous injections in patients with acute lung injury due to H7N9 bird flu infections [52]. Effects of the cells upon lung injury are the main outcomes.

A phase I/II clinical trial investigating the effect of escalating doses of ERCs on limb ischemia has also been launched in 2012, aiming to enroll 15 patients with critical limb ischemia not eligible for invasive interventions [53]. Data on the study status, however, are not available. Effects of ERCs on type 1 diabetes are also under investigation in a phase I/II open label trial launched in 2012 [54]. In this study, multiple doses of ERCs are injected through the pancreatic artery or intravenously. Adverse events and effects on glucose metabolism and pancreatic beta cell function will be evaluated. Updated status of the study is also not available. Finally, the effects of ERCs on embryo implantation in the uterine cavity during *in vitro* fertilization will be investigated [55].

9.6 Conclusions

Menstrual blood-derived cells have great potential in regenerative medicine [56]. Ease of access, availability and safety are considered their main key to future clinical studies. *In vitro* investigations have shown the superiority of MenSC in proliferation and differentiation assays, as well as immunomodulatory potential, when compared with more traditionally used sources such as the bone marrow and cord blood. Moreover, preclinical studies have successfully shown their potential to modulate inflammation, prevent neuronal death and possibly increase angiogenesis, *in vivo*. Importantly, cell fusion may be a therapeutic mechanism in muscle and cardiac tissue injuries. These effects may be result of the more immature behavior of these cells, which although bringing them closer to embryonic stem cells, still retain a safe profile. Although clinical studies have already been started, further experimental studies are still required, as several questions about the full therapeutic potential and specific mechanisms of action persist. In particular, cryopreservation of MenSCs while feasible, may require further validation of the frozen-and-thawed stem cells, as well the cost-limiting ethical concerns that may be associated with such cell banking approach. These investigations may help to explore benefits of such cells and expand applications. Clinical practice bears specific challenges, such as the low yield and difficulty in expansion of ample supply of stem cells from this source, mainly low replication rate and risk of contamination. Additionally, stem cells derived from menstrual blood would only be applicable to the pre-menopausal female population, thereby limiting the target patient population when contemplating autologous menstrual blood cell therapy. A feasible solution would be to educate the female pre-menopausal population about the potential of the menstrual cells and, therefore, encourage this young adult population towards the anticipated harvesting and cryopreservation of the cells, for future autologous use. When such pre-menopausal collection and banking is not pursued in time, the use of stem cells derived from a post-menopausal endometrium may be a solution, though more invasive. For the male population, there remains the alternative of using allogeneic cells and of searching for alternative autologous sources.

Acknowledgements The authors thank Mr. Igor Passioura for excellent technical assistance in the manuscript preparation.

Conflict of Interest PRS and CVB are founders and/or consultants of Saneron-CCEL, and hold patents and patent applications on stem cells and their applications.

References

1. Rammal H, Harmouch C, Lataillade JJ, Laurent-Maquin D, Labrude P, Menu P, Kerdjoudj H (2014) Stem cells: a promising source for vascular regenerative medicine. Stem Cells Dev 23:2931–2949
2. Hass R, Kasper C, Bohm S, Jacobs R (2011) Different populations and sources of human mesenchymal stem cells (MSC): a comparison of adult and neonatal tissue-derived MSC. Cell Commun Signal 9:12
3. Crisan M, Yap S, Casteilla L, Chen CW, Corselli M, Park TS, Andriolo G, Sun B, Zheng B, Zhang L, Norotte C, Teng PN, Traas J, Schugar R, Deasy BM, Badylak S, Buhring HJ, Giacobino JP, Lazzari L, Huard J, Péault B (2008) A perivascular origin for mesenchymal stem cells in multiple human organs. Cell Stem Cell 3:301–313
4. Gronthos S, Mankani M, Brahim J, Robey PG, Shi S (2000) Postnatal human dental pulp stem cells (DPSCs) in vitro and in vivo. Proc Natl Acad Sci U S A 97:13625–13630
5. Brooke G, Rossetti T, Pelekanos R, Ilic N, Murray P, Hancock S, Antonenas V, Huang G, Gottlieb D, Bradstock K, Atkinson K (2009) Manufacturing of human placenta-derived mesenchymal stem cells for clinical trials. Br J Haematol 144:571–579
6. Beane OS, Fonseca VC, Cooper LL, Koren G, Darling EM (2014) Impact of aging on the regenerative properties of bone marrow-, muscle-, and adipose-derived mesenchymal stem/stromal cells. PLoS One 9, e115963
7. Duan B, Hockaday LA, Das S, Xu CY, Butcher JT (2015) Comparison of mesenchymal stem cell source differentiation towards human pediatric aortic valve interstitial cells within 3D engineered matrices. Tissue Eng Part C Methods 21(8):795–807
8. Lotfy A, Salama M, Zahran F, Jones E, Badawy A, Sobh M (2014) Characterization of mesenchymal stem cells derived from rat bone marrow and adipose tissue: a comparative study. Int J Stem Cells 7:135–142
9. Prianishnikov VA (1978) A functional model of the structure of the epithelium of normal, hyperplastic

and malignant human endometrium: a review. Gynecol Oncol 6:420–428
10. Gargett CE (2007) Uterine stem cells: what is the evidence? Hum Reprod Update 13:87–101
11. Gargett CE, Masuda H (2010) Adult stem cells in the endometrium. Mol Hum Reprod 16:818–834
12. Padykula HA (1991) Regeneration in the primate uterus: the role of stem cells. Ann N Y Acad Sci 622:47–56
13. Spencer TE, Hayashi K, Hu J, Carpenter KD (2005) Comparative developmental biology of the mammalian uterus. Curr Top Dev Biol 68:85–122
14. Nguyen HP, Sprung CN, Gargett CE (2012) Differential expression of Wnt signaling molecules between pre- and postmenopausal endometrial epithelial cells suggests a population of putative epithelial stem/progenitor cells reside in the basalis layer. Endocrinology 153:2870–2883
15. Musina R, Belyavski A, Tarusova O, Solovyova E, Sukhikh G (2008) Endometrial mesenchymal stem cells isolated from the menstrual blood. Bull Exp Biol Med 145:539–543
16. Verdi J, Tan A, Shoae-Hassani A, Seifalian AM (2014) Endometrial stem cells in regenerative medicine. J Biol Eng 8:20
17. Chan RW, Schwab KE, Gargett CE (2004) Clonogenicity of human endometrial epithelial and stromal cells. Biol Reprod 70:1738–1750
18. Meng X, Ichim TE, Zhong J, Rogers A, Yin Z, Jackson J, Wang H, Ge W, Bogin V, Chan KW, Thébaud B, Riordan NH (2007) Endometrial regenerative cells: a novel stem cell population. J Transl Med 5:57
19. Van der Molen RG, Schutten JH, van Cranenbroek B, ter Meer M, Donckers J, Scholten RR, van der Heijden OW, Spaanderman ME, Joosten I (2014) Menstrual blood closely resembles the uterine immune microenvironment and is clearly distinct from peripheral blood. Hum Reprod 29:303–314
20. Rossignoli F, Caselli A, Grisendi G, Piccinno S, Burns JS, Murgia A, Veronesi E, Loschi P, Masini C, Conte P, Paolucci P, Horwiz EM, Dominici M (2013) Isolation, characterization, and transduction of endometrial decidual tissue multipotent mesenchymal stromal/stem cells from menstrual blood. Biomed Res Int 2013:901821
21. Patel AN, Silva F (2008) Menstrual blood stromal cells: the potential for regenerative medicine. Regen Med 3:443–444
22. Cui CH, Uyama T, Miyado K, Terai M, Kyo S, Kiyono T, Umezawa A (2007) Menstrual blood-derived cells confer human dystrophin expression in the murinemodel of Duchenne muscular dystrophy via cell fusion and myogenic transdifferentiation. Mol Biol Cell 18:1586–1594
23. Khanjani S, Khanmohammadi M, Zarnani AH, Akhondi MM, Ahani A, Ghaempanah Z, Naderi MM, Eghtesad S, Kazemnejad S (2014) Comparative evaluation of differentiation potential of menstrual blood- versus bone marrow-derived stem cells into hepatocyte-like cells. PLoS One 9, e86075
24. Mou XZ, Lin J, Chen JY, Li YF, Wu XX, Xiang BY, Li CY, Ma JM, Xiang C (2013) Menstrual blood-derived mesenchymal stem cells differentiate into functional hepatocyte-like cells. J Zhejiang Univ Sci B 14:961–972
25. Hu X, Zhou Y, Zheng X, Tian N, Xu C, Wu W, Li F, Zhu S, Zheng Y, Xue E, Yu Y, Zhang X, Xu H (2014) Differentiation of menstrual blood-derived stem cells toward nucleus pulposus-like cells in a coculture system with nucleus pulposus cells. Spine 39:754–760
26. Azedi F, Kazemnejad S, Zarnani AH, Behzadi G, Vasei M, Khanmohammadi M, Khanjani S, Edalatkhah H, Lakpour N (2014) Differentiation potential of menstrual blood- versus bone marrow-stem cells into glial-like cells. Cell Biol Int 38:615–624
27. Rahimi M, Zarnani AH, Mohseni-Kouchesfehani H, Soltanghoraei H, Akhondi MM, Kazemnejad S (2014) Comparative evaluation of cardiac markers in differentiated cells from menstrual blood and bone marrow-derived stem cells in vitro. Mol Biotechnol 56:1151–1162
28. Hida N, Nishiyama N, Miyoshi S, Kira S, Segawa K, Uyama T, Mori T, Miyado K, Ikegami Y, Cui C, Kiyono T, Kyo S, Shimizu T, Okano T, Sakamoto M, Ogawa S, Umezawa A (2008) Novel cardiac precursor-like cells from human menstrual blood-derived mesenchymal cells. Stem Cells 26:1695–1704
29. Kazemnejad S, Zarnani AH, Khanmohammadi M, Mobini S (2013) Chondrogenic differentiation of menstrual blood-derived stem cells on nanofibrous scaffolds. Methods Mol Biol 1058:149–169
30. De Carvalho Rodrigues D, Asensi KD, Vairo L, Azevedo-Pereira RL, Silva R, Rondinelli E, Goldenberg RC, Campos de Carvalho AC, Urményi TP (2012) Human menstrual blood-derived mesenchymal cells as a cell source of rapid and efficient nuclear reprogramming. Cell Transplant 21:2215–2224
31. Li Y, Li X, Zhao H, Feng R, Zhang X, Tai D, An G, Wen J, Tan J (2013) Efficient induction of pluripotent stem cells from menstrual blood. Stem Cells Dev 22:1147–1158
32. Cervelló I, Gil-Sanchis C, Mas A, Faus A, Sanz J, Moscardó F, Higueras G, Sanz MA, Pellicer A, Simón C (2012) Bone marrow-derived cells from male donors do not contribute to the endometrial side population of the recipient. PLoS One 7, e30260
33. Masuda H, Matsuzaki Y, Hiratsu E, Ono M, Nagashima T, Kajitani T, Arase T, Oda H, Uchida H, Asada H, Ito M, Yoshimura Y, Maruyama T, Okano H (2010) Stem cell-like properties of the endometrial side population: implication in endometrial regeneration. PLoS One 5, e10387
34. Murphy MP, Wang H, Patel AN, Kambhampati S, Angle N, Chan K, Marleau AM, Pyszniak A, Carrier E, Ichim TE, Riordan NH (2008) Allogeneic endome-

trial regenerative cells: an "Off the shelf solution" for critical limb ischemia? J Transl Med 6:45
35. Borlongan CV, Kaneko Y, Maki M, Yu SJ, Ali M, Allickson JG, Sanberg CD, Kuzmin-Nichols N, Sanberg PR (2010) Menstrual blood cells display stem cell-like phenotypic markers and exert neuroprotection following transplantation in experimental stroke. Stem Cells Dev 19:439–452
36. Drago H, Marin GH, Sturla F, Roque G, Mártire K, Díaz Aquino V, Lamonega R, Gardiner C, Ichim T, Riordan N, Raimondi JC, Bossi S, Samadikuchaksaraei A, van Leeuwen M, Tau JM, Núñez L, Larsen G, Spretz R, Mansilla E (2010) The next generation of burns treatment: intelligent films and matrix, controlled enzymatic debridement, and adult stem cells. Transplant Proc 42:345–349
37. Wolff EF, Gao XB, Yao KV, Andrews ZB, Du H, Elsworth JD, Taylor HS (2011) Endometrial stem cell transplantation restores dopamine production in a Parkinson's disease model. J Cell Mol Med 15:747–755
38. Taylor HS (2004) Endometrial cells derived from donor stem cells in bone marrow transplant recipients. JAMA 292:81–85
39. Wang H, Jin P, Sabatino M, Ren J, Civini S, Bogin V, Ichim TE, Stroncek DF (2012) Comparison of endometrial regenerative cells and bone marrow stromal cells. J Transl Med 10:207
40. Bratincsák A, Brownstein MJ, Cassiani-Ingoni R, Pastorino S, Szalayova I, Tóth ZE, Key S, Németh K, Pickel J, Mezey E (2007) CD45-positive blood cells give rise to uterine epithelial cells in mice. Stem Cells 25:2820–2826
41. Peron JP, Jazedje T, Brandão WN, Perin PM, Maluf M, Evangelista LP, Halpern S, Nisenbaum MG, Czeresnia CE, Zatz M, Câmara NO, Rizzo LV (2012) Human endometrial-derived mesenchymal stem cells suppress inflammation in the central nervous system of EAE mice. Stem Cell Rev 8:940–952
42. Ulrich D, Edwards SL, Su K, Tan KS, White JF, Ramshaw JA, Lo C, Rosamilia A, Werkmeister JA, Gargett CE (2014) Human endometrial mesenchymal stem cells modulate the tissue response and mechanical behavior of polyamide mesh implants for pelvic organ prolapse repair. Tissue Eng Part A 20:785–798
43. Han X, Meng X, Yin Z, Rogers A, Zhong J, Rillema P, Jackson JA, Ichim TE, Minev B, Carrier E, Patel AN, Murphy MP, Min WP, Riordan NH (2009) Inhibition of intracranial glioma growth by endometrial regenerative cells. Cell Cycle 8:606–610
44. Cho NH, Park YK, Kim YT, Yang H, Kim SK (2004) Lifetime expression of stem cell markers in the uterine endometrium. Fertil Steril 81:403–407
45. Zemel'ko VI, Kozhukharova IV, Kovaleva ZV, Domnina AP, Pugovkina NA, Fridlianskaia II, Puzanov MV, Anisimov SV, Grinchuk TM, Nikol'skiĭ NN (2014) BDNF secretion in human mesenchymal stem cells isolated from bone marrow, endometrium and adipose tissue. Tsitologiia 56:204–211
46. Lv Y, Xu X, Zhang B, Zhou G, Li H, Du C, Han H, Wang H (2014) Endometrial regenerative cells as a novel cell therapy attenuate experimental colitis in mice. J Transl Med 12:344
47. Wu X, Luo Y, Chen J, Pan R, Xiang B, Du X, Xiang L, Shao J, Xiang C (2014) Transplantation of human menstrual blood progenitor cells improves hyperglycemia by promoting endogenous progenitor differentiation in type 1 diabetic mice. Stem Cells Dev 23:1245–1257
48. Zhong Z, Patel AN, Ichim TE, Riordan NH, Wang H, Min WP, Woods EJ, Reid M, Mansilla E, Marin GH, Drago H, Murphy MP, Minev B (2009) Feasibility investigation of allogeneic endometrial regenerative cells. J Transl Med 7:15
49. Bockeria L, Bogin V, Bockeria O, Le T, Alekyan B, Woods EJ, Brown AA, Ichim TE, Patel AN (2013) Endometrial regenerative cells for treatment of heart failure: a new stem cell enters the clinic. J Transl Med 11:56
50. Tuma J, Fernández-Viña R, Carrasco A, Castillo J, Cruz C, Carrillo A, Ercilla J, Yarleque C, Cunza J, Henry TD, Patel AN (2011) Safety and feasibility of percutaneous retrograde coronary sinus delivery of autologous bone marrow mononuclear cell transplantation in patients with chronic refractory angina. J Transl Med 9:183
51. (2012) Human menstrual blood-derived mesenchymal stem cells for patients with liver cirrhosis. Clinicaltrials.gov. https://clinicaltrials.gov/ct2/show/NCT01483248. Accessed 22 Feb 2015
52. (2014) Using human menstrual blood cells to treat acute lung injury caused by H7N9 bird flu virus infection. Clinicaltrials.gov. https://www.clinicaltrials.gov/ct2/show/NCT02095444. Accessed 22 Feb 2015
53. (2012) Phase I/II trial of endometrial regenerative cells (ERC) in patients with critical limb ischemia. Clinicaltrials.gov. https://clinicaltrials.gov/ct2/show/NCT01558908. Accessed 22 Feb 2015
54. (2012) Human menstrual blood-derived mesenchymal stem cells transplantation in treating type 1 diabetic patients. Clinicaltrials.gov https://clinicaltrials.gov/ct2/show/NCT01496339. Accessed 22 Feb 2015
55. (2012) Role of stem cells in improving implantation rates in ICSI patients. Clinicaltrials.gov. https://clinicaltrials.gov/ct2/show/NCT01649752. Accessed 22 Feb 2015
56. Allickson JG, Sanchez A, Yefimenko N, Borlongan CV, Sanberg PR (2011) Recent studies assessing the proliferative capability of a novel adult stem cell identified in menstrual blood. Open Stem Cell J 3:4–10

Cryopreservation of Human Pluripotent Stem Cell-Derived Cardiomyocytes: Strategies, Challenges, and Future Directions

10

Marcela K. Preininger, Monalisa Singh, and Chunhui Xu

Abstract

In recent years, human pluripotent stem cell-derived cardiomyocytes (hPSC-CMs) have emerged as a vital cell source for *in vitro* modeling of genetic cardiovascular disorders, drug screening, and *in vivo* cardiac regeneration research. Looking forward, the ability to efficiently cryopreserve hPSC-CMs without compromising their normal biochemical and physiologic functions will dramatically facilitate their various biomedical applications. Although working protocols for freezing, storing, and thawing hPSC-CMs have been established, the question remains as to whether they are optimal. In this chapter, we discuss our current understanding of cryopreservation appertaining to hPSC-CMs, and proffer key questions regarding the mechanical, contractile, and regenerative properties of cryopreserved hPSC-CMs.

Keywords

Human pluripotent stem cell-derived cardiomyocytes • Cryopreservation • Thawing • Freezing • Contraction • Beating • Cell viability • Electrical coupling • Engraftment • Transplantation

M.K. Preininger • C. Xu (✉)
Division of Pediatric Cardiology, Department of Pediatrics, Emory University School of Medicine and Children's Healthcare of Atlanta, 2015 Uppergate Drive, Atlanta, GA 30322, USA

Wallace H. Coulter Department of Biomedical Engineering, Georgia Institute of Technology and Emory University, Atlanta, GA, USA
e-mail: chunhui.xu@emory.edu

M. Singh
Division of Pediatric Cardiology, Department of Pediatrics, Emory University School of Medicine and Children's Healthcare of Atlanta, 2015 Uppergate Drive, Atlanta, GA 30322, USA

F. Karimi-Busheri, M. Weinfeld (eds.), *Biobanking and Cryopreservation of Stem Cells*, Advances in Experimental Medicine and Biology 951, DOI 10.1007/978-3-319-45457-3_10

Abbreviations

hPSC-CMs	Human pluripotent stem cell-derived cardiomyocytes
CPA	Cryoprotective agent
DMSO	Dimethyl sulfoxide
FBS	Fetal bovine serum
ROCK	Rho-associated kinase

10.1 Introduction

Despite advances in prevention and management, cardiovascular disease remains a leading cause of morbidity and mortality worldwide—underscoring the need for novel and improved therapies. While still critically important, animal models fail to recapitulate several key aspects of human cardiovascular physiology. Thus, the development of new cardiac therapies has been partly hindered by a lack of human cardiac tissues and isolated cardiomyocytes to study *in vitro*. Primary human cardiomyocytes are difficult to obtain, fail to thrive in culture, and are especially refractory to genetic manipulation. In this regard, human pluripotent stem cells (hPSCs), including human induced pluripotent stem cells (hiPSCs) and human embryonic stem cells (hESCs), have been vaunted as a promising platform to effect progress. Due to their unparalleled self-renewal capacity and differentiation potential *in vitro*, hPSCs represent a consummate and theoretically unlimited source of cardiomyocytes for disease modeling, stem cell research, drug discovery, and cell-based regenerative therapies. In contrast to their primary counterparts, cardiomyocytes derived from hPSCs are readily amenable to long-term culture and can be additionally manipulated for genetic modifications and tissue engineering purposes. Considering the time and cost of continuous culture, cryopreservation of hPSC-derived cardiomyocytes (hPSC-CMs) will greatly facilitate their various applications.

In this chapter, we further enumerate the uses of hPSC-CMs in both clinical and research settings and explain the value of their cryopreservation. We additionally discuss considerations and strategies for freezing, storing, and thawing methods as well as our current understanding of the phenotype and function of cryopreserved hPSC-CMs. Finally, we highlight key questions that must be rejoined before cryopreserved hPSC-CMs can be credibly regarded as an equivalent cell source when compared to their freshly-derived counterparts for *in vitro* modeling, drug discovery, and *in vivo* cardiac regeneration.

10.2 Applications of Pluripotent Stem Cell-Derived Cardiomyocytes

hPSC-CMs have been used broadly for cardiac tissue engineering [1, 2] and developmental biology [3, 4], but the applications garnering the most attention are in disease modeling, drug development, and cardiac regeneration therapy. Therefore, in this section, we will provide a brief introduction into these areas of investigation to emphasize the importance of hPSC-CMs in biomedical research.

10.2.1 Disease Modeling

In recent years, hPSC-CMs have emerged as a fruitful tool for modeling genetic cardiac disorders *in vitro* [5–7]. Although valuable, animal models cannot accurately model many aspects of human cardiac syndromes due to substantial differences in cardiac electrophysiology. While seemingly a better alternative, primary human cardiomyocytes are difficult to obtain and genetically manipulate *in vitro*. Furthermore, they cannot proliferate and rapidly dedifferentiate in culture exhibiting dramatic, time-dependent phenotypic changes [8]. These hurdles limit their use for disease pathway studies and drug discovery applications.

In contrast, hiPSCs derived from afflicted patient somatic cells are readily expandable in culture and can be differentiated into cardiomyocytes at any time, providing a theoretically limitless source of human diseased cells for *in vitro* studies. Furthermore, hPSCs are amenable to genetic manipulation prior to differentiation, enabling the use of classic molecular biology techniques to interrogate cardiac biology and disease pathophysiology. Numerous reports have established that hiPSC-CMs can recapitulate the main pathognomonic characteristics of genetic cardiac disorders including inherited arrhythmias at the single cell level, and can thus subsequently be used to gain insights into disease mechanisms [9]. Such disease models have been successfully generated from patients afflicted with disorders including long QT syndrome (LQTS) [10–12], Timothy Syndrome [13], and catecholaminergic polymorphic ventricular tachycardia (CPVT) [14, 15], to name a few. Moreover, in the emerging field of precision medicine, this technology could 1 day be used to tailor new or existing therapies to patient-specific needs [16].

10.2.2 Drug Discovery and Safety Testing

The fact that hPSC-CMs exhibit key cardiac electrophysiological responses to external electrical and pharmacological stimuli at the cellular level makes them suitable for drug discovery and safety testing. Moreover, they are permissive to physiologically-relevant tissue engineering and high throughput screening platforms. While the prospect of drug discovery using hPSC-CMs is currently being explored, these cells have already been successfully applied in safety pharmacology testing to identify drug-induced cardiotoxicity [17, 18]. Numerous drugs make it through animal studies, but are withdrawn during human trials due to proarrhythmic side effects such as QT interval prolongation and torsades de pointes [19–21]. A prevailing expectation is that incorporating hPSC-CMs into early drug screens will minimize the risk of cardiotoxic events in clinical subjects, and thus reduce the rates of late-stage drug attrition [22].

10.2.3 Myocardial Repair

Perhaps the most revolutionary prospect of hPSC-CMs is for use in cardiac regeneration therapy. Due to the heart's limited regenerative capacity, cells lost during ischemic heart disease are not replaced intrinsically—resulting in fibrosis, reduced cardiac function, and overall morbidity [23]. Therefore, proposed strategies to restore the damaged myocardium have focused on external interventions such as cell, tissue, or full organ replacement [24]. In contrast to other non-cardiomyogenic stem cell types such as mesenchymal precursor cells that confer benefits through paracrine effects [25], transplanted hPSC-CMs also have the potential to achieve true contractile tissue regeneration [26]. Indeed, experiments in small and large animal models have demonstrated that hPSC-CMs transplanted to sites of ischemic injury survive and can electrically couple to the host myocardium [27–34]. In the case of human myocardial infarction, it is estimated that up to one billion hPSC-CMs may need to be transplanted for sufficient contractile tissue repair to be accomplished [35, 36], emphasizing the demand for efficient and scalable methods of procuring these cells for medical applications, including forthcoming clinical trials [37].

10.3 Considerations for the Cryopreservation of Pluripotent Stem Cell-Derived Cardiomyocytes

As the usage of hPSC-CMs increases, so does the demand for their long-term storage. Cryopreservation enables the storage of hPSC-CMs with desired phenotypes for later investigation and supports the establishment of clinical large-scale biobanking initiatives. Freezing may also be used as an effective strategy to attain clinically-relevant quantities of hPSC-CMs, since they can be generated in multiple batches and amassed over time, and allow sufficient time to perform quality control for cell preparations. Moreover, frozen hPSC-CMs can be used by researchers with minimal knowledge of hPSC culture or differentiation procedures, since the maintenance of pre-made hPSC-CMs is relatively inexpensive and simple compared to the time-consuming and resource-intensive process of cardiac differentiation from hPSCs. Current protocols for cryopreserving hPSC-CMs are similar to those for other cultured cell types, and thus aim to minimize the formation of damaging ice crystals. These procedures involve cell dissociation, centrifugation, re-suspension in the presence of a cryoprotective agent (CPA), and cooling in cryovials. A working protocol for the cryopreservation of hPSC-CMs has already been established [38], and thus will not be re-iterated here. However, elements of the cryopreservation procedure may differ slightly between laboratories—varying in cell handling, freezing rate, storage temperature, and choice of CPA(s). Therefore, in this section, we review the literature available regarding these variables, and compare methods for successfully freezing and storing hPSC-CMs.

10.3.1 Cardiac Maturity

The optimal differentiation day at which to freeze hPSC-CMs has been explored, with the general consensus being that younger cells exhibit enhanced recovery from cryopreservation compared to older hPSC-CMs. This may be due, at least in part, to the fact that immature cardiomyocytes are smaller, rounder, and less structured than their mature counterparts [39–41]. Another possibility is that immature hPSC-CMs are protected by their distinct metabolic profile, which is more glycolytic in nature than mature hPSC-CMs [3, 39, 42]. Studies have reported high recovery percentages of beating hPSC-CMs frozen at days 12–30 after cardiac induction [28, 29, 43] with diminished survival observed in hPSC-CMs frozen later than 30 days [44]. In a direct comparison between day 12- and day 16-frozen cells (pre-beating and post-beating), day 12-frozen cells exhibited significantly higher post-thaw viability and fewer ultrastructural alterations than day 16-frozen hPSC-CMs [43]. Since hPSC-CMs typically initiate beating after differentiation day 14, these results suggest that improved cryopreservation outcomes may be achieved when cells are frozen at a pre-contraction stage. If this is true, the post-thaw cardiomyogenic capacity of hPSC-CMs frozen at even earlier progenitor stages should be explored.

10.3.2 Pro-survival Treatments

hPSC-CMs that are cryopreserved for eventual cardiac transplantation are often subjected to a pro-survival protocol. The method was initially established by Laflamme and colleagues, and has been shown to enhance engraftment and survival after transplantation [27]. The protocol consists of two parts, the first of which is performed prior to cryopreservation. Approximately 24 h prior to cryopreservation, cells are heat-shocked for 30 min by incubation at 43 °C, then returned to 37 °C in medium supplemented with 100 ng/mL IGF1 (Peprotech) and 0.2 μM cyclosporine A (Sandimmune, Novartis). The next day, hPSC-CMs are dissociated and cryopreserved as usual [29, 33, 45].

10.3.3 Cell Dissociation

The first step in cryopreserving hPSC-CMs (excluding the pro-survival pre-treatment) involves their removal from any culture surface and enzymatic dissociation (Fig. 10.1). Dissociation of hPSC-CMs into a single-cell suspension is imperative for minimizing the formation of damaging ice crystals, since it allows for maximal exposure of the cells to CPAs. Approximately 1 h prior to enzymatic dispersion, cells may be treated with 10 μM of Y-2763, an inhibitor of the pro-apoptotic protein Rho-associated kinase (ROCK), which has been shown to promote post-cryopreservation survival of hPSC-CMs and stem cells [43, 46–49]. Efficient dissociation of hPSC-CMs can be achieved by incubation at 37 °C for 5–10 min with Gibco® 0.25 % Trypsin-EDTA, Accutase®, or the xeno-

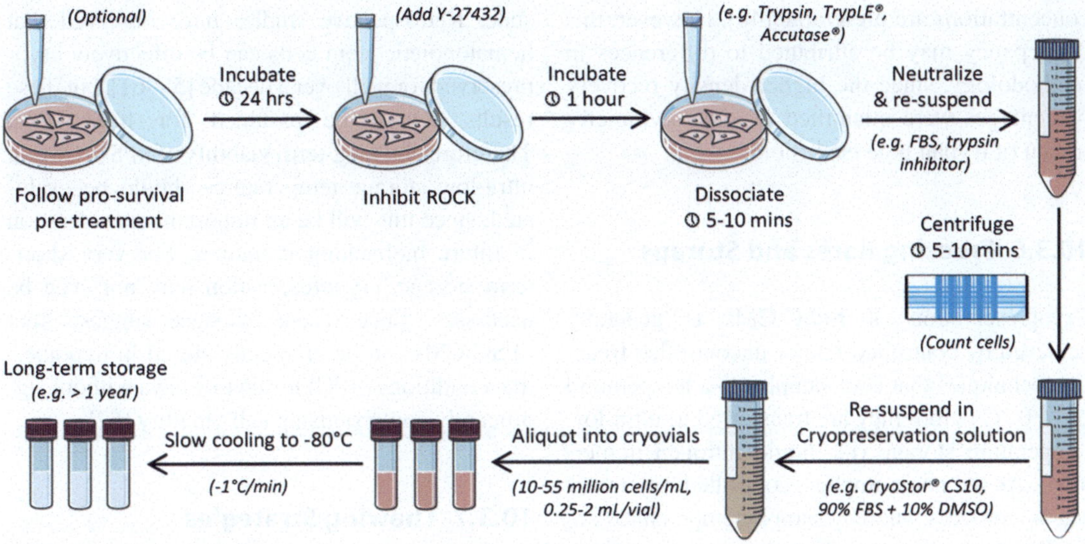

Fig. 10.1 General procedure for the cryopreservation of hPSC-CMs

free TrypLE™ family of products (Thermo Fisher). Gentle dissociation reagents are not advised, since they require extended incubation times and are markedly less effective. If using trypsin, the enzyme should be neutralized after incubation either with serum-containing media or a chemically-defined trypsin inhibitor. Neutralized cells are then washed via centrifugation (300 g, 5 min) and re-suspended directly in a cryoprotective solution in preparation for freezing.

10.3.4 Cryoprotective Solutions

Successful freezing of hPSC-CMs requires cell concentrates to be suspended in a cryopreservation solution. Depending on the eventual application of the cells, the solution usually consists of a xeno-free, chemically-defined cryopreservation solution (e.g. CryoStor® CS10) [38, 45, 48, 49] or fetal bovine serum (FBS) containing a CPA [43, 44]. In principle, CPAs are compounds that reduce cellular damage during freezing by lowering the probability of intracellular and extracellular ice crystal formation [50–53]. A variety of cell-penetrating and membrane non-permeable CPAs are available, but dimethyl sulfoxide (DMSO) at a concentration of 10 % is the most widely-used for hPSC-CM cryopreservation. Careful adherence to optimized protocols is recommended, since prolonged exposure to DMSO is assumed to be harmful to hPSC-CMs based on its clinical cardiotoxicity profile [54, 55]. The percentage of FBS in cryoprotective solutions can range from 30 % to 90 %, with improved recovery observed in preparations containing higher levels of FBS [43, 44]. If a lower percentage of FBS is used, the remaining portion is completed with hPSC-CM culture media, such as RPMI 1640 containing B27 supplement (Life Technologies).

10.3.5 Cell Density and Solution Volume

Cell density and attendant solution volumes for the cryopreservation of hPSC-CMs have also been examined (Table 10.1). Recovery percentages of hPSC-CM preparations with cell densities of approximately 27–53 million cells per mL in CryoStor® CS10 volumes of 0.25–1.5 mL per vial were reported to be comparable [45]. Interestingly, recovery percentages in CryoStor® CS10 preparations containing 10–15 million cells per mL in volumes of 1–2 mL per vial were reported to be higher, suggesting that lower cell concentrations are preferable [48]. However, this discrepancy may be attributed to differences in methodology, since the higher density recovery percentages were quantified via flow cytometry in lieu of trypan blue exclusion.

10.3.6 Freezing Rates and Storage

Cryopreservation of hPSC-CMs is generally achieved by controlled-rate or uncontrolled freezing techniques that cool samples to a temperature of −80 °C before they are transferred to ultra-low temperature storage (i.e. liquid nitrogen immersion). As the name implies, controlled-rate freezing is a process whereby sample temperatures are continuously monitored by a programmable device and adjusted internally to maintain a predetermined cooling rate. Uncontrolled freezing techniques utilize containers in which samples are surrounded by an insulating substance that is placed into a −80 °C freezer to achieve an approximate cooling profile of 1 °C per minute. The obvious advantage of controlled-rate freezing is that multiple freezing rate steps can be incorporated to minimize the liberation of fusion heat, which is cell-injurious [52, 56]. Studies have optimized freezing protocols for other cell types [57], but no consensus currently exists regarding the best strategy for freezing hPSC-CMs. However, a few groups have published their controlled-rate freeing protocols, which we have enumerated here (Table 10.1). Another aspect that remains undetermined is the durability of hPSC-CMs in a cryopreserved state. Retrospective studies have observed that hematopoietic stem cells can be effectively cryopreserved for well over a decade [58–61], but these results cannot be assumed for hPSC-CMs. Therefore, the long-term viability of hPSC-CMs at ultra-low storage temperatures should be evaluated, since this will be an important consideration in future biobanking initiatives. For very short-term storage, cryopreservation may not even be necessary, since recent evidence suggests that hPSC-CMs can be efficiently stored in hypothermic conditions (4 °C) for up to 7 days without significantly compromising cell viability [62].

10.3.7 Thawing Strategies

Rapid thawing of frozen samples minimizes recrystallization of ice, and thus provides improved cell viability compared to slow thawing [63–66]. Once samples are removed from cryogenic storage, they are dipped immediately in a warm 37 °C water bath. If there is distance between the cryogenic storage vessel and the water bath, the frozen samples are transported on dry ice (−80 °C) until immediately before thawing. Samples are swirled or gently shaken in the bath to ensure the entire sample thaws at a consistent rate, and removed when the last visible ice crystal has melted. To remove the cryoprotective agent from the sample, the thawed cells are thoroughly dispersed in an excess of culture media, and washed once via centrifugation (300 g, 5 min) before re-suspension and plating. To enhance via-

Table 10.1 Cell densities, suspension volumes, and freezing rates for hPSC-CM cryopreservation

Reference	Cell density (cells/mL)	Freezing volume (mL)	Recovery rate (%), assay	Freezing rate (Δ°C/min)
Xu et al. [45]	~27 million	1.5	70–77, cell counting and flow cytometry	1 until −40 °C
	~53 million	1.5		5 from −40 °C to −80 °C
	40 million	0.25		
Zhu et al [38]	40 million	0.25	Not reported	1 from 0 °C to −7 °C
				0.75 from −7 °C to −10 °C
				1 from −10 °C to −80 °C
Chen et al. [48]	10 million	1	84.3 ± 5.2, trypan blue exclusion	Controlled rate program not specified
	15 million	2		

bility, cells may be thawed in media containing Y-2763 [46–48]. hPSC-CMs that are thawed for continued culture may be seeded onto matrigel-coated plates. In some instances, hPSC-CMs may form aggregates during the freeze/thaw cycle. Presumably, cell aggregation occurs as a result of adhesive DNA molecules released from dying cells. In these cases, addition of the endonuclease deoxyribonuclease I (DNase I) may be added to curtail the presence of free DNA fragments and cell clumps [49]. DNase I Solution (Stem Cell Technologies) is added to the media during the initial dispersal and washing step to achieve a final concentration of 0.1 mg per mL (i.e. 200 Kunitz units per mL), followed by incubation for 15 min at room temperature prior to proceeding with downstream applications (Fig. 10.2).

The percentage of cell recovery after cryopreservation of hPSC-CMs ranges from 55 % to more than 90 %, with an average of ~83 % [45, 48]. The method of cardiac differentiation does not appear to affect the cryopreservation outcomes of resultant hPSC-CMs, since both growth factor and small molecule-induced hPSC-CMs recover equally well [44]. The percentage of cardiomyocytes in a mixed population of hPSC-CMs also does not appear to affect post-thaw recovery, since viability has been found to be similar across cell preparations with varying degrees of cardiac purity [45].

10.4 Functional Recovery of Stem Cell-Derived Cardiomyocytes After Cryopreservation

After rapid thawing of ≤day 30 hPSC-CMs, high cell viability and recovery of spontaneous beating can be expected. Several studies have reported data regarding post-thaw cell viability, cardiac cell purity, engraftment efficiency, and contractility parameters from frozen hPSC-CM preparations [29, 38, 43–45, 48, 67]. In general, these studies suggest that the form and function of cryopreserved hPSC-CMs are similar to their freshly-derived counterparts in the short term, but one study has reported that accelerated, irregular, and arrhythmic beating occurs during extended post-thaw culture [43]. In-depth characterization of the transcriptomic, proteomic, metabolic, epigenetic, and therapeutic properties of cryopreserved hPSC-CMs is lacking, and will be necessary to determine if these cells can be used for transplantation comparable to freshly-derived cells. In this section, we review previously reported cell recovery parameters.

10.4.1 Cell Structure and Cycling

Micro- and ultrastructural alterations in cryopreserved hPSC-CMs have been reported, but not yet linked to functional consequences. Post-thaw hPSC-CMs retain their cardiogenic features, but do exhibit features of organelle damage. Transition electron microscopic analysis of cellular components in cryopreserved hPSC-CMs revealed partially disrupted nuclei, damaged cell membranes, and loss of mitochondrial integrity [43]. However, the maturation stage of the cell appears to be critical in structural recovery, since hPSC-CMs frozen at a pre-contraction differentiation stage (day 12) exhibited less structural damage than cells frozen at a post-contraction stage (day 16). In terms of cell cycling, cryopreservation does not appear to arrest cells that are in active phases, since the expression of Ki-67 [45] and BrdU incorporation [43] were both detectable in post-thaw cryopreserved hPSC-CMs.

10.4.2 Mechanical and Contractile Properties

The majority of cryopreserved hPSC-CMs recover contractility 1–5 days post-thawing [43, 45, 48], but the question remains as to whether they recover 100 % of their pre-frozen contractile capacity. An early study used perforated patch clamp to show that cryopreserved hPSC-CMs retained cardiac action potential characteristics and expected electrophysiological responses to pharmacological modulators [48]. A later study performed a comprehensive physiological analysis of cryopreserved hPSC-CMs and acutely-

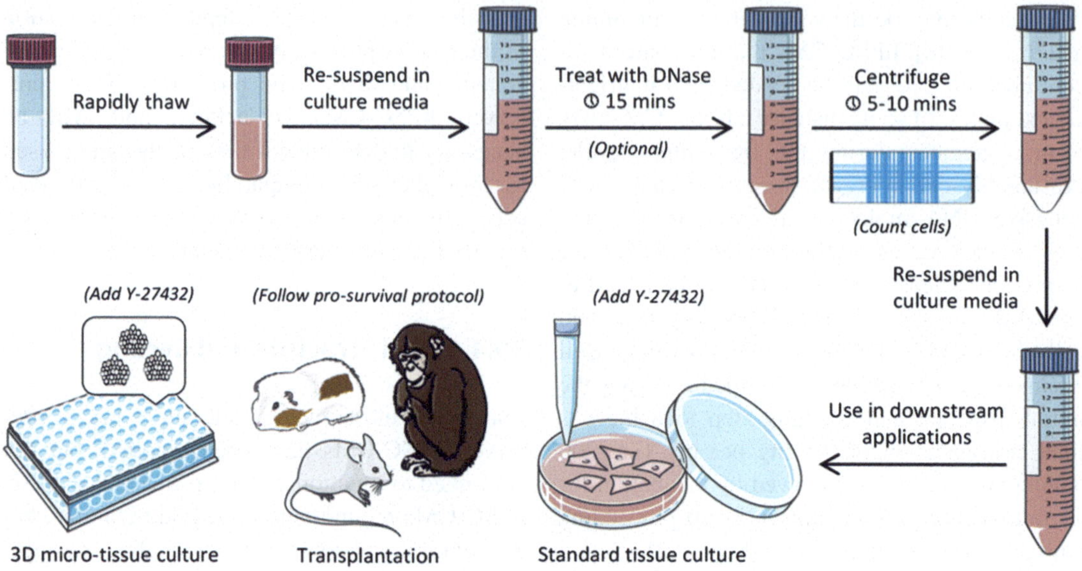

Fig. 10.2 General procedure for thawing cryopreserved hPSC-CMs

isolated adult rabbit ventricular cardiomyocytes, but did not directly compare the contractile properties of thawed hPSC-CMs to freshly-derived hPSC-CMs [44]. In this study, experiments were performed 3–5 days post-thaw, and demonstrated that hPSC-CMs retained a functional sarcoplasmic reticulum (SR) and robust intracellular Ca^{2+} handling after cryopreservation. Moreover, the study showed that cytosolic Ca^{2+} buffering, twitch transient amplitude, resting diastolic Ca^{2+} content, SR Ca^{2+} content, and relative contribution of SR Ca^{2+} flux were comparable between cryopreserved hPSC-CMs and acutely-isolated adult rabbit cardiomyocytes. Collectively, these results imply that the contractile parameters of cryopreserved hPSC-CMs are mostly uncompromised during short-term post-thaw culture. However, a separate study that examined thawed hPSC-CMs in extended culture observed profound contractile alterations that progressively deteriorated until ultimate arrest [43]. In this study, spontaneous beating frequency of post-thaw hPSC-CMs was initially similar to that of non-frozen cells, but began to accelerate 6 days after thawing to unprecedented rates of 108 beats per minute compared to 20–40 beats per minute in non-frozen cells. Furthermore, the beating became irregular 8 days after thawing, and completely arrested at post-thaw day 12.

In addition to apoptosis and anoikis, cryopreservation is known to disrupt cell-cell and cell-matrix adhesion [68–70], increase production of reactive oxygen species [71], and activate various cellular stress pathways [51, 52]. In other proliferating cell types, any losses of function may be recovered in subsequent generations [72, 73]. However, the opportunity for hPSC-CMs to recover during extended culture is forfeited, since the majority of fully differentiated cardiomyocytes have negligible cell cycle activity [74]. Therefore, it is not surprising that post-thaw functionality may be compromised during prolonged post-thaw culture. This again brings the question of whether freezing hPSC-CMs during their still-proliferative cardiac progenitor phase would result in improved post-thaw outcomes, assuming that the ultimate cardiomyogenic capacity of the cells is unaffected by the cryopreservation procedure. This is an intriguing notion, and should be further explored.

Another interesting concept is that the post-thaw environment may influence the functional recovery of cryopreserved hPSC-CMs. A study performed in standard two-dimensional (2D) culture conditions reported progressive functional decline during extended (≥6 days) post-thaw maintenance, while a study that maintained hPSC-CMs in a three-dimensional (3D) micro-tissue culture system did not report any beating abnormalities up to a week post-thaw [49]. Furthermore, studies that have transplanted cryo-preserved hPSC-CMs into rodents presented no evidence of arrhythmias after host-graft electrical coupling [33, 49]. These results suggest that thawed hPSC-CMs are especially vulnerable to 2D culture, but may demonstrate improved long-term outcomes in host tissues or under tissue-like conditions. This inference is corroborated by a recent study that demonstrated hPSC-CMs cultured in 3D micro-tissue aggregates also are more resistant to prolonged hypothermic (4 °C) storage-induced cell injury compared to their counterparts cultured in 2D monolayers [62]. However, differences in cryopreservation methodology may have also contributed to the observed differences, so these results must be interpreted with caution.

10.4.3 Post-transplantation Engraftment and Electrical Coupling

Experiments in rodents and large animal models have demonstrated that transplanted cryopreserved hPSC-CMs partially re-muscularize injured hearts and can beat synchronously with the host myocardium [28, 29, 33, 45, 49, 75]. Xu et al. compared post-transplantation cell survival of freshly-derived and cryopreserved hPSC-CMs using an athymic rat model of acute myocardial ischemia-reperfusion. Immunohistological analysis of the injured myocardial tissue 1 week and 4 weeks after transplantation revealed similar overall graft sizes between the two hPSC-CM preparations, demonstrating comparable short and long-term engraftment efficiency at small scales [45]. Using a large-scale animal model, Chong et al. also found no adverse impact of cryopreservation on hPSC-CM graft size in pig-tail macaques [29]. Efficient host-graft electro-mechanical integration of cryopreserved hPSC-CMs has also been reported, with 1:1 coupling between *in vivo* hPSC-CM calcium fluorescence transients and the host ECG in rodents [33, 49] and non-human primates [29]. The fact that the monkey study utilized approximately one billion cells suggests cryopreservation of hPSC-CMs may be a sound strategy for scaling-up to transplantation in humans. However, in contrast to rodents, non-fatal ventricular arrhythmias were observed transiently in primates that received cryopreserved hPSC-CMs over the course of a 3-month period—although this might be due to species-specific differences in heart size and beating rate, and may not be attributed to cryopreservation per se [29].

10.5 Conclusions and Future Directions

Although current cryopreservation protocols are viable, the question still remains as to whether or not they are optimal. To adequately address this question, more studies that directly compare the physiological properties of freshly-derived and cryopreserved hiPSC-CMs and hESC-CMs are warranted. As mentioned earlier, the differentiation stage at which hPSC-CMs should be frozen for maximal post-thaw functional recovery remains to be established, and the potential for freezing cardiac progenitors should be further explored. Also imperative are studies that investigate and characterize the biology of cryopreservation-associated physiological sequelae in hPSC-CMs, since there is evidence from both *in vitro* [43] and *in vivo* [29] studies that cryopreserved hPSC-CMs acquire pro-arrhythmic tendencies during extended post-thaw time periods. Another important concern stems from evidence showing that freeze/thawed mesenchymal stromal cells [76] and natural killer cells [77, 78] show impaired immunomodulatory and therapeutic properties compared to freshly-derived cells, despite exhibiting high post-thaw viability. In light of this, a side-by-side comparison of transplanted non-frozen and cryopreserved hPSC-CMs' ability to improve cardiac function in recipient hearts would be very informative.

Finally, since frozen cells often show variable viability upon thawing, standardized or automated methods of thawing hPSC-CMs should be established.

In conclusion, the ability to successfully cryopreserve and thaw hPSC-CMs with their therapeutic potential intact would greatly facilitate their availability and distribution for preclinical studies and clinical trials of myocardial repair. While progress has been made in developing methods and characterizing freeze-thawed hPSC-CMs, further studies are needed before cryopreserved hPSC-CMs can be regarded as functionally and therapeutically equivalent to their freshly-derived counterparts for their multiple *in vitro* and *in vivo* applications.

Acknowledgements This work was supported in part by grants GA-2014-126 from the Center for the Advancement of Science in Space and R21 HL123928 from the NIH. M.K.P. was supported by the Center for Pediatric Nanomedicine at Emory/Georgia Tech. Figures were prepared using Servier Medical Art.

References

1. Turnbull IC, Lieu DK, Li RA, Costa KD (2012) Cardiac tissue engineering using human stem cell-derived cardiomyocytes for disease modeling and drug discovery. Drug Discov Today Dis Models 9(4):e219–e227
2. Feric NT, Radisic M (2016) Maturing human pluripotent stem cell-derived cardiomyocytes in human engineered cardiac tissues. Adv Drug Deliv Rev 96:110–134
3. Birket Matthew J, Casini S, Kosmidis G, Elliott David A, Gerencser Akos A, Baartscheer A, Schumacher C, Mastroberardino Pier G, Elefanty Andrew G, Stanley Ed G, Mummery CL (2013) PGC-1α and reactive oxygen species regulate human embryonic stem cell-derived cardiomyocyte function. Stem Cell Reports 1(6):560–574
4. Birket MJ, Ribeiro MC, Verkerk AO, Ward D, Leitoguinho AR, den Hartogh SC, Orlova VV, Devalla HD, Schwach V, Bellin M, Passier R, Mummery CL (2015) Expansion and patterning of cardiovascular progenitors derived from human pluripotent stem cells. Nat Biotechnol 33(9):970–979
5. Lui KO, Stachel MW, Lieu DK, Li RA, Bu L (2012) Induced pluripotent stem cells as a disease model for studying inherited arrhythmias: promises and hurdles. Drug Discov Today Dis Models 9(4):e199–e207
6. Li RA (2012) The use of induced pluripotent stem cells for disease modeling: what are the promises and hurdles? Drug Discov Today Dis Models 9(4):e143–e145
7. Davis RP, van den Berg CW, Casini S, Braam SR, Mummery CL (2011) Pluripotent stem cell models of cardiac disease and their implication for drug discovery and development. Trends Mol Med 17(9):475–484
8. Li RK, Mickle DA, Weisel RD, Carson S, Omar SA, Tumiati LC, Wilson GJ, Williams WG (1996) Human pediatric and adult ventricular cardiomyocytes in culture: assessment of phenotypic changes with passaging. Cardiovasc Res 32(2):362–373
9. Lieu DK, Turnbull IC, Costa KD, Li RA (2012) Engineered human pluripotent stem cell-derived cardiac cells and tissues for electrophysiological studies. Drug Discov Today Dis Models 9(4):e209–e217
10. Itzhaki I, Maizels L, Huber I, Zwi-Dantsis L, Caspi O, Winterstern A, Feldman O, Gepstein A, Arbel G, Hammerman H, Boulos M, Gepstein L (2011) Modelling the long QT syndrome with induced pluripotent stem cells. Nature 471(7337):225–229
11. Lahti AL, Kujala VJ, Chapman H, Koivisto A-P, Pekkanen-Mattila M, Kerkelä E, Hyttinen J, Kontula K, Swan H, Conklin BR, Yamanaka S, Silvennoinen O, Aalto-Setälä K (2012) Model for long QT syndrome type 2 using human iPS cells demonstrates arrhythmogenic characteristics in cell culture. Dis Model Mech 5(2):220–230
12. Moretti A, Bellin M, Welling A, Jung CB, Lam JT, Bott-Flügel L, Dorn T, Goedel A, Höhnke C, Hofmann F, Seyfarth M, Sinnecker D, Schömig A, Laugwitz K-L (2010) Patient-specific induced pluripotent stem-cell models for long-QT syndrome. N Engl J Med 363(15):1397–1409
13. Yazawa M, Hsueh B, Jia X, Pasca AM, Bernstein JA, Hallmayer J, Dolmetsch RE (2011) Using induced pluripotent stem cells to investigate cardiac phenotypes in Timothy syndrome. Nature 471(7337):230–234
14. Jung CB, Moretti A, Mederos y Schnitzler M, Lop L, Storch U, Bellin M, Dorn T, Ruppenthal S, Pfeiffer S, Goedel A, Dirschinger RJ, Seyfarth M, Lam JT, Sinnecker D, Gudermann T, Lipp P, Laugwitz KL (2012) Dantrolene rescues arrhythmogenic RYR2 defect in a patient-specific stem cell model of catecholaminergic polymorphic ventricular tachycardia. EMBO Mol Med 4(3):180–191
15. Itzhaki I, Maizels L, Huber I, Gepstein A, Arbel G, Caspi O, Miller L, Belhassen B, Nof E, Glikson M, Gepstein L (2012) Modeling of catecholaminergic polymorphic ventricular tachycardia with patient-specific human-induced pluripotent stem cells. J Am Coll Cardiol 60(11):990–1000
16. Chen IY, Matsa E, Wu JC (2016) Induced pluripotent stem cells: at the heart of cardiovascular precision medicine. Nat Rev Cardiol. doi:10.1038/nrcardio.2016.36
17. Mandenius CF, Steel D, Noor F, Meyer T, Heinzle E, Asp J, Arain S, Kraushaar U, Bremer S, Class R,

Sartipy P (2011) Cardiotoxicity testing using pluripotent stem cell-derived human cardiomyocytes and state-of-the-art bioanalytics: a review. J Appl Toxicol 31(3):191–205
18. Navarrete EG, Liang P, Lan F, Sanchez-Freire V, Simmons C, Gong T, Sharma A, Burridge PW, Patlolla B, Lee AS, Wu H, Beygui RE, Wu SM, Robbins RC, Bers DM, Wu JC (2013) Screening drug-induced arrhythmia using human induced pluripotent stem cell–derived cardiomyocytes and low-impedance microelectrode arrays. Circulation 128(11 suppl 1):S3–S13
19. MacDonald JS, Robertson RT (2009) Toxicity testing in the 21st century: a view from the pharmaceutical industry. Toxicol Sci 110(1):40–46
20. Fermini B, Fossa AA (2003) The impact of drug-induced QT interval prolongation on drug discovery and development. Nat Rev Drug Discov 2(6):439–447
21. Frothingham R (2001) Rates of torsades de pointes associated with ciprofloxacin, ofloxacin, levofloxacin, gatifloxacin, and moxifloxacin. Pharmacotherapy 21(12):1468–1472
22. Laustriat D, Gide J, Peschanski M (2010) Human pluripotent stem cells in drug discovery and predictive toxicology. Biochem Soc Trans 38(4):1051–1057
23. Jessup M, Brozena S (2003) Heart failure. N Engl J Med 348(20):2007–2018
24. Lin Z, Pu WT (2014) Strategies for cardiac regeneration and repair. Sci Transl Med 6(239):239rv1
25. Perin EC, Borow KM, Silva GV, DeMaria AN, Marroquin OC, Huang PP, Traverse JH, Krum H, Skerrett D, Zheng Y, Willerson JT, Itescu S, Henry TD (2015) A phase II dose-escalation study of allogeneic mesenchymal precursor cells in patients with ischemic or nonischemic heart failure. Circ Res 117(6):576–584
26. Kuraitis D, Suuronen EJ, Sellke FW, Ruel M (2010) The future of regenerating the myocardium. Curr Opin Cardiol 25(6):575–582
27. Laflamme MA, Chen KY, Naumova AV, Muskheli V, Fugate JA, Dupras SK, Reinecke H, Xu C, Hassanipour M, Police S, O'Sullivan C, Collins L, Chen Y, Minami E, Gill EA, Ueno S, Yuan C, Gold J, Murry CE (2007) Cardiomyocytes derived from human embryonic stem cells in pro-survival factors enhance function of infarcted rat hearts. Nat Biotechnol 25(9):1015–1024
28. Chong JJH, Murry CE (2014) Cardiac regeneration using pluripotent stem cells—progression to large animal models. Stem Cell Res 13(3):654–665, Part B
29. Chong JJ, Yang X, Don CW, Minami E, Liu YW, Weyers JJ, Mahoney WM, Van Biber B, Cook SM, Palpant NJ, Gantz JA, Fugate JA, Muskheli V, Gough GM, Vogel KW, Astley CA, Hotchkiss CE, Baldessari A, Pabon L, Reinecke H, Gill EA, Nelson V, Kiem HP, Laflamme MA, Murry CE (2014) Human embryonic-stem-cell-derived cardiomyocytes regenerate non-human primate hearts. Nature 510(7504):273–277
30. Caspi O, Huber I, Kehat I, Habib M, Arbel G, Gepstein A, Yankelson L, Aronson D, Beyar R, Gepstein L (2007) Transplantation of human embryonic stem cell-derived cardiomyocytes improves myocardial performance in infarcted rat hearts. J Am Coll Cardiol 50(19):1884–1893
31. van Laake LW, Passier R, Monshouwer-Kloots J, Verkleij AJ, Lips DJ, Freund C, den Ouden K, Ward-van Oostwaard D, Korving J, Tertoolen LG, van Echteld CJ, Doevendans PA, Mummery CL (2007) Human embryonic stem cell-derived cardiomyocytes survive and mature in the mouse heart and transiently improve function after myocardial infarction. Stem Cell Res 1(1):9–24
32. Laflamme MA, Gold J, Xu C, Hassanipour M, Rosler E, Police S, Muskheli V, Murry CE (2005) Formation of human myocardium in the rat heart from human embryonic stem cells. Am J Pathol 167(3):663–671
33. Shiba Y, Fernandes S, Zhu WZ, Filice D, Muskheli V, Kim J, Palpant NJ, Gantz J, Moyes KW, Reinecke H, Van Biber B, Dardas T, Mignone JL, Izawa A, Hanna R, Viswanathan M, Gold JD, Kotlikoff MI, Sarvazyan N, Kay MW, Murry CE, Laflamme MA (2012) Human ES-cell-derived cardiomyocytes electrically couple and suppress arrhythmias in injured hearts. Nature 489(7415):322–325
34. Blin G, Nury D, Stefanovic S, Neri T, Guillevic O, Brinon B, Bellamy V, Rucker-Martin C, Barbry P, Bel A, Bruneval P, Cowan C, Pouly J, Mitalipov S, Gouadon E, Binder P, Hagege A, Desnos M, Renaud JF, Menasche P, Puceat M (2010) A purified population of multipotent cardiovascular progenitors derived from primate pluripotent stem cells engrafts in post-myocardial infarcted nonhuman primates. J Clin Invest 120(4):1125–1139
35. Laflamme MA, Murry CE (2005) Regenerating the heart. Nat Biotechnol 23(7):845–856
36. Murry CE, Reinecke H, Pabon LM (2006) Regeneration gaps: observations on stem cells and cardiac repair. J Am Coll Cardiol 47(9):1777–1785
37. Menasche P, Vanneaux V, Fabreguettes JR, Bel A, Tosca L, Garcia S, Bellamy V, Farouz Y, Pouly J, Damour O, Perier MC, Desnos M, Hagege A, Agbulut O, Bruneval P, Tachdjian G, Trouvin JH, Larghero J (2015) Towards a clinical use of human embryonic stem cell-derived cardiac progenitors: a translational experience. Eur Heart J 36(12):743–750
38. Zhu W-Z, Van Biber B, Laflamme MA (2011) Methods for the derivation and use of cardiomyocytes from human pluripotent stem cells. Methods Mol Biol (Clifton, NJ) 767:419–431
39. Lundy SD, Zhu WZ, Regnier M, Laflamme MA (2013) Structural and functional maturation of cardiomyocytes derived from human pluripotent stem cells. Stem Cells Dev 22(14):1991–2002
40. Robertson C, Tran DD, George SC (2013) Concise review: maturation phases of human pluripotent stem cell-derived cardiomyocytes. Stem Cells 31(5):829–837

41. Yang X, Pabon L, Murry CE (2014) Engineering adolescence: maturation of human pluripotent stem cell-derived cardiomyocytes. Circ Res 114(3):511–523
42. Poon E, Keung W, Liang Y, Ramalingam R, Yan B, Zhang S, Chopra A, Moore J, Herren A, Lieu DK, Wong HS, Weng Z, Wong OT, Lam YW, Tomaselli GF, Chen C, Boheler KR, Li RA (2015) Proteomic analysis of human pluripotent stem cell-derived, fetal, and adult ventricular cardiomyocytes reveals pathways crucial for cardiac metabolism and maturation. Circ Cardiovasc Genet 8(3):427–436
43. Kim YY, Ku SY, Liu HC, Cho HJ, Oh SK, Moon SY, Choi YM (2011) Cryopreservation of human embryonic stem cells derived-cardiomyocytes induced by BMP2 in serum-free condition. Reprod Sci 18(3):252–260
44. Hwang HS, Kryshtal DO, Feaster TK, Sanchez-Freire V, Zhang J, Kamp TJ, Hong CC, Wu JC, Knollmann BC (2015) Comparable calcium handling of human iPSC-derived cardiomyocytes generated by multiple laboratories. J Mol Cell Cardiol 85:79–88
45. Xu C, Police S, Hassanipour M, Li Y, Chen Y, Priest C, O'Sullivan C, Laflamme MA, Zhu WZ, Van Biber B, Hegerova L, Yang J, Delavan-Boorsma K, Davies A, Lebkowski J, Gold JD (2011) Efficient generation and cryopreservation of cardiomyocytes derived from human embryonic stem cells. Regen Med 6(1):53–66
46. Heng BC (2009) Effect of Rho-associated kinase (ROCK) inhibitor Y-27632 on the post-thaw viability of cryopreserved human bone marrow-derived mesenchymal stem cells. Tissue Cell 41(5):376–380
47. Martin-Ibanez R, Unger C, Stromberg A, Baker D, Canals JM, Hovatta O (2008) Novel cryopreservation method for dissociated human embryonic stem cells in the presence of a ROCK inhibitor. Hum Reprod 23(12):2744–2754
48. Chen VC, Ye J, Shukla P, Hua G, Chen D, Lin Z, Liu JC, Chai J, Gold J, Wu J, Hsu D, Couture LA (2015) Development of a scalable suspension culture for cardiac differentiation from human pluripotent stem cells. Stem Cell Res 15(2):365–375
49. Gerbin KA, Yang X, Murry CE, Coulombe KL (2015) Enhanced electrical integration of engineered human myocardium via intramyocardial versus epicardial delivery in infarcted rat hearts. PLoS One 10(7):e0131446
50. Gosden RG (2014) General principles of cryopreservation. Methods Mol Biol 1154:261–268
51. Hubel A (1997) Parameters of cell freezing: implications for the cryopreservation of stem cells. Transfus Med Rev 11(3):224–233
52. Pegg DE (2015) Principles of cryopreservation. Methods Mol Biol 1257:3–19
53. Meryman HT (2007) Cryopreservation of living cells: principles and practice. Transfusion 47(5):935–945
54. Davis JM, Rowley SD, Braine HG, Piantadosi S, Santos GW (1990) Clinical toxicity of cryopreserved bone marrow graft infusion. Blood 75(3):781–786
55. Zenhausern R, Tobler A, Leoncini L, Hess OM, Ferrari P (2000) Fatal cardiac arrhythmia after infusion of dimethyl sulfoxide-cryopreserved hematopoietic stem cells in a patient with severe primary cardiac amyloidosis and end-stage renal failure. Ann Hematol 79(9):523–526
56. De Santis L, Coticchio G (2011) Theoretical and experimental basis of slow freezing. Reprod Biomed Online 22(2):125–132
57. Balint B, Ivanovic Z, Petakov M, Taseski J, Jovcic G, Stojanovic N, Milenkovic P (1999) The cryopreservation protocol optimal for progenitor recovery is not optimal for preservation of marrow repopulating ability. Bone Marrow Transplant 23(6):613–619
58. Mugishima H, Harada K, Chin M, Suzuki T, Takagi K, Hayakawa S, Sato K, Klein JP, Gale RP (1999) Effects of long-term cryopreservation on hematopoietic progenitor cells in umbilical cord blood. Bone Marrow Transplant 23(4):395–396
59. Aird W, Labopin M, Gorin NC, Antin JH (1992) Long-term cryopreservation of human stem cells. Bone Marrow Transplant 9(6):487–490
60. Attarian H, Feng Z, Buckner CD, MacLeod B, Rowley SD (1996) Long-term cryopreservation of bone marrow for autologous transplantation. Bone Marrow Transplant 17(3):425–430
61. Veeraputhiran M, Theus JW, Pesek G, Barlogie B, Cottler-Fox M (2010) Viability and engraftment of hematopoietic progenitor cells after long-term cryopreservation: effect of diagnosis and percentage dimethyl sulfoxide concentration. Cytotherapy 12(6):764–766
62. Correia C, Koshkin A, Carido M, Espinha N, Šarić T, Lima PA, Serra M, Alves PM (2016) Effective hypothermic storage of human pluripotent stem cell-derived cardiomyocytes compatible with global distribution of cells for clinical applications and toxicology testing. Stem Cells Transl Med 5:658–669, pii: sctm.2015-0238
63. Seki S, Mazur P (2008) Effect of warming rate on the survival of vitrified mouse oocytes and on the recrystallization of intracellular ice. Biol Reprod 79(4):727–737
64. Tao J, Du J, Kleinhans FW, Critser ES, Mazur P, Critser JK (1995) The effect of collection temperature, cooling rate and warming rate on chilling injury and cryopreservation of mouse spermatozoa. J Reprod Fertil 104(2):231–236
65. El-Naggar MM, Al-Mashat FM, Elayat AA, Sibiany AR, Ardawi MS, Badawoud MH (2006) Effect of thawing rate and post-thaw culture on the cryopreserved fetal rat islets: functional and morphological correlation. Life Sci 78(17):1925–1932
66. Hochi S, Semple E, Leibo SP (1996) Effect of cooling and warming rates during cryopreservation on survival of in vitro-produced bovine embryos. Theriogenology 46(5):837–847
67. Moriguchi H, Madson J (2013) Autologous human cardiac stem cells transplantation for the treatment of

ischaemic cardiomyopathy: first study of human-induced pluripotent stem (iPS) cell-derived cardiomyocytes transplantation. BMJ Case Rep. doi:10.1136/bcr-2013-008960

68. Wong RC, Dottori M, Koh KL, Nguyen LT, Pera MF, Pebay A (2006) Gap junctions modulate apoptosis and colony growth of human embryonic stem cells maintained in a serum-free system. Biochem Biophys Res Commun 344(1):181–188
69. Wong RC, Pebay A, Nguyen LT, Koh KL, Pera MF (2004) Presence of functional gap junctions in human embryonic stem cells. Stem Cells 22(6):883–889
70. Ichikawa H, Yoshie S, Shirasawa S, Yokoyama T, Yue F, Tomotsune D, Sasaki K (2011) Freeze-thawing single human embryonic stem cells induce e-cadherin and actin filament network disruption via g13 signaling. Cryo Letters 32(6):516–524
71. Xu X, Cowley S, Flaim CJ, James W, Seymour L, Cui Z (2010) The roles of apoptotic pathways in the low recovery rate after cryopreservation of dissociated human embryonic stem cells. Biotechnol Prog 26(3):827–837
72. Sugimachi K, Sosef MN, Baust JM, Fowler A, Tompkins RG, Toner M (2004) Long-term function of cryopreserved rat hepatocytes in a coculture system. Cell Transplant 13(2):187–195
73. Sosef MN, Baust JM, Sugimachi K, Fowler A, Tompkins RG, Toner M (2005) Cryopreservation of isolated primary rat hepatocytes: enhanced survival and long-term hepatospecific function. Ann Surg 241(1):125–133
74. Snir M, Kehat I, Gepstein A, Coleman R, Itskovitz-Eldor J, Livne E, Gepstein L (2003) Assessment of the ultrastructural and proliferative properties of human embryonic stem cell-derived cardiomyocytes. Am J Physiol Heart Circ Physiol 285(6):H2355–H2363
75. Riegler J, Ebert A, Qin X, Shen Q, Wang M, Ameen M, Kodo K, Ong SG, Lee WH, Lee G, Neofytou E, Gold JD, Connolly AJ, Wu JC (2016) Comparison of magnetic resonance imaging and serum biomarkers for detection of human pluripotent stem cell-derived teratomas. Stem Cell Reports 6(2):176–187
76. Moll G, Alm JJ, Davies LC, von Bahr L, Heldring N, Stenbeck-Funke L, Hamad OA, Hinsch R, Ignatowicz L, Locke M, Lönnies H, Lambris JD, Teramura Y, Nilsson-Ekdahl K, Nilsson B, Le Blanc K (2014) Do cryopreserved mesenchymal stromal cells display impaired immunomodulatory and therapeutic properties? Stem Cells 32(9):2430–2442
77. Mata MM, Mahmood F, Sowell RT, Baum LL (2014) Effects of cryopreservation on effector cells for antibody dependent cell-mediated cytotoxicity (ADCC) and natural killer cell (NK) activity in (51)Cr-release and CD107a assays. J Immunol Methods 406:1–9
78. Dominguez E, Lowdell MW, Perez-Cruz I, Madrigal A, Cohen SBA (1997) Natural killer cell function is altered by freezing in DMSO. Biochem Soc Trans 25(2):175S–S

Cryopreserved Adipose Tissue-Derived Stromal/Stem Cells: Potential for Applications in Clinic and Therapy

11

Fabiana Zanata, Shahensha Shaik, Ram V. Devireddy, Xiying Wu, Lydia Masako Ferreira, and Jeffrey M. Gimble

Abstract

Adipose-Derived Stromal/Stem Cells (ASC) have considerable potential for regenerative medicine due to their abilities to proliferate, differentiate into multiple cell lineages, high cell yield, relative ease of acquisition, and almost no ethical concerns since they are derived from adult tissue. Storage of ASC by cryopreservation has been well described that maintains high cell yield and viability, stable immunophenotype, and robust differentiation potential post-thaw. This ability is crucial for banking research and for clinical therapeutic purposes that avoid the morbidity related to repetitive liposuction tissue harvests. ASC secrete various biomolecules such as cytokines which are reported to have immunomodulatory properties and therapeutic potential to reverse symptoms of multiple degenerative diseases/disorders. Nevertheless, safety regarding the use of these cells clinically is still under investigation. This chapter focuses on the different aspects of cryopreserved ASC and the methods to evaluate their functionality for future clinical use.

F. Zanata
Federal University of Sao Paulo,
Sao Paulo, SP, Brazil

Center for Stem Cell Research & Regenerative Medicine, Tulane University, New Orleans, LA, USA

S. Shaik • R.V. Devireddy
Department of Mechanical Engineering, Louisiana State University, Baton Rouge, LA, USA

X. Wu
La Cell LLC, New Orleans BioInnovation Center, Suite 304, 1441 Canal Street, New Orleans, LA 70112, USA

L.M. Ferreira
Federal University of Sao Paulo,
Sao Paulo, SP, Brazil

J.M. Gimble (✉)
Center for Stem Cell Research & Regenerative Medicine, Tulane University, New Orleans, LA, USA

La Cell LLC, New Orleans BioInnovation Center, Suite 304, 1441 Canal Street, New Orleans, LA 70112, USA
e-mail: jeffrey.gimble@lacell-usa.com

F. Karimi-Busheri, M. Weinfeld (eds.), *Biobanking and Cryopreservation of Stem Cells*, Advances in Experimental Medicine and Biology 951, DOI 10.1007/978-3-319-45457-3_11

Keywords
Adipose-derived stromal/stem cells • Regenerative medicine • Adipose tissue • Human cells • Cell differentiation • Immunophenotype • Cryopreservation • Clinical trials

Abbreviations

ASC	Adipose-derived stromal/stem cells
ANGPT1	Angiopoietin 1
CAL	Cell Assisted Lipotranfer
CPA	Cryoprotectant agents
cGMP	Current Good Manufacturing Practice
DS	Dextran sulfate
DMSO	Dimethyl sulfoxide
EG	Ethylene glycol
EAE	Experimental autoimmune encephalitis
FDA	U.S. Food and Drug Administration
GLUT4	Glucose transporter
HGF	Hepatocyte growth factor
HES	Hydroxyethylstarch
INSR	Insulin receptor
IL-1β	Interleukin 1 beta
IL-6	Interleukin 6
IL-12	Interleukin 12
IFN-γ	Interferon gamma
MSC	Mesenchymal stem cell
SVF	Stromal vascular fraction
MC	Methylcellulose
PD	Parkinson's disease
PVP	Polyvinylpyrollidone
TNF-α	Tumor necrosis factor alpha
VEGF	Vascular endothelial growth factor

11.1 Introduction

Regenerative medicine aims to restore tissue or organ form and function with the use of cell therapy, organ transplant, tissue engineering among other techniques and the combination of biological tissues with natural or synthetic materials [1]. Mesenchymal stem cells (MSC) including Adipose-Derived Stromal/Stem Cells (ASC) are multipotent with the ability to differentiate into tissues of mesodermal origin (bone, cartilage and fat). This property, associated to other mechanisms such as the release of paracrine factors, scavenging of reactive oxygen species, immunomodulatory function are ASC mechanisms contribute for tissue regeneration and possible translational applications [2–5]. Adipose tissue is a useful source for stromal/stem cells with no ethical restriction given its adult tissue origin and accessibility, targeting, ease of collection, ability to expand, maintenance in culture, and ability to contribute to tissue regeneration after implantation [6]. The Stromal Vascular Fraction (SVF), the uncultured heterogeneous cell population obtained after adipose tissue digestion, have also increasing research interest due to less manipulation related to cell culture and immunomodulatory effects [7, 8].

The potential challenge to the clinical translation of MSC-based products is the diversity in how MSC are defined. This can be based on multiple criteria, including the tissue source, methods for processing/manufacturing, cell surface marker expression and other *in vitro* and *in vivo* characteristics as described in regulatory submissions to the FDA [9]. Current Good Manufacturing Practice (cGMP) protocols for generating clinical-grade cells are necessary before beginning a clinical trial based on guidance documents [10]. Donor tissue and cell products must be screened for viral agents (such as HIV, Hepatitis C in case of allogenic use), aerobic and anaerobic bacteria, endotoxin, and mycoplasma. Cell biological characterization and product quality control must be included. The manufacturing facility must meet and be routinely monitored based on rigorous standards which include documentation of all processes based on established and validated standard operating procedures [11]. In this chapter we discuss the critical aspects for the clinical use of ASC for regenerative/reconstruction purposes.

11.2 Adipose-Derived Stromal/Stem Cells Cryopreservation

Long-term cryopreservation is fundamental for a reliable product supply for research or clinical purposes. Current cryopreservation techniques already allow easier access and use of the products for clinical use such as blood transfusion, bone marrow transplantation, *in vitro* fertilization, vascular grafts, bone grafts and skin grafts [12]. This knowledge can be transferred for ASC cryopreservation and some principles can be kept such as addition of cryoprotective agents (CPA) before controlled freezing at which the cells are stored, rapidly thaw cells and remove the CPA for use [13]. Simply freezing a cell or tissue is not enough to maintain its long-term viability since ice crystal formation will damage the cells by direct disruption or due to the increase in solute concentrations as water within the cytoplasm is sequestered [14]. Permeable CPA prevent this outcome by entering the cell and equilibrating across cell membranes where it replaces the lost water and maintains an acceptable volume allowing cell survival [15]. Dimethyl sulfoxide (DMSO) is the most commonly employed permeable CPA. Bone marrow and blood cell product cryopreservation routinely combines the use of albumin and DMSO; however, DMSO can be clinically associated with toxicity including genotypic changes, impaired differentiation potentiality, and symptomatic side effects in the recipient, thus making dilution or washing steps to reduce or remove DMSO fundamental to protect against such complications [16, 17]. Alternative permeable CPA include glycerol, ethylene glycol and propylene glycol. Although these do not cause the same degree of toxicity as DMSO, none has displayed equivalent or better cell viability post-thaw after cryopreservation. Impermeable CPA act extracellularly to prevent ice crystal formation and protect against plasma membrane damage. Impermeable CPAs include both low molecular weight compounds (glucose, sucrose, trehalose, hydroxyethylstarch/HES) and high molecular weight macromolecular polymers (dextran sulfate/DS, methylcellulose/MC, Polyvinylpyrollidone/PVP) [18, 19].

ASC isolation, expansion and cryopreservation methods typically use animal or human products such as serum [20]. Historically, serum is employed as a source of proteins and chemicals promoting cell growth, attachment, and neutralization of toxins and oxidants. Bovine or human serum derived products added to DMSO or other low molecular weight chemical CPA serves as a source of additional high molecular weight impermeable colloid CPA; however, even with rigorous screening for common infectious agents (such as HIV, hepatitis B or C), human serum proteins remain a potential source of bacterial, prion, or viral contaminants [21]. Consequently, it is important to establish xeno-free defined protocols capable of preserving cell function post-thaw to levels comparable to that seen in freshly isolated tissue at the various steps during ASC processing in order to avoid contamination risks associated with human or animal cell culture and/or cryopreservation components [4, 9–11, 22, 23]. While autologous donor serum or platelet-derived supplements have potential as acceptable alternatives, an optimal human serum reagent would need to have any antibodies or complement proteins depleted to remove any risk of cell damage [24].

Systematic analyses combining different CPA to optimize a CPA solution's function could minimize cytotoxicity and provide either an allogenic or xenogeneic serum-free cryopreservation solution. Published studies have begun to pursue such an approach [24]. After storage in liquid nitrogen, 10 % PVP in the absence of serum or DMSO at concentrations as low as 2 % maintained post-thaw cell viability at levels close to those displayed by freshly isolated cells [17]. The favorable outcomes using DMSO and/or PVP were not observed with the use of glycerol or MC as CPA, which resulted in sub-optimal cryopreservation based on *in vitro* assays [25, 26]. Trehalose has been used to preserve not only cells but also intact tissues (including pancreatic islets, skin, oocytes, testis, and lipoaspirate fragments) that displayed functional viability when transplanted *in vivo* [27]. With the aim to avoid the use of xenogenic product, a CPA solution with antioxidants (reduced glutathione and ascorbic acid), polymers (PVA and ficoll), permeating CPA (ethylene glycol/EG and DMSO), a disaccharide (trehalose), and a calcium chelator (EGTA) were added to HEPES-buffered DMEM/F12 and ASC retained their multipotency and chromosomal normality [20].

The optimization of CPA solution has so far focused on the clinical translation of adipose tissue and cells for regenerative medical applications and the complete characterization of the cells is of fundamental importance for product development, regulatory approval and patient safety [17, 25, 26, 28–31]. Extensive published studies have addressed optimized approaches to SVF and ASC cryopreservation based on *in vitro* assays. *In vitro* assays include adipogenic, osteogenic and chondrogenic differentiation, cell apoptosis, glycerophosphate dehydrogenase (GPDH) enzyme activity, cell proliferation rate, surface immunophenotype (FACS) and Magnetic Cell Sorting (MACS), adipose-derived cell recovery rates, viability (MTT), karyotype and gene expression [20, 28, 29, 31, 32]. The vast majority of the protocols utilize storage in liquid nitrogen after controlled rate cooling [20, 28, 31–34] (Table 11.1).

11.3 Why ASC Cryopreservation Remains a Relevant Research Question

Recent studies have found that the age, body mass index, and overall health condition of donors impacts the functionality of isolated ASC. The ASC from elderly donors expressed elevated levels of senescence markers, oxidative stress, lower antioxidant enzyme (superoxide dimutase) activity, increased cell doubling times and reduced clonogenic potential [35, 36]. The differentiation potential of ASC especially towards osteogenic and chondrogenic pathways was compromised from the aged donors whereas the adipogenic ability tended to increase [35, 36]. The Angiogenic differentiation potential and the ability to secrete pro-angiogenic growth factors such as ANGPT1, HGF, and VEGF, was also hampered in ASC isolated from older donors [37, 38]. Furthermore, ASC isolated from the obese individuals display impaired functionality. The therapeutic efficacy of ASC from obese donors to treat multiple sclerosis was decreased in comparison to ASC from lean donors [39]. The osteogenic potential of ASC from obese donors was also reduced relative to ASC from lean donors [39]. Collectively, these published findings suggest that ASC from young, healthy, lean donors will have the greatest potential for efficacy in clinical applications. However, most of the patients who benefit from

Table 11.1 Adipose-derived stem/stromal cells (ASC) cryopreservation literature

Author	CPA	Time
De Rosa et al.	4 % DMSO/6 % trehalose	12 months
Gonda et al.	Proprietary + bovine serum	6 months
Devireddy, Gimble et al.	DMSO/glycerol/MC/PVP	2 weeks
Lopez et al.	Antioxidants + polymers + EG/DMSO + trehalose + EGTA	Not specified
Hoogduijn et al.	20 % DMSO + 20 % FBS	Not specified

ASC therapies are elderly with co-morbidities associated with obesity. Thus, their autologous ASC may not display the desired therapeutic benefits, making such patients dependent on the use of allogeneic ASC. Consequently, cryopreservation of ASC from young and healthy individuals may emerge as an opportunity to create a new cell product, similar to type O erythrocytes that can be transplanted to multiple recipients with minimal risk and maximal benefit. For these reasons, it remains important to develop cryopreservation protocols that retain the post thaw functionality of ASC while minimizing risks of infectious agent contamination or transmission.

11.4 Pre-clinical Models and Clinical Use

Despite variation in processing methods, isolated ASC have displayed regenerative, anti-inflammatory and immunomodulatory potential. Studies in animal models have determined the beneficial effects of ASC in the healing of Acute Myocardial Infarction, Chronic Ulcers, Peripheral Vascular Disease, and Bone defects among other conditions [11, 40–44]. There remain challenges to the development of clinical grade quality cell products without biological contaminants, with high cell viability, low product variability and constant and reproducible potency [45]. In addition to the impact of cryopreservation and thawing on the cell characteristics, studies must consider the final success of the ASC when administered to the recipient in vivo [32]. For example, the intravenous infusion of MSC has been found to modulate the cytokines released by resident lung immune cells [46]. Furthermore, the gene expression profile can change after MSC have been exposed to the host/recipient [32]. Thus, pre-clinical animal models are critically necessary to define the mechanisms of action and to identify the ideal method for cell transplantation [45].

The existing body of basic science investigations has contributed to an increased number of clinical trials using ASC for Soft Tissue Reconstruction, Osteoarticular Defects, and Cardiovascular, Pulmonary and Neurological diseases, among others (www.clinicaltrials.org); however, at present, there are no U.S. FDA-approved ASC-based products (Table 11.2).

11.5 Potential Reconstructive and Chronic Disease Targets for ASC Therapies

11.5.1 Soft Tissue Reconstruction

Fat grafting is a useful tool for soft tissue reconstruction to repair volume loss or improve shape. Recently, the use of Cell Assisted Lipotransfer (CAL), which combines autologous SVF cells with fat grafts, has gained considerable interest among plastic surgeons internationally. The SVF cell enriched fat graft contains increased numbers of ASC and associated cells which have been reported to improve breast reconstruction

Table 11.2 Clinical studies using ASC/SVF

Target tissue	Soft tissue reconstruction	Breast reconstruction and augmentation; pressure ulcer; facial atrophy; enterocutaneous/recto-vaginal fistulas
	Hard tissue reconstruction	Osteoarticular (osteoartritis, degenerative disc disease; osteoporotic fractures; genu varum)
Underlying disease	Neurological	Parkinson's disease; multiple sclerosis; Autism
	Circulatory and cardiovascular	Stroke; chronic ischemic heart disease; acute myocardial infarction; critical limb ischemia; ischemic nephropathy
	Methabolic and auto-immune	Diabetes type I and II; liver cirrhosis

outcomes by increasing volume retention and reducing irregularities or retractions [47, 48]. Facial atrophy [49] and other craniofacial malformations [50] classically treated with fat grafts have also displayed improved outcomes when treated with CAL.

11.5.2 Hard Tissue Reconstruction

Although autologous or allogeneic bone graft remains the bone regeneration "gold standard" for treatment, this approach is flawed due to volume limitation and morbidity associated with the tissue harvesting. Thus, ongoing efforts are underway to seek an improved solution [42]. Published studies have achieved the successful reconstruction of large mandibular defects using a tissue-engineered construct containing beta-tricalcium phosphate, recombinant bone morphogenetic protein and Good Manufacturing Practice-level autologous ASC allowing for full functional recover and patient rehabilitation [51]. Likewise, similar approaches have used autologous ASC to repair calvarian defects through local ossification, thereby improving the patient's quality of life [52].

11.5.3 Multiple Sclerosis

Administration of SVF cells has ameliorated the severity of experimental autoimmune encephalitis (EAE) in mice (an animal model of Multiple Sclerosis) by decreasing the magnitude of the pro-inflammatory cytokines interleukin (IL-12) and interferon gamma (IFN-γ) which contribute to the progression of the disease [53]. Furthermore, administration of culture expanded ASC after the establishment of the disease could still alleviate the signs and symptoms associated with chronic EAE [54]. In contrast, bone marrow-derived mesenchymal stem cells (BM-MSCs) were only able cure the disease when they were injected at the initiation stage of the EAE model [55]. Currently, at least one company is sponsoring a registered clinical trial to treat multiple sclerosis using autologous SVF with an estimated enrolment of 100 participants [ClinicalTrials.gov Identifier: NCT02157064]. Their goal is to complete the study by May 2017.

11.5.4 Osteoarthritis

A phase I/II clinical study designed to test the effect of injected ASC in to intra-articular cartilage of knee in patients suffering with osteoarthritis has so far shown reported promising results [56]. This trial has enrolled 18 patients with osteoarthiritis divided into groups who received autologous ASC at one of three different cell concentrations: low (10^7 cells), intermediate (5×10^7 cells), and high (10^8 cells) respectively [56]. In the group that received high dose of ASC hyaline-like cartilage was regenerated at cartilage defects resulting in improved function and relief of knee pain without any severe side effects [56]. Another clinical trial has reported that the 14 out of 16 elderly osteoarthritis patients injected intra-articularly with ASC suspended in platelet rich plasma showed significant cartilage regeneration, reduced pain relief, and improved ambulatory function [57].

11.5.5 Crohn's Disease

Crohn's disease is an inflammatory disease that affects the entire digestive tract and commonly is associated with the development of fistulas. The use of ASC to treat Crohn's disease has shown some success in a phase I clinical study [58]. Patients suffering from a perianal fistula received 2×10^7 ASC which resulted in complete healing of fistula 8 weeks post injection with no adverse events [58, 59]. There were no signs of recurrence of fistula 8 months after the ASC injection [58]. In a phase II clinical study, the same group demonstrated that administration of ASC improved recovery of fistulas in 27 out of 33 [59]. A follow up phase II trial investigation conducted in 43 patients showed the long-term sustainability of ASC treatment since 75 % of the patients maintained complete closure of the fistula without recurrence [60]. Other groups have reported similar beneficial results in phase I and II clinical trials [61, 62]. Anterogen Co., Ltd. a South Korean based company, is currently marketing an ASC based product, (Cupistem®) as an injection to treat Crohn's disease.

11.5.6 Parkinson's Disease

Parkinson's disease (PD) is a progressive central nervous system disorder that results in the chronic degeneration of dopamine producing neurons in the substantia nigra region of brain [63] and the accumulation of alpha-synuclein protein aggregates called Lewy Bodies [64]. ASC transplanted into the subventicular region of the brain in a rat PD model significantly improved neurogenesis in comparison to controls [65]. Furthermore, there was an increase in the levels of brain-derived neurotrophic factor and the number of tyrosine hydroxylase and glial fibrillary acidic secreting cells resulting in betterment of motor function in the PD model rats transplanted with ASC [66]. At present a clinical study sponsored by StemGenex medical group is recruiting participants with PD to study the safety and efficacy of autologous SVF transplantation for up to 12 months [ClinicalTrials.gov Identifier: NCT02184546]. The study is expected to be complete by June 2017.

11.5.7 Diabetes

Diabetes is a chronic disease that affects the insulin secretion and glucose metabolism leading to hyperglycemia. An intravenous infusion of autologous ASC in diabetic rats significantly reduced hyperglycemia [67, 68]. The mechanism involved the regeneration of pancreatic β cells and increased insulin secretion [67]. Additionally, ASC infusion significantly lowered the level of pro-inflammatory cytokines such as TNF-α, IL-1β, and IL-6 in comparison to untreated diabetic controls [67, 68]. Moreover, cells in the recipient animal's liver and adipose tissues showed improved insulin sensitivity based on enhanced expression levels of both the glucose transporter (GLUT4) and the insulin receptor (INSR) [67]. In addition to ameliorating the major diabetic symptoms such as hyperglycemia, impaired insulin secretion, and insulin resistance ASC infusion reduced other adverse effects such as diabetic nephropathy and diabetic wounds [68, 69]. Adistem Ltd is conducting phase I and II clinical studies to treat diabetes type 1 and 2 using intravenous administration of ASC [ClinicalTrials.gov Identifiers: NCT00703599; NCT00703612].

11.6 Summary

Although ASC are relatively easy to harvest, ASC cryopreservation will allow for their use as an "of the shelf" product immediately available at the bedside. Although clinical applications for ASC and SVF cells remains experimental, their future success will rely in part on the development of fully tested protocols relating to tissue acquisition, processing, cryopreservation, thawing, CPA dilution and infusion techniques. These may need to be modified depending on the target organ or tissue to be treated. These advances will require the use of internationally harmonized assays predictive of both functionality and potency.

Disclosures Fabiana Zanata received financial support from CAPES, Brazil (Process BEX 1524/15-1). Jeffrey Gimble and Xiying Wu are co-owners and employees of LaCell LLC, a biotechnology company focusing on research and clinical translation involving ASC and SVF cells.

References

1. Jenkins DD, Yang GP, Lorenz HP et al (2003) Tissue engineering and regenerative medicine. Clin Plast Surg 30:581
2. Horwitz EM, Le Blanc K, Dominici M et al (2005) Clarification of the nomenclature for MSC: the International Society for Cellular Therapy position statement. Cytotherapy 7:393–395
3. Dominici M, Le Blanc K, Mueller I et al (2006) Minimal criteria for defining multipotent mesenchymal stromal cells. The International Society for Cellular Therapy position statement. Cytotherapy 8:315–317

4. McIntosh KR, Frazier T, Rowan BG et al (2013) Evolution and future prospects of adipose-derived immunomodulatory cell therapeutics. Expert Rev Clin Immunol 9:175–184
5. Davis TA, Anam K, Lazdun Y et al (2014) Adipose-derived stromal cells promote allograft tolerance induction. Stem Cells Transl Med 3:1444–1450
6. Gimble JM, Bunnell BA, Frazier T et al (2013) Adipose-derived stromal/stem cells: a primer. Organogenesis 9:3–10, United States
7. Cawthorn WP, Scheller EL, MacDougald OA (2012) Adipose tissue stem cells meet preadipocyte commitment: going back to the future. J Lipid Res 53:227–246
8. McIntosh K, Zvonic S, Garrett S et al (2006) The immunogenicity of human adipose-derived cells: temporal changes in vitro. Stem Cells 24:1246–1253
9. Mendicino M, Bailey AM, Wonnacott K et al (2014) MSC-based product characterization for clinical trials: an FDA perspective. Cell Stem Cell 14:141–145
10. Halme DG, Kessler DA (2006) FDA regulation of stem-cell-based therapies. N Engl J Med 355:1730–1735
11. Gimble JM, Guilak F, Bunnell BA (2010) Clinical and preclinical translation of cell-based therapies using adipose tissue-derived cells. Stem Cell Res Ther 1:19
12. Hunt CJ (2011) Cryopreservation of human stem cells for clinical application: a review. Transfus Med Hemother 38:107–123
13. Pegg DE (2015) Principles of cryopreservation. Methods Mol Biol 1257:3–19
14. Mazur P (1984) Freezing of living cells: mechanisms and implications. Am J Physiol 247:C125–C142
15. Saragusty J (2015) Directional freezing for large volume cryopreservation. Methods Mol Biol 1257:381–397
16. Gorin NC (1986) Collection, manipulation and freezing of haemopoietic stem cells. Clin Haematol 15:19–48
17. Thirumala S, Gimble JM, Devireddy RV (2010) Cryopreservation of stromal vascular fraction of adipose tissue in a serum-free freezing medium. J Tissue Eng Regen Med 4:224–232
18. Balci D, Can A (2013) The assessment of cryopreservation conditions for human umbilical cord stroma-derived mesenchymal stem cells towards a potential use for stem cell banking. Curr Stem Cell Res Ther 8:60–72
19. Pegg DE (2002) The history and principles of cryopreservation. Semin Reprod Med 20:5–13
20. Lopez M, Bollag RJ, Yu JC et al (2016) Chemically defined and xeno-free cryopreservation of human adipose-derived stem cells. PLoS One 11:e0152161
21. Brockbank KG, Heacox AE, Schenke-Layland K (2011) Guidance for removal of fetal bovine serum from cryopreserved heart valve processing. Cells Tissues Organs 193:264–273
22. Carvalho PP, Wu X, Yu G et al (2011) Use of animal protein-free products for passaging adherent human adipose-derived stromal/stem cells. Cytotherapy 13:594–597
23. Dos Santos F, Campbell A, Fernandes-Platzgummer A et al (2014) A xenogeneic-free bioreactor system for the clinical-scale expansion of human mesenchymal stem/stromal cells. Biotechnol Bioeng 111:1116–1127
24. Gimble JM, Bunnell BA, Chiu ES et al (2011) Concise review: adipose-derived stromal vascular fraction cells and stem cells: let's not get lost in translation. Stem Cells 29:749–754
25. Thirumala S, Wu X, Gimble JM et al (2010) Evaluation of polyvinylpyrollidone as a cryoprotectant for adipose tissue-derived adult stem cells. Tissue Eng Part C Methods 16:783–792
26. Thirumala S, Gimble JM, Devireddy RV (2010) Evaluation of methylcellulose and dimethyl sulfoxide as the cryoprotectants in a serum-free freezing media for cryopreservation of adipose-derived adult stem cells. Stem Cells Dev 19:513–522
27. Pu LL, Cui X, Fink BF et al (2005) Cryopreservation of adipose tissues: the role of trehalose. Aesthet Surg J 25:126–131
28. Devireddy RV, Thirumala S, Gimble JM (2005) Cellular response of adipose derived passage-4 adult stem cells to freezing stress. J Biomech Eng 127:1081–1086
29. Goh BC, Thirumala S, Kilroy G et al (2007) Cryopreservation characteristics of adipose-derived stem cells: maintenance of differentiation potential and viability. J Tissue Eng Regen Med 1:322–324
30. Thirumala S, Gimble JM, Devireddy RV (2005) Transport phenomena during freezing of adipose tissue derived adult stem cells. Biotechnol Bioeng 92:372–383
31. Thirumala S, Zvonic S, Floyd E et al (2005) Effect of various freezing parameters on the immediate post-thaw membrane integrity of adipose tissue derived adult stem cells. Biotechnol Prog 21:1511–1524
32. Hoogduijn MJ, de Witte SF, Luk F et al (2016) Effects of freeze-thawing and intravenous infusion on mesenchymal stromal cell gene expression. Stem Cells Dev 25:586–597
33. De Rosa A, De Francesco F, Tirino V et al (2009) A new method for cryopreserving adipose-derived stem cells: an attractive and suitable large-scale and long-term cell banking technology. Tissue Eng Part C Methods 15:659–667
34. Gonda K, Shigeura T, Sato T et al (2008) Preserved proliferative capacity and multipotency of human adipose-derived stem cells after long-term cryopreservation. Plast Reconstr Surg 121:401–410
35. Choudhery MS, Badowski M, Muise A et al (2014) Donor age negatively impacts adipose tissue-derived mesenchymal stem cell expansion and differentiation. J Transl Med 12:8
36. Kornicka K, Marycz K, Tomaszewski KA et al (2015) The effect of age on osteogenic and adipogenic differentiation potential of human adipose derived stromal stem cells (hASCs) and the impact of stress factors in the course of the differentiation process. Oxid Med Cell Longev 2015:309169

37. Efimenko A, Dzhoyashvili N, Kalinina N et al (2014) Adipose-derived mesenchymal stromal cells from aged patients with coronary artery disease keep mesenchymal stromal cell properties but exhibit characteristics of aging and have impaired angiogenic potential. Stem Cells Transl Med 3:32–41
38. Duscher D, Rennert RC, Januszyk M et al (2014) Aging disrupts cell subpopulation dynamics and diminishes the function of mesenchymal stem cells. Sci Rep 4:7144
39. Strong AL, Bowles AC, Wise RM et al (2016) Human adipose stromal/stem cells from obese donors show reduced efficacy in halting disease progression in the experimental autoimmune encephalomyelitis model of multiple sclerosis. Stem Cells 34:614–626
40. Gauglitz GG, Jeschke MG (2011) Combined gene and stem cell therapy for cutaneous wound healing. Mol Pharm 8:1471–1479
41. Gimble JM, Grayson W, Guilak F et al (2011) Adipose tissue as a stem cell source for musculoskeletal regeneration. Front Biosci (Schol Ed) 3:69–81
42. Grayson WL, Bunnell BA, Martin E et al (2015) Stromal cells and stem cells in clinical bone regeneration. Nat Rev Endocrinol 11:140–150
43. Sultan SM, Stern CS, Allen RJ Jr et al (2011) Human fat grafting alleviates radiation skin damage in a murine model. Plast Reconstr Surg 128:363–372
44. Cowan CM, Shi YY, Aalami OO et al (2004) Adipose-derived adult stromal cells heal critical-size mouse calvarial defects. Nat Biotechnol 22:560–567
45. Galipeau J, Krampera M, Barrett J et al (2016) International Society for Cellular Therapy perspective on immune functional assays for mesenchymal stromal cells as potency release criterion for advanced phase clinical trials. Cytotherapy 18:151–159
46. Fischer UM, Harting MT, Jimenez F et al (2009) Pulmonary passage is a major obstacle for intravenous stem cell delivery: the pulmonary first-pass effect. Stem Cells Dev 18:683–692
47. Yoshimura K, Sato K, Aoi N et al (2008) Cell-assisted lipotransfer for cosmetic breast augmentation: supportive use of adipose-derived stem/stromal cells. Aesthetic Plast Surg 32:48–55, discussion 56–47
48. Tissiani LA, Alonso N (2016) A prospective and controlled clinical trial on stromal vascular fraction enriched fat grafts in secondary breast reconstruction. Stem Cells Int 2016:2636454
49. Yoshimura K, Sato K, Aoi N et al (2008) Cell-assisted lipotransfer for facial lipoatrophy: efficacy of clinical use of adipose-derived stem cells. Dermatol Surg 34:1178–1185
50. Tanikawa DY, Aguena M, Bueno DF et al (2013) Fat grafts supplemented with adipose-derived stromal cells in the rehabilitation of patients with craniofacial microsomia. Plast Reconstr Surg 132:141–152
51. Sandor GK, Tuovinen VJ, Wolff J et al (2013) Adipose stem cell tissue-engineered construct used to treat large anterior mandibular defect: a case report and review of the clinical application of good manufacturing practice-level adipose stem cells for bone regeneration. J Oral Maxillofac Surg 71:938–950
52. Lendeckel S, Jodicke A, Christophis P et al (2004) Autologous stem cells (adipose) and fibrin glue used to treat widespread traumatic calvarial defects: case report. J Craniomaxillofac Surg 32:370–373
53. Semon JA, Maness C, Zhang X et al (2014) Comparison of human adult stem cells from adipose tissue and bone marrow in the treatment of experimental autoimmune encephalomyelitis. Stem Cell Res Ther 5:2
54. Constantin G, Marconi S, Rossi B et al (2009) Adipose-derived mesenchymal stem cells ameliorate chronic experimental autoimmune encephalomyelitis. Stem Cells 27:2624–2635
55. Zappia E, Casazza S, Pedemonte E et al (2005) Mesenchymal stem cells ameliorate experimental autoimmune encephalomyelitis inducing T-cell anergy. Blood 106:1755–1761
56. Jo CH, Lee YG, Shin WH et al (2014) Intra-articular injection of mesenchymal stem cells for the treatment of osteoarthritis of the knee: a proof-of-concept clinical trial. Stem Cells 32:1254–1266
57. Koh YG, Choi YJ, Kwon SK et al (2015) Clinical results and second-look arthroscopic findings after treatment with adipose-derived stem cells for knee osteoarthritis. Knee Surg Sports Traumatol Arthrosc 23:1308–1316
58. Cho YB, Lee WY, Park KJ et al (2013) Autologous adipose tissue-derived stem cells for the treatment of Crohn's fistula: a phase I clinical study. Cell Transplant 22:279–285
59. Lee WY, Park KJ, Cho YB et al (2013) Autologous adipose tissue-derived stem cells treatment demonstrated favorable and sustainable therapeutic effect for Crohn's fistula. Stem Cells 31:2575–2581
60. Cho YB, Park KJ, Yoon SN et al (2015) Long-term results of adipose-derived stem cell therapy for the treatment of Crohn's fistula. Stem Cells Transl Med 4:532–537
61. Garcia-Olmo D, Garcia-Arranz M, Herreros D et al (2005) A phase I clinical trial of the treatment of Crohn's fistula by adipose mesenchymal stem cell transplantation. Dis Colon Rectum 48:1416–1423
62. Garcia-Olmo D, Herreros D, Pascual I et al (2009) Expanded adipose-derived stem cells for the treatment of complex perianal fistula: a phase II clinical trial. Dis Colon Rectum 52:79–86
63. Braak H, Del Tredici K, Rub U et al (2003) Staging of brain pathology related to sporadic Parkinson's disease. Neurobiol Aging 24:197–211
64. Ubeda-Banon I, Saiz-Sanchez D, de la Rosa-Prieto C et al (2010) alpha-synucleinopathy in the human olfactory system in Parkinson's disease: involvement of calcium-binding protein- and substance P-positive cells. Acta Neuropathol 119:723–735
65. Schwerk A, Altschuler J, Roch M et al (2015) Human adipose-derived mesenchymal stromal cells increase endogenous neurogenesis in the rat subventricular zone acutely after 6-hydroxydopamine lesioning. Cytotherapy 17:199–214

66. Berg J, Roch M, Altschuler J et al (2015) Human adipose-derived mesenchymal stem cells improve motor functions and are neuroprotective in the 6-hydroxydopamine-rat model for Parkinson's disease when cultured in monolayer cultures but suppress hippocampal neurogenesis and hippocampal memory function when cultured in spheroids. Stem Cell Rev 11:133–149
67. Hu J, Fu Z, Chen Y et al (2015) Effects of autologous adipose-derived stem cell infusion on type 2 diabetic rats. Endocr J 62:339–352
68. Fang Y, Tian X, Bai S et al (2012) Autologous transplantation of adipose-derived mesenchymal stem cells ameliorates streptozotocin-induced diabetic nephropathy in rats by inhibiting oxidative stress, pro-inflammatory cytokines and the p38 MAPK signaling pathway. Int J Mol Med 30:85–92
69. Zografou A, Papadopoulos O, Tsigris C et al (2013) Autologous transplantation of adipose-derived stem cells enhances skin graft survival and wound healing in diabetic rats. Ann Plast Surg 71:225–232

Banking of Adipose- and Cord Tissue-Derived Stem Cells: Technical and Regulatory Issues

12

David T. Harris

Abstract

Stem cells are found in all multicellular organisms and are defined as cells that can differentiate into specialized mature cells as well as divide to produce more stem cells. Mesenchymal stem cells (MSC) were among the first stem cell types to be utilized for regenerative medicine. Although initially isolated from bone marrow, based on ease and costs of procurement, MSC derived from adipose tissue (AT-MSC) and umbilical cord tissue (CT-MSC) are now preferred stem cell sources for these applications. Both adipose tissues and cord tissue present unique problems for biobanking however, in that these are whole tissues, not cellular suspensions. Although the tissues could be processed to facilitate the biobanking process, by doing so additional regulatory issues arise that must be addressed. This review will discuss the technical issues associated with biobanking of these tissues, as well as regulatory concerns when banking of utilizing MSC derived from these sources in the clinic.

Keywords

Stem cells • Cord tissue • Adipose tissue • FDA • Regulations • Biobanking • Regenerative medicine

D.T. Harris (✉)
University of Arizona Health Sciences Biorepository, Tucson, AZ, USA

University of Arizona, Tucson, AZ, USA

Division of Translational Medicine, University of Arizona, Tucson, AZ, USA

GMP Laboratory at the University of Arizona, Tucson, AZ, USA

Celebration Stem Cell Centre, Gilbert, AZ, USA
e-mail: davidh@email.arizona.edu

F. Karimi-Busheri, M. Weinfeld (eds.), *Biobanking and Cryopreservation of Stem Cells*, Advances in Experimental Medicine and Biology 951, DOI 10.1007/978-3-319-45457-3_12

Abbreviations

MSCs	Mesenchymal stem cells
AT-MSC	Adipose tissue-MSC
CT-MSC	Cord tissue-MSC
BM	Bone marrow
MS	Multiple sclerosis
TNC	Total nucleated cells
SVF	Stromal vascular fraction
IND	Investigational new drug
BB	Obstetrician

12.1 Introduction

Stem cells are found in all multicellular organisms and are defined as cells that can differentiate into specialized mature cells as well as divide to produce more stem cells. Stem cells can be divided into embryonic/fetal stem cells, and adult stem cells, based on their origin. Stem cells can be further classified as totipotental capable of giving rise to all tissues in the organism including the organism itself; as pluripotential able to give rise to multiple lineages of tissues and cells from different germ lineages; as multipotential which can give rise to different cell types generally within the same germ lineage; or as progenitor cells which only give rise to more lineage-restricted cells and tissues from a single germ layer origin. This appreciation for the numbers, types and potentials of stem cells has given rise to the fields of regenerative medicine and tissue engineering which encompasses a variety of cellular therapies. Mesenchymal stem cells (MSC) were among the first stem cell types to be utilized for such applications. Initially, all MSC were bone marrow (BM) derived. However, based on the ease and costs of procurement, as well as the possibility of obtaining autologous MSC from most patients using these sources, MSC derived from adipose tissue (AT-MSC) and umbilical cord tissue (CT-MSC) are now beginning to replace BM as the preferred source of clinically applicable MSC. In fact, as of March 2016 there were 176 clinical studies listed on www.clinicaltrials.gov for AT-MSC and an additional 342 studies listed for CT-MSC, with thousands of patients having received such treatments. These studies encompassed such clinical applications as osteoarthritis, orthopedic reconstructions, autism, multiple sclerosis (MS), spinal cord injury, diabetes, wound healing, cardiovascular disease and pulmonary disease.

Work done over the past decade has demonstrated that subcutaneous adipose tissue (AT) is the richest source of MSC in the human body, containing 100–1000× more MSC/g or cc of tissue that either BM or CT [1, 2]. In fact, as many as 1 % of all total nucleated cells (TNC) contained within AT may be MSC [1, 2]. However, AT is a tissue, not a cellular suspension, posing unique constraints on its use. In fact, lipoaspirate, which is often the starting point for AT-MSC collection, is a viscous gelatinous tissue that is difficult to manipulate, even at room temperatures (see Fig. 12.1). Although the lipoaspirate can be digested and processed, such a procedure would necessarily require imposition of additional and somewhat onerous regulatory guidelines (see below). The physical characteristics of the biospecimen also present unique requirements when it comes time to thaw, wash and clinically use the banked sample. Although MSC in general have long been studied as a clinically relevant source of stem cells, and have been extensively studied in multiple regenerative medicine and tissue engineering applications [3–5], it was not until recently that the umbilical cord (specifically the Wharton's jelly contained within the tissue; CT) itself was recognized as an economical and readily available source of large numbers of MSC [6, 7]. Similar to adipose tissues, CT presents unique problems for biobanking in that it is a whole tissue, not a cellular suspension (see Fig. 12.2). Although the tissue could be processed to facilitate the biobanking process, by doing so additional regulatory issues arise that must be addressed. However, as one of the youngest sources of MSC available its inclusion into any stem cell banking program is worthwhile.

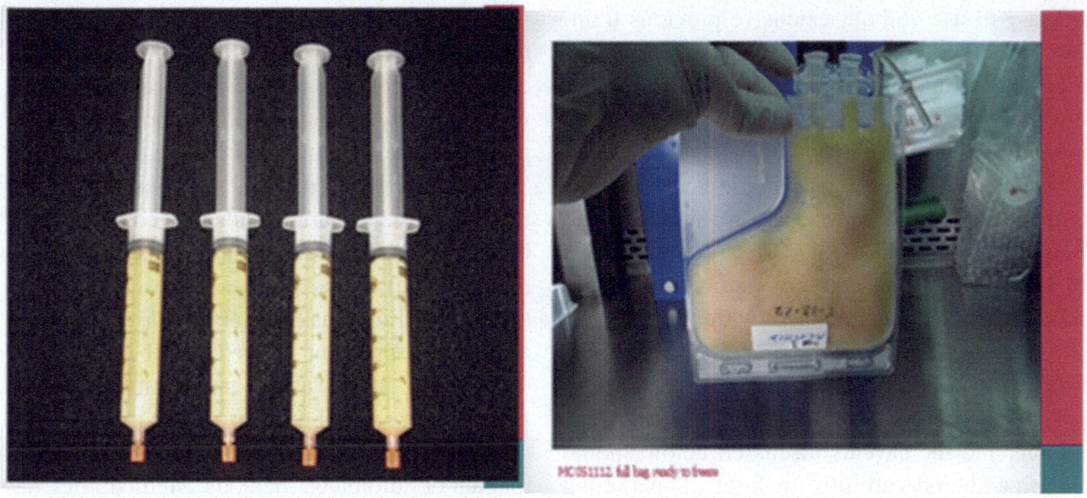

Fig. 12.1 Adipose tissue collection (*left*) and storage (*right*) using 60 cc syringes and 60 cc bags, respectively. Note the gelatinous nature of the stem cell source

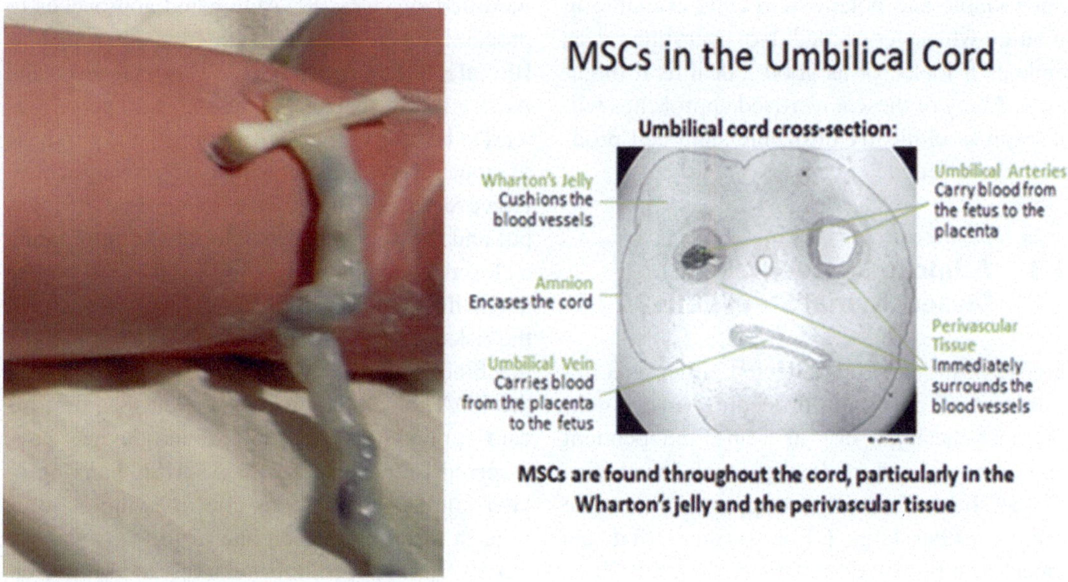

Fig. 12.2 The umbilical cord of a newborn (*left*) and a cross-sectional representation of the anatomy (*right*) demonstrating the unique structure of the stem cell source

12.2 Technical Issues in Tissue Stem Cell Banking

Whenever possible it is always preferable to utilize a closed and sterile system for stem cell collection and processing, which simplifies and/or eliminates many regulatory requirements. If the stem cell collection device and the cell processing device can be combined into a single system design, then the methodological approach becomes more attractive to the end-user and more likely to be utilized. As most stem cell collections (and eventually their utilizations) will be performed by clinicians who may or may not have extensive experience in stem cell collection or cell manipulations, the employed systems should be easy to use without extensive previous training. Although most collection systems are generally sterile, the commercially available options are constrained by the particulars of the stem cell source.

Neither CT nor AT are single cell solutions similar to blood that are easily manipulated and banked. Rather, both AT and CT are intact tissues that require extensive manipulation to obtain/retrieve the MSC contained within the tissue. This observation is true whether the tissues are to be used immediately in the clinic (e.g., point of care use) or banked for future use. These constraints and the caveats discussed below should be remembered carefully in light of marketing efforts by numerous commercial entities that entreat parent and patients to pay for AT-MSC and CT-MSC biobanking with the promise of having viable and potent stem cells available in the future whenever needed. Not everything is as simple as it looks, or as good a deal as it might appear. Many of these advertised approaches will not result in clinically utilizable stem cell products if ever needed.

12.3 Adipose-Derived Mesenchymal Stem Cells

Adipose tissue may be collected either as a by-product of a liposuction procedure (under general anesthesia), or as an independent stand-alone procedure (using local anesthetics such as lidocaine). Both approaches utilize syringes (either large 60 cc syringes that can connect to a liposuction canister or small 20 cc syringes with a cannula for the manual procedure). Both collection systems can be considered "closed". The syringes can also be sterilely connected in a closed fashion to large (commercially available) bags for processing and storage (depending on volumes desired). As an add-on to a liposuction procedure it is possible to obtain several liters of lipoaspirate (although most adipose banks collect and store no more than 2000 cc). As a stand-alone procedure done under local anesthetic, harvests are normally no larger than 20–100 cc.

AT-MSC have unique processing requirements in that the MSC are not obtained from a cellular suspension like BM, but are contained within a fairly viscous tissue composed of multiple cell types. Initial processing (often called "enhancement") removes as much tumescent fluid as possible along with any contaminating blood via the use of several low speed centrifugations (300×g). This process leaves the adipose tissue containing the MSC amenable for immediate cryopreservation or for clinical use, or as a source of MSC after enzymatic digestion. However, in each of these examples it can be processed sterilely in a closed system using either manual or automated methods. Methodology has been developed that allows for economical, closed system processing that meets FDA requirements for minimal manipulation utilizing modified syringes [8]. Automated approaches to processing can be found with companies such as BioSafe, GDP Inc., Cytori, TissueGenesis and American Cryostem. The automated approaches tend to be more expensive, requiring the purchase of machinery and/or expensive consumables. However, these approaches do increase throughput and are very reproducible. In addition, being a closed system these methodologies are compliant with (some) FDA regulations and minimize the risks of sample contamination. Generally, the machines have received a 510 k registration with the FDA for clinical use, and even for point-of-care applications, but none of the devices have received FDA approval for AT-MSC banking or later clinical applications after thawing at some time in the future. Each one of these systems is closed from start to finish, harvest to cryopreservation, including the thawing of the sample after storage. However, the major and significant difference between the manual systems and the automated systems is that the automated systems produce what is referred to as a stromal vascular fraction (SVF) after enzymatic digestion of the "enhanced fat" which by definition is "more than minimally manipulated" and requires an Investigational New Drug (IND) application from the FDA before clinical use (see www.fda.gov and guidance on adipose tissues). The manual system discussed in [8] only produces "enhanced fat" that is minimally manipulated and can be immediately used for cosmetic and reconstructive purposes, or processed further under IND for regenerative medicine applications. Oftentimes, the SVF must be cultured in vitro to purify the AT-MSC before clinical use in order to remove the contaminating cell types. In vitro expansion (although not as often needed with AT-MSC as with other MSC sources) requires adherence to another and different set of FDA regulations. As AT contains the greatest number of MSC per gm or cc of tissue, and since most individuals have plenty of AT to harvest, it is generally more a matter of MSC purification or enrichment through in vitro culture than designed MSC expansion by culturing. A recent presentation has described a short term panning procedure that allows for almost 100 % enrichment with less than one to two cell doublings over the course of 3–5 days [9] that may be applicable to most clinical settings where MSC numbers are sufficient but non-MSC cells need to be removed. However, this methodology would still be considered a manipulation under FDA law requiring IND clearance before clinical use. It should be noted that AT, unlike CT as discussed below, is the only one of these two MSC sources that can be processed in a closed system, biobanked without manipulation (by FDA definitions), and later thawed with 80 % or greater cellular recovery for research or clinical use [8].

12.4 Umbilical Cord Tissue-Derived Mesenchymal Stem Cells

Collection of umbilical cord tissue (CT) is a recent development in the stem cell field (within the last 5 years or so). By its nature, CT is not sterile (at least the exterior which recently exited the birth canal and is generally covered with blood and mucus from the birth) and cannot be collected in a closed system, as its connected on one end to the baby and on the other end to the placenta. Normally, the physician, obstetrician (OB), or other caregiver will cut a 6–10 in. segment of the umbilical cord after birth of the child and ligation of the cord. The segment of umbilical cord is then placed into a sterile capped cup (e.g. sterile, screw-capped urine specimen cup) that contains a transport buffer (nutrients as well as antibiotics and antifungals). Care must be taken to insure that the CT is not exposed to excessive air and is kept "wet" (i.e., submerged) during transport.

Due to the nature by which CT is harvested and its structural composition, CT must be processed in an "open" fashion. Generally this means that it is either stored as whole minced tissue or the CT is enzymatically processed to its cellular components before use or banking (this latter approach would be classified as "more than minimally manipulated" for regulatory purposes). Therefore, extensive sterility testing (bacterial, fungal and mycobacterial) is necessary, as is testing of the donor for a variety of viral infections. The stem cells of interest, the MSC, lie within the cord tissue either as perivascular cells or contained within the Wharton's jelly [3, 4], and this anatomical distribution must be taken into consideration when processing and banking the stem cells. If one attempts to cryopreserve the whole, intact unprocessed tissue the MSC will be damaged and will not survive the procedure. Thus, one must either finely mince the tissue before cryopreservation or enzymatically digest the tissue and then freeze the isolated MSC as is typically done for a single cell suspension [6, 7]. Finely minced tissue requires the slow infusion of DMSO over prolonged periods of time to insure homogenous distribution of the cryoprotectant throughout the tissue [10]. The finely minced tissue may also be used for explant cultures although the time required to obtain sufficient numbers of MSC for use can be several weeks or more. Isolated MSC obtained from an enzymatic digest (or from in vitro expansion) can be frozen as for any single cell suspension. Two problems present themselves however, with either methodology. If one enzymatically digests the cord tissue before cryopreservation it is now considered more than minimally manipulated by the FDA, requiring an IND prior to clinical use.

Cryopreserved whole, minced CT meets the regulatory definition of minimal manipulation, but one generally only recovers approx. 10 % of starting MSC population, requiring extensive ex vivo expansion prior to clinical use (again requiring an IND prior to clinical utilization), as compared to 80 % or higher recoveries for frozen adipose tissues [11]. CT is problematic to thaw as the tissue itself serves as a "sink: for the cryoprotectant DMSO, making it difficult to thoroughly remove which results in a loss of viable cells. Thus, CT is a less than optimal stem cell source for clinical applications due to the regulatory oversight that is necessary for its clinical use.

Oftentimes thawing of whole minced CT results in few viable MSC being obtained, and many fewer than needed for most clinical applications [10]. However, thawed tissue can be grown in vitro as explants rather than digested with much more successful results. Unfortunately, in vitro culture and expansion is also considered a manipulation and requires extensive regulatory oversight along with an IND application. Although fresh CT can be easily digested with combinations of fairly benign enzymes (such as collagenase and hyaluronidase), the numbers of MSC obtained per gram of tissue is often quite low (often 50–100 times fewer MSC per gm of tissue for CT as compared to AT; [10, 11]). Thus, CT-MSC will most likely always require some in vitro expansion before clinical use, mandating federal oversight and need to obtain an IND.

12.5 Regulatory Issues

As with any type of cellular therapy, regenerative medicine application or tissue engineering procedure, there are always concerns and confusions about which guidelines mandated by the FDA (see 21CFR regulations) must be followed. Stem cell therapies are more in the spotlight these days than ever before with the seemingly nonstop construction of stem cell clinics and the rapid commercialization of this medical field (endeavor) by individuals that are neither qualified by training nor experience. Generally, the FDA is concerned with several basic issues such as sample identity (does the stem cell sample come from the person that is going to use it in an autologous setting, or is it suitably matched if unrelated?); sample purity (has it been adulterated by foreign substances which may be harmful or which may obliterate its usefulness?); the potency of the sample (is there a standard against which the sample can be compared to measure its potential biological function?); is the sample efficacious (will the sample have any beneficial effects after administration?), and is the sample free of exogenous disease causing organisms (i.e., what is the risk of disease transmission?). In order to address these issues and to determine the level of oversight required the FDA has instituted about 10 years ago two sets of guidelines termed "351" and "361" regulations. Generally the "361" guidelines are less onerous and sometimes apply to therapies carried out under the practice of medicine but often requiring an IRB approval. However, the "351" guidelines require an IND prior to clinical use (21 CFR Part 1271, human cells, tissues, and cellular and tissue-based products or HCT/Ps) in that cells and tissues (including stem cells) falling under these guidelines are considered "biological drugs". For a procedure to qualify under the "361" regulations one must use autologous tissues and cells, do no more than minimally manipulate the cells and tissues, and then use the cells and tissues in a homologous fashion. Under strict FDA interpretations, it is difficult to qualify under the less strenuous "361" guidelines.

The FDA's view on AT-MSC procedures has been clearly enunciated as can be found at (http://www.fda.gov/biologicsbloodvaccines/guidancecomplianceregulatoryinformation/guidances/cellularandgenetherapy/ucm427692.htm). However, there is surprisingly little guidance when it comes to CT-MSC. Autologous use of cells and tissues is fairly straightforward although sometimes the definition has been expanded to include first and second degree relatives (e.g., parents, siblings, aunts and uncles, and cousins). In terms of

AT-MSC the FDA defines "minimal manipulation" for structural tissue such as AT as "processing that does not alter the original relevant characteristics of the tissue relating to the tissue's utility for reconstruction, repair, or replacement". By definition then, enzymatically digested AT to generate SVF would be considered "more than minimally manipulated" and would require an IND. An IND is undesirable not just because of regulatory paperwork and inspections that accompany it, but also because non-academic establishments are not allowed to charge a profit for any cellular endeavor until it is proven safe and efficacious which may take up to a decade or more at an investment that is often in the hundreds of millions of dollars range. In addition, with regard to AT the federal guidelines have sometimes been interpreted to mean that MSC isolated from various anatomical locations throughout the body cannot be used at any other dissimilar location in the body, although there is no data to support that assumption [2]. In regard to CT these guidelines could be interpreted to mean that CT-MSC will always require an IND in that the umbilical cord and those MSC contained within only exist during gestation whose only purpose is to protect the fetal blood supply during pregnancy. Homologous use of cells and tissues seems to pose the greatest confusion for practitioners. A strict definition would imply use only in the same context as the cells and tissues were originally found. For example, AT-MSC used to reconstruct atrophied facial fat after chemotherapy or used to reconstruct structural tissue in a different part of the body or a limb would be homologous use. However, utilization of AT-MSC or CT-MSC for treatment of stroke, myocardial infarction, or Multiple Sclerosis most probably would be considered non-homologous use. Many stem cell clinics and physicians claim to be exempt from such guidelines under the practice of medicine. However, nothing could be further from reality, as the practice of medicine does not allow for the routine application of unproven or unsafe medical practices, particularly for profit. It should be noted that several such stem cell clinics have been closed and physicians have been arrested and their medical license suspended for continuing to perform such unauthorized and unproven practices, particularly when they are not qualified by training or experience (see for example, http://www.ipscell.com/2015/01/stem-humanexperiment/; http://www.nejm.org/doi/full/10.1056/NEJMp1504560 and http://www.ipscell.com/tag/celltex/). As an example, wouldn't you prefer a licensed professional to perform your liposuction and do you really want an everyday GP to be injecting your joints or spinal cord (see http://www.scripps.edu/friedlander/docs/Graefe's%20Arch%20Clin%20Exp%20Ophthalmol.pdf versus https://www.ipscell.com/2016/01/us-stem-cell-clinic-sued-for-injection-into-patients-eyes-landmark-case/ for example).

For some reason CT has really not been regulated or even discussed in this context at all despite a large and successful commercial endeavor on the part of numerous cord blood banks. Many consumers are misled by these evangelists, opportunists and charlatans to believe that their CT-MSC will survive banking and be immediately available for clinical use when needed in the future. Those banks freezing minced whole tissue will recover few cells upon thaw that require extensive expansion before use which will require weeks to months (see above and reference [11]). Those banks that digest the CT before banking in order to cryopreserve the isolated MSC may not be able to use the samples upon thaw due to improper methodology that runs afoul of regulations. In both cases the numbers of cells obtained is generally below clinically useful levels and will require expansion before use. Perhaps the FDA has concluded that since there will always be a requirement for in vitro expansion before clinical use they will wait to have regulatory oversight at that time. Or perhaps the cord blood banking industry is more influential than the AT-MSC industry (e.g., lobbying).

12.6 Conclusions

Stem cells are found in different locations throughout the body, with each anatomical site generally containing a mixture of stem cell types. However, the most frequently utilized sources due to ease of accessibility and reduced costs are those stem cells found in adipose tissue, bone marrow (similar to mobilized peripheral blood), umbilical cord blood, and umbilical cord tissue. Each of these stem cell sources has different requirements when it comes to collection, processing, cryopreservation and storage. Cord tissue and adipose tissue are unique in that both are (semi) solid tissues that require enzymatic digestion before the MSC contained within can be obtained, purified and stored or utilized. CT is a preferred MSC source in that it represents the youngest gestational source of MSC for regenerative medicine and tissue engineering, having demonstrated superior proliferative capacity to other MSC types. AT is a preferred source of MSC for many applications in that it generally can be obtained from all potential subjects (autologous) removing concerns of immune rejection, and AT is the richest source of MSC in the body with more than 500,000 MSC/g of tissue [2, 10, 11], eliminating the need to expand MSC before use, which can induce cellular senescence [12]. However, for both CT and AT the need to use enzymatic digestion to obtain the MSC prior to banking or utilization falls under the "more than minimally manipulated" category, even in an autologous setting, mandating FDA oversight and the need to obtain an IND approval before clinical use. This point cannot be emphasized or overstated enough in that although both CT and AT are sources of MSC with significant potential for use in regenerative medicine and tissue engineering, without adherence to the regulatory guidelines (including oversight and an understanding of proper methodology) that have been instated for such tissues and their clinical use, it is only a matter of time before a patient is injured or is killed. If that were to happen, the fields of regenerative medicine and tissue engineering could be set back years with patients in need not having access to lifesaving and life-altering medical advances.

References

1. Bunnell BA, Flaat M, Gagliardi C, Patel B, Ripoll C (2008) Adipose-derived stem cells: isolation, expansion and differentiation. Methods 45(2):115–120. doi:10.1016/j.ymeth.2008.03.006, Published online 2008 May 29. PMCID: PMC3668445, NIHMSID: NIHMS59647
2. Choudhery MS, Badowski MS, Muise A, Pierce J, T Harris DT (2015) Adipose drived stem cell utility is independent of anatomical harvest site. BioRes Open Access 4(1):131–145. Published online 2015 Feb 1. doi 10.1089/biores.2014.0059
3. Caplan AI (1991) Mesenchymal stem cells. J Orthop Res 9:641–650
4. Caplan A (2007) Adult mesenchymal stem cells for tissue engineering versus regenerative medicine. J Cell Physiol 213(2):341–347
5. Choudhery MS, Badowski M, Muise A, Harris DT (2013) Utility of cryopreserved umbilical cord tissue for regenerative medicine. Curr Stem Cell Res Ther 8(5):370–380
6. Seshareddy K, Troyer D, Weiss ML (2008) Method to isolate mesenchymal-like cells from Wharton's jelly of umbilical cord. Methods Cell Biol 86:101–119
7. Wang HS, Hung SC, Peng ST, Huang CC, Wei HM, Guo YJ, Fu YS, Lai MC, Chen CC (2004) Mesenchymal stem cells in the Wharton's jelly of the human umbilical cord. Stem Cells 22(7):1330–1337
8. Choudhery MS, Badowski M, Muise A, Pierce J, Harris DT (2014) Cryopreservation of whole adipose tissue for future use in regenerative medicine. J Surg Res 187:24–35
9. Harris DT, Muise A, White L, Badowski M, Rapid purification of adipose-derived MSC for use in cellular therapies by short-term panning. Abstract, ISCT 2016 Annual Meeting, Singapore, May 2016
10. Choudhery MS, Badowski M, Muise A, Harris DT (2013) Comparison of the regenerative potential of human adipose and cord tissue derived mesenchymal stem cells. Cytotherapy 15:330–343
11. Badowski M, Muise A, Harris DT (2014) Mixed effects of long term frozen storage on cord tissue stem cells. Cytotherapy 16(9):1313–1321
12. Choudhery MS, Badowski M, Muise A, Harris DT (2015) Effect of mild heat stress on the regenerative potential of human mesenchymal stem cells. Cytotherapy 17:359–368

Mature Oocyte Cryopreservation for Fertility Preservation

13

Tina Liang and Tarek Motan

Abstract

In recent decades, advances in cancer treatment have led to a dramatic improvement in long term survival. This has led to an increasing focus on quality of life after surviving cancer treatment, with fertility being an important aspect. Given the known reproductive risks of cancer therapies, there has been a growing interest in the field of fertility preservation (also referred to as oncofertility). Mature oocyte cryopreservation is no longer considered experimental and has become a realistic option for reproductive aged women prior to undergoing cancer treatment. Additionally, as cryopreservation techniques improve, mature oocyte cryopreservation is increasing being marketed to healthy women without cancer wishing to delay child bearing, also termed "social egg freezing". This chapter provides a review of the current technology, use, and outcomes of mature oocyte cryopreservation. It also outlines the ethical debate surrounding social egg freezing and directions for future research in female fertility preservation.

Keywords

Oocyte cryopreservation • Vitrification • Egg freezing • Fertility preservation

T. Liang (✉)
Department of Obstetrics and Gynecology, Faculty of Medicine and Dentistry, University of Alberta, Edmonton, AB, Canada
e-mail: liang3@ualberta.ca

T. Motan
Division of Reproductive Endocrinology & Infertility, Department of Obstetrics and Gynecology, Faculty of Medicine and Dentistry, University of Alberta, Edmonton, AB, Canada

Abbreviations

ART	Assisted reproductive technologies
IVF	In-vitro fertilization
DMSO	Dimethylsulfoxide
ICSI	Intracytoplasmic sperm injection
OHSS	Ovarian hyperstimulation syndrome
AMA	Advanced maternal age

F. Karimi-Busheri, M. Weinfeld (eds.), *Biobanking and Cryopreservation of Stem Cells*, Advances in Experimental Medicine and Biology 951, DOI 10.1007/978-3-319-45457-3_13

13.1 Introduction

The correlation of increasing age with decreasing female fertility is due to the natural decline in ovarian reserve throughout all stages of a woman's life. Women are born with a finite number of oocytes, decreasing from a peak of 6–7 million in a 20-week gestation fetus to 1–2 million at birth [1]. This number continues to decline from 300,000–500,000 at puberty to less than 1000 at age 51, the average age of menopause in the United States [1]. In addition to the decrease in number, there is a decrease in the quality of remaining oocytes. Epidemiological studies have consistently demonstrated a decline in female fertility as early as beginning of the thirties to a more dramatic decline after the mid-thirties [2]. Although outcomes have significantly improved in assisted reproductive technologies (ART), the factor of advanced age or decreased ovarian reserve still cannot be completely overcome with the exception of using donor eggs [2–4]. As a result, the advent of mature oocyte cryopreservation offers much hope for fertility preservation in women facing anticipated fertility decline. These include women undergoing gonadotoxic chemo or radiation therapies for cancer, women with a genetic predisposition to primary ovarian insufficiency, or healthy women who wish to delay child-bearing.

Among current existing options for fertility preservation, embryo cryopreservation is a well-established technology that is widely used in in-vitro fertilization (IVF) treatments [5]. However, it requires a male partner or the use of donor sperm in order to fertilize oocytes retrieved. Mature oocyte cryopreservation is a newer technology but no longer considered experimental. Its advantages include lack of need for male partner or donor sperm, lack of ethical, personal, religious or legal concerns for embryo storage and disposition, and greater control of disposition of gametes in the future [4–7].

Table 13.1 Current uses for mature oocyte cryopreservation

Fertility preservation prior to gonadotoxic cancer therapies
Fertility preservation prior to surgeries with risk of damage to ovary
Fertility preservation due to risk of primary ovarian insufficiency
Fertility preservation for elective reasons to delay child bearing
Storage and banking of donor eggs
Failure to obtain sperm on day of oocyte retrieval in IVF treatment for infertility
Excess oocytes obtained with legal or moral restrictions to embryo cryopreservation

13.2 Current Uses for Mature Oocyte Cryopreservation

Indications for mature oocyte cryopreservation are summarized in Table 13.1 [6, 8]. Mature oocyte cryopreservation can be used for women facing anticipated fertility decline for various reasons. Certain gonadotoxic cancer therapies involving chemotherapy or radiation destroy growing oocytes in the ovary, therefore mature oocytes may be retrieved and preserved prior to undergoing such treatments. Women may also wish to preserve their fertility prior to undergoing surgeries involving risk of damage to the ovaries or oophorectomies, for example in the case of genetic conditions like BRCA for cancer prophylaxis. Women with an increased risk of primary ovarian insufficiency such as those with a family history may also wish to preserve their fertility at an earlier age. Other medical indications include storage of excessive numbers of oocytes obtained during hyper stimulation of patients with polycystic ovarian syndrome. Recently women with severe progressive endometriosis have requested oocyte cryopreservation [9]. A growing number of women in developed countries where there is an increasing trend to delay child bearing also undergo elective mature oocyte cryopreservation in hopes of circumventing the age-related fertility decline (also termed as "social egg freezing"). In addition, there are many indications for mature oocyte cryopreservation other than anticipated fertility decline. Much of the research in the early days of oocyte cryopreservation was related to preservation of donor eggs. Today, storage and use of mature oocytes in donor egg banks is available worldwide. Certain countries (eg. Italy) have legal restrictions on embryo freezing and storage. Therefore, mature oocyte cryopreservation is used instead of embryo cryopreservation in IVF treatments to minimize the number of controlled ovarian stimulation cycles a woman may need to undergo. For couples undergoing IVF treatments for infertility, mature oocyte cryopreservation is also used for instances where sperm cannot be obtained on the day of oocyte retrieval or if couples have ethical or moral problems with embryo cryopreservation.

13.3 Oocyte Cryopreservation Technology

Advances in cryopreservation have allowed major breakthroughs in assisted reproduction. The first major turning point came in 1949 with the discovery of glycerol as a cryoprotectant for human sperm [10], which quickly led to the first reported human birth from frozen sperm in 1953 [11]. This was followed by extensive work on the physiology of cellular water movement and mechanisms of cell freezing [12]. Other cryoprotectants that minimized cellular damage were also identified, such as propanediol, ethylene glycol and dimethylsulfoxide (DMSO) [6]. In 1984, the first birth from a human frozen embryo was reported. Since then, embryo cryopreservation has been widely and successfully applied [6].

In 1986, Chen reported the first pregnancy from a frozen oocyte [13]. However, technical difficulties due to unique properties of the mature oocyte prevented consistent success initially. In the following decade, only a handful of pregnancies and no births from frozen oocytes were reported [12]. Challenges to oocyte freezing include its relatively large size, high water content, unique chromosomal arrangement and meiotic spindle [6]. These properties make the oocyte especially vulnerable to damage from ice crystal formation during the freezing and thawing process. Two main freezing techniques have been developed to overcome these challenges. In the older slow-freeze technique, cells are gradually dehydrated in the presence of cryoprotectants while programmable freezers lower the temperature at a very slow rate to minimize ice crystal formation. However, cells exposed to low temperatures for long periods were still prone to chilling injury [14]. Subsequently, an alternative to slow-freeze, vitrification has been developed. Vitrification uses high initial concentrations of cryoprotectants and ultra-rapid cooling to minimize crystallization. Various vitrification protocols using different devices have been developed. Most devices are known as "open systems" as they expose samples directly to liquid nitrogen to maximize the cooling and thawing rate [14]. The Cryotop method is one such widely used open system developed in Japan which has been pivotal to success in vitrification [15]. Most recent studies suggest superior post thaw oocyte survival and pregnancy rates of vitrification compared to the slow-freeze technique [6, 14].

In most cryopreservation protocols, oocytes are denuded by removing the outer layer of cumulus cells in order to assess for oocyte maturity, as only mature metaphase II oocytes are selected for cryopreservation. The removal of cumulus cells raises concern for reduced fertilization following standard insemination. In addition, hardening of the zona pellucida of the oocyte has been seen following rewarming of cryopreserved oocytes which may also affect fertilization. Therefore, intracytoplasmic sperm injection (ICSI) is generally performed for fertilization of cryopreserved oocytes to improve fertilization rates [6].

13.4 Clinical Outcomes

Many studies comparing clinical outcomes of fresh versus frozen oocytes involve donor oocytes. One of the largest published prospective randomized controlled trials included 600 patients randomized to receive vitrified or fresh donor oocytes, with over 3000 oocytes used in each arm [16]. Survival of vitrified oocytes was reported at 92.5 % and there were no differences in fertilization rate (74.2 % vs. 73.3 %), clinical (50.2 % vs. 49.8 %) or ongoing pregnancy rates per transfer cycle (44.4 % vs. 43.3 %) between vitrified and fresh oocytes. Similarly, 2 prospective randomized studies using non-donor sibling oocytes also showed comparable high survival rates and no statistical differences in fertilization rates between vitrified and fresh oocytes [17, 18]. In a 2014 meta-analysis including 21 prospective studies involving vitrified oocytes, the ongoing pregnancy rate per warmed oocyte or "oocyte efficiency" was found to be 7 %, similar to efficiency estimates using fresh oocytes [19].

The effect of storage time on outcomes is demonstrated by a recent large retrospective study with over 42,000 vitrified donor oocytes [20]. No differences in oocyte survival, clinical and ongoing pregnancy rates were found between 8 categories of storage time ranging from less than 6 months to over 5 years. Currently, the longest reported storage period of a cryopreserved oocyte resulting in a live birth is 14 years [21].

Numerous studies have also shown that outcome is dependent on the age at which oocytes are cryopreserved [22–24]. In an individual patient data meta-analysis including 2265 cycles from 1805 patients, live birth rates declined with increasing age at freezing regardless of the freezing technique used [22]. An age cut-off point of 36 years was found to provide the highest discrimination between success and failure. Similarly, a more recent retrospective study of 1468 women who underwent elective oocyte cryopreservation showed a live birth rate of 50 % for women 35 years or younger at vitrification, compared to 23 % for women 36 years or older at vitrification [24]. In addition, freezing 10–15 oocytes yielded a 60.5–85.2 % chance of live birth in women 35 years or younger, thus providing a "number needed to freeze" for a reasonable chance of success [24]. It is evident that to achieve a high probability of a live birth, multiple stimulation cycles may be required to attain the optimal number of oocytes.

Relatively few studies focus specifically on cancer patients undergoing fertility preservation [25, 26]. A recent study of 176 cancer patients who underwent oocyte cryopreservation found similar oocyte survival rate (86 %) and fertilization rate (72 %) to previous reports [25]. There were 4 live births from 10 patients who returned for oocyte thaw.

13.5 Risks

A theoretical concern with oocyte cryopreservation is possible damage to the meiotic spindle from the freeze/thaw process could lead to greater risk of aneuploidy [6]. Two well designed studies where all resulting embryos were genetically tested showed reassuring outcomes with no difference in rate of embryo aneuploidy comparing cryopreserved to fresh oocytes [27, 28]. Studies looking at rates of congenital anomalies have also been reassuring. Two larger studies including 200 and over 900 infants born from cryopreserved oocytes showed incidences of congenital anomalies to be 2.5 % and 1.3 %, respectively, which are no higher than incidences of congenital anomalies in infants born from IVF using fresh oocytes [29, 30]. Additionally, a large retrospective study including over 1000 infants born from vitrified oocytes compared to over 1200 infants born from fresh oocytes showed no difference in rates of various adverse obstetrical and perinatal outcomes [31]. Although short term safety data appear reassuring, there is still a lack of long term data such as developmental outcomes in children born from cryopreserved oocytes [6].

Risks associated with ovarian stimulation and oocyte retrieval, all of which have been previously described at less than 1 %, also apply to oocyte cryopreservation [7]. However, since the majority of women undergoing oocyte cryopreservation do not undergo embryo transfer immediately following oocyte retrieval, the risk of ovarian hyperstimulation syndrome (OHSS) is low [6]. An overall complication rate of 0.8 %

and severe OHSS of 0.25 % has been reported [32]. It is important to bear in mind the well described risks associated with pregnancies from ART compared to pregnancies conceived naturally. These include multiple pregnancies, as well as various adverse obstetrical and perinatal outcomes which are independently associated with ART and advanced maternal age (AMA) [2, 33].

13.6 Ethical Considerations

While oocyte cryopreservation for fertility preservation was initially introduced for women facing cancer treatment related fertility loss, the procedure is increasingly being marketed to women who wish to delay child bearing for various social reasons. The major reasons for postponing parenthood include desire to achieve higher education, career or financial goals, and not having met a suitable partner [7]. Proponents of elective oocyte cryopreservation argue that it offers women the reproductive freedom to choose parenthood at a more suitable stage of life. It allows more personal autonomy in the ability to retain the legal right to use one's own oocytes in case of a change in partner. However, the increased marketing of oocyte cryopreservation has also been criticized for encouraging women to delay motherhood while providing them with a false sense of security. Women must be aware that successful live birth cannot be guaranteed with the use of cryopreserved oocytes and that it is by no means an insurance policy, but rather a second option in case natural conception fails [7, 34–37].

13.7 Future Directions

Mature oocyte cryopreservation is no longer considered an experimental technique for fertility preservation. However, it does require a patient to be post pubertal in order to undergo controlled ovarian stimulation, which takes a time period ranging from 1 to 2 weeks [5]. The supraphysiologic estrogen levels achieved with ovarian stimulation may also be a concern for hormone sensitive malignancies such as breast cancer [5, 38]. For pre-pubertal adolescents or those unable to undergo ovarian stimulation, there are experimental options which are directions for future research.

Ovarian tissue cryopreservation is one option where ovarian cortical tissue is surgically obtained and cryopreserved either by slow freezing or vitrification [4, 39]. The ovarian tissue is then thawed and transplanted back to a pelvic (orthotopic) site or extrapelvic (heterotopic) site. To date, at least 60 live births from natural conception or IVF have been achieved with re-implantation of cryopreserved ovarian tissue, with one case involving tissue cryopreserved pre-pubertally as an adolescent [5]. However, an important disadvantage is risk of cancer recurrence or malignant transformation of the transplanted tissue [5, 39].

Alternatively, immature oocytes can be retrieved, not requiring ovarian stimulation, and may be used in combination with ovarian tissue cryopreservation [5, 38, 40, 41]. Previous studies have shown poor outcomes of vitrification of immature oocytes compared to mature oocytes, therefore in-vitro maturation (IVM) is carried out prior to vitrification [42]. To date, only a few live births have been reported from cryopreserved in-vitro matured oocytes [43].

13.8 Summary

Mature oocyte cryopreservation is an established method for female fertility preservation. It is currently being used as fertility preservation for women facing anticipated fertility loss due to various medical reasons or due to age related fertility decline. It is also widely used in donor egg banking and IVF treatments for infertility. Recent advances in freezing technology have resulted in outcomes comparable to fresh oocytes allowing women the opportunity of motherhood by saving a small number of oocytes for future utilisation. Although short-term safety data appears reassuring, long-term data and studies on children born from cryopreserved oocytes is ongoing. There is a lack of awareness of fertility preservation strategies which prevents oncology patients from benefiting from the available technologies. Proactive counseling by health professionals (oncologists, surgeons, nurses, etc.), public awareness campaigns presented in the media and education of patients, is required to improve utilisation of fertility preservation options. However healthy women choosing to freeze oocytes for elective reasons in order to delay child bearing must be cautioned that the procedure is not a guarantee for future live birth.

References

1. American College of Obstetricians and Gynecologists Committee on Gynecologic Practice and Practice Committee (2014) Female age-related fertility decline. Committee Opinion No. 589. Fertil Steril 101(3):633–634
2. Johnson J-A, Tough S, Society of Obstetricians and Gynaecologists of Canada (2012) Delayed childbearing. J Obstet Gynaecol Can 34(1):80–93
3. Centers for Disease Control and Prevention, American Society for Reproductive Medicine, Society for Assisted Reproductive Technology (2015) 2013 Assisted Reproductive Technology National Summary Report. US Dept of Health and Human Services, Atlanta
4. Stoop D, Cobo A, Silber S (2014) Fertility preservation for age-related fertility decline. Lancet 384(9950):1311–1319
5. Kim SY, Kim SK, Lee JR, Woodruff TK (2016) Toward precision medicine for preserving fertility in cancer patients: existing and emerging fertility preservation options for women. J Gynecol Oncol 27(2):e22
6. Practice Committees of American Society for Reproductive Medicine, Society for Assisted Reproductive Technology (2013) Mature oocyte cryopreservation: a guideline. Fertil Steril 99(1):37–43
7. Cil AY, Turkgeldi L, Seli E (2015) Oocyte cryopreservation as a preventive measure for age-related fertility loss. Semin Reprod Med 33(6):429–435
8. Schattman GL (2015) Clinical practice – cryopreservation of oocytes. N Engl J Med 373(18):1755–1760
9. Elizur SE, Chian RC, Holzer HE, Gidoni Y, Tulandi T, Tan SL (2009) Cryopreservation of oocytes in a young woman with severe and symptomatic endometriosis: a new indication for fertility preservation. Fertil Steril 91(1):293.e1–3
10. Polge C, Smith AU, Parkes AS (1949) Revival of spermatozoa after vitrification and dehydration at low temperatures. Nature 164(4172):666
11. Bunge RG, Keettel WC, Sherman JK (1954) Clinical use of frozen semen: report of four cases. Fertil Steril 5(6):520–529
12. Gook DA (2011) History of oocyte cryopreservation. Reprod Biomed Online 23(3):281–289
13. Chen C (1986) Pregnancy after human oocyte cryopreservation. Lancet 1(8486):884–886
14. Cobo A, Garcia-Velasco JA, Domingo J, Remohí J, Pellicer A (2013) Is vitrification of oocytes useful for fertility preservation for age-related fertility decline and in cancer patients? Fertil Steril 99(6):1485–1495
15. Kuwayama M, Vajta G, Kato O, Leibo SP (2005) Highly efficient vitrification method for cryopreservation of human oocytes. Reprod Biomed Online 11(3):300–308
16. Cobo A, Meseguer M, Remohí J, Pellicer A (2010) Use of cryo-banked oocytes in an ovum donation programme: a prospective, randomized, controlled, clinical trial. Hum Reprod 25(9):2239–2246
17. Rienzi L, Romano S, Albricci L, Maggiulli R, Capalbo A, Baroni E, Colamaria S, Sapienza F, Ubaldi F (2010) Embryo development of fresh "versus" vitrified metaphase II oocytes after ICSI: a prospective randomized sibling-oocyte study. Hum Reprod 25(1):66–73
18. Parmegiani L, Cognigni GE, Bernardi S, Cuomo S, Ciampaglia W, Infante FE, Tabarelli de Fatis C, Arnone A, Maccarini AM, Filicori M (2011) Efficiency of aseptic open vitrification and hermetical cryostorage of human oocytes. Reprod Biomed Online 23(4):505–512
19. Potdar N, Gelbaya TA, Nardo LG (2014) Oocyte vitrification in the 21st century and post-warming fertility outcomes: a systematic review and meta-analysis. Reprod Biomed Online 29(2):159–176
20. Cobo A, Garrido N, Pellicer A, Remohí J (2015) Six years' experience in ovum donation using vitrified oocytes: report of cumulative outcomes, impact of storage time, and development of a predictive model for oocyte survival rate. Fertil Steril 104(6):1426–1434.e1428
21. Urquiza MF, Carretero I, Cano Carabajal PR, Pasqualini RA, Felici MM, Pasqualini RS, Quintans CJ (2014) Successful live birth from oocytes after more than 14 years of cryopreservation. J Assist Reprod Genet 31(11):1553–1555
22. Cil AP, Bang H, Oktay K (2013) Age-specific probability of live birth with oocyte cryopreservation: an individual patient data meta-analysis. Fertil Steril 100(2):492–9.e493
23. Doyle JO, Richter KS, Lim J, Stillman RJ, Graham JR, Tucker MJ (2016) Successful elective and medically indicated oocyte vitrification and warming for autologous in vitro fertilization, with predicted birth probabilities for fertility preservation according to number of cryopreserved oocytes and age at retrieval. Fertil Steril 105(2):459–466.e2
24. Cobo A, Garcia-Velasco JA, Coello A, Domingo J, Pellicer A, Remohí J (2016) Oocytes vitrification as

an efficient option for elective fertility preservation. Fertil Steril 105(3):755–764.e8
25. Druckenmiller S, Goldman KN, Labella PA, Fino ME, Bazzocchi A, Noyes N (2016) Successful oocyte cryopreservation in reproductive-aged cancer survivors. Obstet Gynecol 127(3):474–480
26. Martínez M, Rabadan S, Domingo J, Cobo A, Pellicer A, Garcia-Velasco JA (2014) Obstetric outcome after oocyte vitrification and warming for fertility preservation in women with cancer. Reprod Biomed Online 29(6):722–728
27. Forman EJ, Li X, Ferry KM, Scott K, Treff NR, Scott RT (2012) Oocyte vitrification does not increase the risk of embryonic aneuploidy or diminish the implantation potential of blastocysts created after intracytoplasmic sperm injection: a novel, paired randomized controlled trial using DNA fingerprinting. Fertil Steril 98(3):644–649
28. Goldman KN, Kramer Y, Hodes-Wertz B, Noyes N, McCaffrey C, Grifo JA (2015) Long-term cryopreservation of human oocytes does not increase embryonic aneuploidy. Fertil Steril 103(3):662–668
29. Chian R-C, Huang JYJ, Tan SL, Lucena E, Saa A, Rojas A, Castellón LAR, Amador MIG, Sarmiento JEM (2008) Obstetric and perinatal outcome in 200 infants conceived from vitrified oocytes. Reprod Biomed Online 16(5):608–610
30. Noyes N, Porcu E, Borini A (2009) Over 900 oocyte cryopreservation babies born with no apparent increase in congenital anomalies. Reprod Biomed Online 18(6):769–776
31. Cobo A, Serra V, Garrido N, Olmo I, Pellicer A, Remohí J (2014) Obstetric and perinatal outcome of babies born from vitrified oocytes. Fertil Steril 102(4):1006–1015.e4
32. Mv W, Germeyer A, Nawroth F (2015) Fertility preservation for non-medical reasons: controversial, but increasingly common. Dtsch Arztebl Int 112(3):27–32
33. Okun N, Sierra S, Society of Obstetricians and Gynaecologists of Canada (2014) Pregnancy outcomes after assisted human reproduction. J Obstet Gynaecol Can 36(1):64–83
34. ESHRE Task Force on Ethics and Law, Dondorp W, de Wert G, Pennings G, Shenfield F, Devroey P, Tarlatzis B, Barri P, Diedrich K (2012) Oocyte cryopreservation for age-related fertility loss. Hum Reprod 27(5):1231–1237
35. Baylis F (2015) Left out in the cold: arguments against non-medical oocyte cryopreservation. J Obstet Gynaecol Can 37(1):64–67
36. Linkeviciute A, Peccatori FA, Sanchini V, Boniolo G (2015) Oocyte cryopreservation beyond cancer: tools for ethical reflection. J Assist Reprod Genet 32(8):1211–1220
37. Petropanagos A, Cattapan A, Baylis F, Leader A (2015) Social egg freezing: risk, benefits and other considerations. CMAJ 187(9):666–669
38. Grynberg M, Poulain M, le Parco S, Sifer C, Fanchin R, Frydman N (2016) Similar in vitro maturation rates of oocytes retrieved during the follicular or luteal phase offer flexible options for urgent fertility preservation in breast cancer patients. Hum Reprod 31(3):dev325–dev629
39. Practice Committee of American Society for Reproductive Medicine (2014) Ovarian tissue cryopreservation: a committee opinion. Fertil Steril 101(5):1237–1243
40. Hourvitz A, Yerushalmi GM, Maman E, Raanani H, Elizur S, Brengauz M, Orvieto R, Dor J, Meirow D (2015) Combination of ovarian tissue harvesting and immature oocyte collection for fertility preservation increases preservation yield. Reprod Biomed Online 31(4):497–505
41. Abir R, Ben-Aharon I, Garor R, Yaniv I, Ash S, Stemmer SM, Ben-Haroush A, Freud E, Kravarusic D, Sapir O, Fisch B (2016) Cryopreservation of in vitro matured oocytes in addition to ovarian tissue freezing for fertility preservation in paediatric female cancer patients before and after cancer therapy. Hum Reprod 31(4):750–762
42. Borini A, Bianchi V (2010) Cryopreservation of mature and immature oocytes. Clin Obstet Gynecol 53(4):763–774
43. Chen H, Lv J-Q, Ge H-S, Ge H-S, Wu X-M, Xi H-T, Chi H-H, Zhu C-F, Huang J-Y (2014) Live birth following vitrification of in vitro matured oocytes derived from sibling smaller follicles at follicle selection phase in the context of in vitro fertilization. Gynecol Endocrinol 30(9):624–626

Stem Cell Banking and Its Impact on Cardiac Regenerative Medicine

14

Silvana Bardelli and Marco Moccetti

Abstract

Cardiovascular diseases, including heart failure, are the most frequent cause of death annually, even higher than any other pathologies. Specifically, patients who suffer from myocardial infarction may encounter adverse remodeling processes of the heart that can ultimately lead to heart failure. Prognosis of patients affected by heart failure is very poor with 5-year mortality close to 50 %. Despite the impressive progress in the clinical treatment of heart failure in recent years, heart transplantation is still required to avoid death as the result of the inexorable decline in cardiac function. Unfortunately, the availability of donor human hearts for transplantation largely fails to cover the number of potential recipient requests. From this urgent unmet clinical need the interest in stem cell applications for heart regeneration made its start, and has rapidly grown in the last decades. Indeed, the discovery and application of stem and progenitor cells as therapeutic agents has raised substantial interest with the objective of reversing these processes, and ultimately inducing cardiac regeneration. In this scenario, the role of biobanking may play a remarkable role to provide cells at the right time according to the patient's clinical needs, mostly for autologous use in the acute setting of myocardial infarction, largely reducing the time needed for cell preparation and expansion before administration.

S. Bardelli (✉)
Swiss Institute for Regenerative Medicine,
Cardiocentro Ticino Foundation,
Via Tesserete 48, 6900 Lugano, Switzerland
e-mail: Silvana.Bardelli@cardiocentro.org

M. Moccetti
Cardiology Department, Cardiocentro Ticino
Foundation, Via Tesserete 48,
6900 Lugano, Switzerland

F. Karimi-Busheri, M. Weinfeld (eds.), *Biobanking and Cryopreservation of Stem Cells*,
Advances in Experimental Medicine and Biology 951, DOI 10.1007/978-3-319-45457-3_14

Keywords
Cardiovascular diseases • Cardiac progenitor cells • Biobanking • Cell storage • Acute setting • Regenerative medicine

Abbreviations

MI	Myocardial Infarction
AMI	Acute Myocardial Infarcion
BMMNCs	Bone Marrow Mononuclear Cells
BMMSCs	Bone Marrow Mesenchymal Stem Cells
CHF	Chronic Heart Failure
SPECT	Single-Photon Emission Computed Tomography
LV	Left Ventricular
MRI	Magnetic Resonance Imaging
PCI	Percutaneous Coronary Intervention

14.1 Heart Regeneration

The human heart is one of the organs which regenerate less in the body, or at least its regenerative potential is clearly lower than the intestine, liver, bone or skin [1]. As a consequence, heart failure and its growing incidence raises many questions from the regenerative medicine point of view, mostly considering the predicition that it will reach epidemic proportions as the population ages.

Unlike humans, many lower vertebrates can regenerate limbs and internal organs after injury. Amphibian heart regeneration is well known. Furthermore, more recently the zebrafish has demonstrated to be a useful experimental model based on its broad regenerative capacity and easiness of genetic manipulation. Of note, the zebrafish heart can fully regenerate following surgical amputation of up to 20 % of the ventricular mass [2, 3]. This remarkable regenerative capacity is missing in the postnatal mammalian hearts, however some degree of cardiomyocyte renewal has to be recognized [4–6]. Despite the fact that proliferative rates are clearly small and quite difficult to detect, they raise the question whether such innate processes could be increased and employed therapeutically. To achieve functional heart recovery, it is critical to address its real cause which is the loss of cardiac tissue. The human left ventricle has 2–4 billion cardiomyocytes, and a myocardial infarction can eliminate up to 25 % of them in a few hours causing cardiomyocyte deficieny in the heart. Disorders of cardiac overload like hypertension or valvular disease slowly eliminate cardiomyocytes over many years, and aging is associated with physiological loss of ~1 g of myocardium per year in the absence of any specific cardiac disease. Given these observations, the main objective of cardiac regenerative medicine is to replace damaged heart cells and, therefore, to restore the physiological structure and function of the organ [7–9]. These innovative cell-based therapeutic applications are of particular interest in the clinical setting of cardiovascular diseases which could significantly benefit from a regenerative approach.

Over the past decades, several stem cell types have been identified for their capacity to regenerate the human heart. Bone marrow, and its subpopulations, including mesenchymal stem cells (MSCs) and endothelial progenitor cells (EPCs), have been thoroughly evaluated [10–14]. In addition, other therapeutically appealing sources are adipose tissue or the umbilical cord blood which seem to harbor a reservoir of suitable stem cells for cardiac regeneration. Furthermore, the more recent identification of endogenous progenitor cells in the human heart pushed the development of next-generation regenerative approaches employing heart-derived stem cells [15–17], as well as stem cells guided toward cardiac lineage *ex vivo* [18–20].

We are on the verge of a remarkable development in the field of cardiac regenerative medicine. Having such valuable biological sources available rapidly moves the question on the opportunity of storing them according to the patient's clinical needs. Hence the opportunity of storing such reservoirs for specific therapeutic use is progressively under evaluation. For successful clinical application in the setting of regenerative medicine or other therapies, the concept of an off-the-shelf product is desirable. One important feature of such cellular product is its instant availability for the intended applications. Thus, biobanking technology applied to cardiac regenerative medicine will further facilitate clinical application since it allows sufficient time for product qualification, and mostly it can favor the optimal timing for transplantation, when the patient is ready to receive the cells.

In this chapter we will review to the best of our knowledge the most updated applications of cryopreserved stem cells sources for human cardiac regeneration.

14.2 Available Sources for Cardiac Regenerative Medicine

In recent years, stem cell biology has become one of the fastest moving areas of biomedical research. Particularly, the heart is one of the human organs on which this type of research is more actively investigating for regenerative purposes. In general and quite simplifying, we can divide the field of cardiac regenerative medicine into two main areas concerning cell type: endogenous and exogenous cells.

Resident stem or progenitor cell populations were identified in postnatal hearts by means of a variety of approaches, including expression of surface markers like c-Kit, Sca-1, Isl-1 or physiological properties like the ability to efflux fluorescent dyes or form multicellular spheres in culture. Initially, little overlap among these progenitor classes was thought to exist. However, recent studies indicate shared markers among cell populations or different stages of maturation of the same cell type. C-kit-positive cells or Cardiac Stem Cells are extensively studied. The tyrosine kinase receptor c-Kit is expressed by hematopoietic stem cells, mast cells and other mature circulating cells, as well as in the thymic epithelium. Small c-Kit-positive cells have been identified in niches throughout the adult heart and, following isolation, they have been reported to give rise to cardiomyocytes, smooth muscle cells and endothelial cells when administered, eventually contributing to the regeneration of the myocardium [21–25]. These observations led to the SCIPIO (Stem Cell Infusion in Patients with Ischemic cardiOmyopathy, ClinicalTrials.gov Indentifier: NCT00474461) Phase I randomised, open-label, single centre trial which tested the safety and feasibility of autologous c-Kit-positive cell administration as an adjunctive treatment to patients undergoing coronary bypass surgery [26, 27].

Cardiosphere-derived cells (CDCs) are the other cardiac progenitor cell population already employed in clinical trials for the treatment of recent myocardial infarctions (CADUCEUS trial, ClinicalTrials.gov Indentifier: NCT00893360). CDCs outgrow from cultured cardiac tissue fragments and form the so-called cardiospheres. These spheres obviously represent a mixture of cells which express stem cell markers like c-Kit, and other markers typical of the stromal-vascular compartment [28, 29]. Patients treated with CDCs showed significant reductions in scar mass, and increase in viable heart mass [30, 31].

Considering stem cell sources outside the heart, considerable interest was focused in the earlier studies on bone marrow-derived cells for cardiac regeneration. This interest was derived by the observation that hematopoietic stem cells could transdifferentiate into cardiomyocytes [10]. However, subsequent reports showed that hematopoietic stem cells do not form cardiomyocytes but become mature blood cells after transplantation [12, 13]. Nevertheless, the administration of hematopoietic stem cells after infarction showed improvement in ventricular function, and paracrine signaling was identified as the principal mechanism of their action. In analogy, bone marrow-derived mesenchymal stromal cells (MSCs) were originally reported to differentiate into cardiomyocytes and, similarly, are now thought to exert their actions through paracrine effect [14]. Clinical trials indicate the safety and feasibility of bone marrow-derived cells through coronary injection, however the observed benefits are quite modest [32–39]. Bone marrow-derived MSCs are also employed in clinical trials (e.g. Prospective Randomized Study of Mesenchymal Stem Cell Therapy in Patients Undergoing Cardiac Surgery, PROMETHEUS, ClinicalTrials.gov Indentifier: NCT00587990) as well as adipose tissue-derived stromal cells (ClinicalTrials.gov Indentifier: NCT02387723). On the overall, the most updated evidence indicates that these cells produce signals that control the response of cells resident in the myocardium, and therefore regulate cardiac healing [40, 41].

Another very interesting approach involves the "cardiopoietic" guidance of multipotent adult stem cells [19, 20, 42]. This is achieved through a specific cardiogenic cocktail which guides human Mesenchymal Stem Cells towards the cardiac lineage through manipulation of their culture environment as was assessed in the C-CURE trial (ClinicalTrials.gov Indentifier: NCT00810238).

For a comprehensive overview of the different adult stem cell types employed in clinical trials for cardiac regeneration, please refer to Table 14.1.

The inherent characteristics of pluripotent stem cells offer a potential solution to the current epidemics of heart failure by providing human cardiomyocytes to support heart regeneration. These properties apply to embryonic stem cells (ESCs) and the more recently developed induced pluripotent stem cells (iPSCs). Human embryonic stem cells (hESCs) derived from the inner cell mass of the pre-implantation embryo appear to be a promising cell source for regenerative medicine [43, 44]. They are capable of extensive proliferation in and self-renewal and, as pluripotent cells, may be induced to generate more differentiated cell types such as functional cardiomyocyte. ESCs have been differentiated in three-dimensional aggregates, the so-called embryoid bodies (EBs), which typically delivers <1 % cardiomyocytes. Further improvement of the methodology are available through Wnt/β-- catenin signaling pathway. Human ESC-derived cardiomyocytes express early cardiac transcription factors like Nkx2.5 and GATA4, and possess functional properties similar to cardiomyocytes in the developing heart [45]. Obviously, also these approaches have their drawbacks: ESCs are derived from the inner cell mass of pre-implantation blastocysts, and this raises the ethical controversy of their use in addition to the potential formation of teratomas [46]. Moreover, ESC-based therapies will be allogeneic and therefore require immunosuppression.

IPSCs were originally generated by the reprogramming of adult somatic cells (mainly dermal fibroblasts) via the expression of up to four embryonic stem cell-related transcription factors (Oct-4, Sox2, Klf4 and c-Myc) [47]. These "first-generation" iPSCs were questionable because viruses were used to introduce reprogramming factors, therefore raising concerns about neoplastic transformation. The use of episomal gene delivery, excisable transgenes, or cell-permeant recombinant proteins have been introduced quite recently to overcome this issue. In the most successful experimental setting, cardiac differentiation was obtained by reprogramming mouse embryonic fibroblasts into cardiomyocytes [48, 49] with the so-called Yamanaka factors (Oct-4, Sox2, Klf4 and c-Myc) to initiate reprogramming, subsequently blocking signaling through the JAK-STAT pathway, which is required for pluripotency in the mouse, and by finally adding the cardiogenic factor BMP-4. This method activated the cardiac lineage program and, within 2 weeks, substantial numbers of beating colonies were observed. Despite some challenges, this is an exciting new avenue of research and could be used in autologous cell therapies.

The emergence of personalized medicine has raised especially in complex diseases such as cancer, cardiovascular and neurodegenerative diseases in which individual responses often complicate the development of therapeutic treatement. In the growing field of research on extracellular vesicles (EVs), exosomes (EXOs), a specific type of EVs, have been proposed as an interesting tool. Both healthy and unhealthy cells secrete vesicles into the extracellular space [50, 51]. EVs entrap lipids, proteins and nucleic acids which can mediate different biological functions. The extracellular vesicles (EVs) are mainly clas-

Table 14.1 First generation stem cell-based studies in ischemic heart disease

Name of the study	Acronym, PI	Year	Clinical setting	Stem cell type	Cell harvesting and manipulation	Cell delivery: Number of cells and route of administration	Effect on Ejection Fraction (EF)
Transplantation of progenitor cells and regeneration enhancement in acute myocardial infarction	TOPCARE-AMI (Assmus, B. et al.)	2002	AMI, Time of cell administration: 3 days after MI	BMMNCs	Density gradient and cell culture	7.3×10^6, intracoronary injection (3 injections of 10 ml each in X-VIVO™ 10 medium)	No change by LV angiography
Bone-marrow-derived cell transfer after ST-elevation myocardial infarction	BOOST (Woller, K. C. et al.)	2004	AMI, Time of cell administration: 5 days after MI	BMMNCs	Density gradient (Ficoll) centrifugation	2.4×10^9, intracoronary injection (4 or five injections in heparinized saline)	Positive by MRI
Reinfusion of enriched progenitor cells and infarct remodeling in acute myocardial infarction	REPAIR-AMI (Schachinger, V. et al.)	2006	AMI, Time of cell administration: 3–6 days after MI	BMMNCs	Density gradient (Ficoll) centrifugation	1.98×10^8, intracoronary injection in X-VIVO™ 10 medium and 20 % autologous serum	Positive by LV angiography
Autologous stem cell transplantation in acute myocardial infarction	ASTAMI (Lunde, K. et al.)	2006	AMI, Time of cell admin–stration: 4–7 days after MI	BMMNCs	Density gradient (Ficoll) centrifugation	6.8×10^7, intracoronary injection in heparin-treated plasma	No change by CT (SPECT)
Effects of intracoronary injection of mononuclear bone marrow cells on left ventricular function, arrhythmia risk profile, and restenosis after thrombolytic therapy of acute myocardial infarction	FINCELL (Lunde, K. et al.)	2008	AMI, Time of cell administration: morning of the day of PCI	BMMNCs	Density gradient (Ficoll) centrifugation	3.6×10^9, intracoronary injection in unspecified medium and 50 % autologous serum	Positive by echocardiography

Table 14.1 (continued)

Name of the study	Acronym, PI	Year	Clinical setting	Stem cell type	Cell harvesting and manipulation	Cell delivery: Number of cells and route of administration	Effect on Ejection Fraction (EF)
Intracoronary infusion of mononuclear cells from bone marrow or peripheral blood compared with standard therapy in patients after acute myocardial infarction treated by primary percutaneous coronary intervention	HEBE (Hirsch, A. et al.)	2011	AMI, Time of cell administration: <8 days after MI	BMMNCs	Density gradient (Lymphoprep™) centrifugation	2.96×10^8, intracoronary injection in sodium heparin and 4 % human serum albumin	No change by MRI
Effect of the use and timing of bone marrow mononuclear cell delivery on left ventricular function after acute myocardial infarction	TIME (Traverse, J. H. et al.)	2012	AMI, Time of cell administration: 3 or 7 days after MI	BMMNCs	Sepax® cell separation system	1.50×10^8, intracoronary injection in normal saline and 5 % Human Serum Albumin	No change by MRI
Effect of intracoronary delivery of autologous bone marrow mononuclear cells 2–3 weeks following acute myocardial infarction on left ventricular function	Late-TIME (Traverse, J. H. et al.)	2011	AMI, Time of cell administration: 2–3 weeks after MI	BMMNCs	Sepax® cell separation system	1.50×10^8, intracoronary injection in normal saline and 5 % Human Serum Albumin	No change by MRI
SWiss multicenter intracoronary stem cells study in acute myocardial infarction	SWISS-AMI (Sürder, D. et al.)	2010	AMI, 5–7 days or 3–4 weeks after MI	BMMNCs	Density gradient (Ficoll) centrifugation	$\geq 5 \times 10^7$, intracoronary injection in 10 ml of X-VIVO 10 with 20 % of autologous serum	No change by Cardiac Magnetic resonance (CMR)

Cardiac stem cells in patients with ischaemic cardiomyopathy	SCIPIO (Bolli, R. et al.)	2011	CHF, infUsed 4 months after CABG surgery	c-kit-positive resident cardiac stem cells	Cell culture and magnetic sorting (MACS)	0.5×10^6–1×10^6, intracoronary injection in PlasmaLyte A	Positive by Echocardiography
Intracoronary cardiosphere-derived cells for heart regeneration after myocardial infarction	CADUCEUS (Makkar, R. R. et al.)	2012		cardiosphere-derived cells	Cell outgrowth from cardiac biopsy, and cardiosphere formation	12.5×10^6–25×10^6, intracoronary injection in a saline solution containing heparin (100 U/mL) and nitroglycerin (50 μg/mL)	Positive by MRI
The prospective randomized study of mesenchymal stem cell therapy in patients undergoing cardiac surgery	PROMETHEUS (Hare, J. M. et al.)	2014	Chronic ischemic left ventricular dysfunction secondary to MI	BMMSCs	Bone marrow aspiration from the iliac crest, isolation and expansion of MSCs	2×10^7–2×10^8 MSCs intramyocardial injection, transepicardial delivery, in PBS buffer or Plasma-Lyte A supplemented with 1 % human serum albumin	Positive by MRI

sified as exosomes (EXOs), microvesicles (MVs), retrovirus-like particles, and apoptotic bodies (APOs) according to their origin. In particular, EXOs are currently defined as cup-shaped nanovesicles about 30–100 nm wide that originate within the endosomal network and can be found in body fluids, including urine, saliva, blood, breast milk, and cerebrospinal fluid. They are of particular interest to the study of complex diseases for their contribution to long-range intercellular communication. Moreover, the experimental observation that the production of EXOs is increased in association with disease, and the fact that their content varies with disease state, led to the evaluation of the potential of EXOs as biomarkers and vehicles to regulate the spread of disease [52, 53]. Current experimental evidence supports a central role of EXOs in cancer development, progression, metastasis, and drug resistance through promotion of carcinogenesis and tumor growth, angiogenesis, modulation of the tumor microenvironment, modulation of immune responses, and induction of mechanisms to acquire therapy resistance. Despite some controversial results found in literature, efforts are being made to resolve the lack of consensus on the methods of isolation and rigorous criteria to characterize them highlighting their potential importance in clinical applications, especially in the cancer field. As our understanding of the biology of EXOs increases, the opportunity of employing EXOs as nanocarriers for immunotherapeutics or vaccines, as angiogenesis modulators and for many other applications is intensely being evaluated.

A critical issue which emerged from the "first generation" of cell therapy-based clinical trials for cardiovascular regenerative medicine is the overall low retention rate after cell delivery to the myocardium. In order to exploit their full therapeutic potential, cells need to be retained at the site of injury, and their long-term survival is a fundamental characteristic to enable them to fully achieve regeneration. To overcome this problem, the concept of 3D cell culture has been proposed to enhance cell engraftment based on the mimicking of the *in-vivo* tissue microenvironment. The growth of 3D cell culture is essential to consider cells in their comprehensive context because in nature, i.e. within the body, a cell is always surrounded by its niche or microenvironment that has profound infuence on its properties. Most commonly, this involves the use of porous, biodegradable scaffolds onto which cells are seeded, but other approaches include the integration of cells into hydrogels or creating scaffold-free tissues composed only of cells and the matrix they secrete. Of great interest is the use of so-called microtissues or spheroids [54, 55]. Microtissues are composed of 500 up to 10,000 cells and were initially used as a model to elucidate tumor biology; they have recently evolved from their known role as *in vitro* models to a novel role as therapeutic agents. In the setting of cardiovascular applications, microtissues may provide an optimal strategy to implant stem cells into the injured heart since they can be transplanted via transcatheter delivery because of their scaffold-free nature. The cardiomyocytes used in tissue engineering are commonly immature cells or stem cells [56]. In order to sustain an adult workload, these cells will need to organize into the cable-like structure of myocardium and increase in size more than ten-fold. There is an ongoing debate whether this maturation should take place before or after transplantation. Certainly, when myocyte-only constructs are transplanted the tissue survives poorly. When vascular endothelial cells and stromal cells are also included, the mixed population creates a synergic effect: the endothelial cells form networks resembling a primitive vascular system, and the stromal cells create a matrix to provide mechanical integrity. Interestingly, the body mostly reacts to the surface topography of the implanted material (whether it is smooth or contains pores and grooves) rather than to its chemical composition. More complex topographies provide less inflammation, less scarring, and increased angiogenesis. Although this rapidly emerging field of tissue engineering has been extensively studied so far mainly in preclinical models, such as constructs of engineered heart tissue from neonatal cardiomyocytes sutured to the surface of infarcted rat hearts, initial studies are promising [57].

14.3 Biobanking: A Clinical Need for Cardiac Regenerative Applications

As previously mentioned, the importance of stem cell biobanking for cardiovascular diseases lays on the opportunity of delivering an "off-the-shelf" product readily available according to the patient's needs. This applies in particular in the acute clinical setting, when the rapid administration of an autologous stem cell therapy product is critical to beneficially affect the clinical outcome.

The intrinsic nature of biobanking activities related to standardization of protocols, evaluation of isolation yields, factors that affect characterization and storage of samples, argues in favor of biobanks playing a central role in the advancement of stem cell-based research and mostly in the development of stem cell-based clinical applications. Biobanks, intended as institutions applying standardized protocols for the collection, processing, storage and release for administration of human stem cells are the ideal infrastructure acting as a service platform to overcome current research hurdles and mainly to expedite the progression of stem cell-based applications in the treatment of cardiovascular diseases for the upcoming era of precision and personalized medicine [58–60].

In general, ice formation during cryopreservation may be a crucial and critical event because it can produce severe cell damage, both during freezing and thawing processes. Vitrification is the transformation of liquids into a solid state. This can be achieved by rapid cooling in combination with a cryoprotective agent with a high osmolarity which prevents cryoinjury due to ice formation. Commonly, the slow-cooling approach is accompanied by a low concentration of the cryoprotective agent to reduce cytotoxicity. Of note, every cell type has its specific optimal cooling rate. Interestingly, gap junctions propagate intercellular ice formation at a given temperature, and may lead to increased intracellular damage and finally to cryoinjury. In addition to the general problems related to the cryopreservation of single cells, such as cryoinjury caused by ice formation, there are additional considerations for the cryopreservation of 3D tissues [54]. Importantly, the cryoprotective agent should be distributed homogeneously within the entire tissue to protect the cells from intracellular ice formation, which can lead to severe cell damage and cell death. For some cell types, such as embryonic stem cells, freezing in aggregates results in higher cell number recovery after thawing compared to single-cell suspensions. Beneficial effects are attributed to the cryoprotective role of the surrounding extracellular matrix.

Historically, cardiac valve cryopreservation has long been used for the treatment of valvular heart disease. Valvular heart disease is still a significant cause of morbidity and mortality worldwide. In the United States, approximately 60,000 valve replacement operations are performed annually, and valve replacement surgery is efficacious [1, 61]. State-of-the-art valves used in the clinical setting include mechanical valves and biological valves such as glutaraldehyde fixed xenografts (derived from animals) or cryopreserved homografts (derived from human donors). Glutaraldehyde-fixed or cryopreserved biological valves do not require anticoagulation treatment, however, they represent non-viable prostheses. They basically lack the ability to grow, to repair, or to remodel. This raises severe problems specifically related to pediatric patients. Approximately 1 % of all newborns have congenital heart defects, and many of them require surgical heart valve replacement. Since currently available valves cannot grow with the young patients, repeated surgical replacement operations have to be performed with exponentially increased morbidity and mortality. Given these crucial observations, the importance of providing patient-specific cryopreserved cardiomyocytes is imperative. For this important reason, biobanking patient-specific stem cells is of utmost importance [62, 63].

Here we report some specific recent examples in which cardiac stem cell banking has been particularly successful in terms of its therapeutic application.

As an important translational step towards creating an effective clinical therapy, Chong et al.

[64] investigated the ability of exogenously delivered human Embryonic Stem Cell-derived Cardiomyocytes (hESC-CMs) to engraft and electrically couple to infarcted host myocardium in a non-human primate (NHP) model of myocardial infarction (MI). This model provides a heart size and rate more comparable to the human and required delivery of about 1 × 10^9 cells. Feasibility of this large-scale hESC-CM delivery required cryopreservation of cells, a method which was previously validated by the authors in an established immunodeficient mouse model of MI. No adverse effects of cryopreservation on hESC-CM were reported Therefore, delivery of cryopreserved hESC-CMs appears to be a sound strategy for large- scale transplantation in large animals or in humans. In particular, seven pigtail macaques (Macaca nemestrina) were used for this study without randomization. MI was created by percutaneous ischemia-reperfusion 2 weeks prior cell administration, and immunosuppression started 5 days prior to hESC-CM delivery. In order to obtain differentiated cultures containing beating cardiomyocytes, Activin A and Bone Morphogenetic Protein 4 (BMP4) were applied to defined, serum-free, monolayer culture conditions. hESC-CMs were collected and cryopreserved after 16–20 days of CM differentiation. One day before collection for delivery, cells were subjected to a pro-survival "cocktail" (PSC) protocol, to enhance engraftment after transplantation. Briefly, cultures were heat-shocked and exposed to RPMI-B27 medium supplemented with IGF1 (100 ng ml^{-1}) and cyclosporine A (0.2 mM). One day later, cultures were collected with Trypsin-EDTA, and neutralized with a Trypsin inhibitor. Cells were washed and resuspended by slowly adding cryopreservation solution CryoStorTM CS-10 (BioLife Solutions Inc.). The final cell concentrations were approximately 5–10 × 10^6 cells in 0.25 ml/vial or 4–8 × 10^7 cells in 1.5 ml/vial. Cells were frozen using a controlled rate freezer at −1 °C/min before the temperature reached −40 °C, and −5 °C/min from −40 °C to −80 °C. Vials were then transferred to a liquid nitrogen tank after reaching −80 °C. To thaw the cells, the vials were incubated in a water bath at 37 °C until no ice crystals were visible. The cell suspension was slowly diluted with RPMI/B27. The cells were subjected to flow cytometric analysis, cultured on Matrigel-coated plates in RPMI/B27 or prepared for transplantation. Immediately before transplantation, cells were suspended in a 1.5 ml volume (per animal) of modified PSC consisting of 50 % (v/v) growth factor–reduced Matrigel, supplemented with Bcl-XL BH4 (50 nM), cyclosporine A (200 nM), IGF1 (100 ng ml^{-1}) and pinacidil (50 mM). hESC-CMs were delivered into the infarct region and surrounding border zones under direct surgical visualization using a method optimized to improve cell retention. No macroscopic or microscopic evidence of teratoma or other tumor was detected. All hESC-CM- treated monkeys showed significant remuscularization of the infarct areas. Graft size, calculated on the basis of green fluorescence protein (GFP) expression, ranged from 0.7 to 5.3 % of the LV, averaging 40 % of the infarct volume. Greater than 98 % of engrafted human cells expressed the sarcomeric protein α-actinin, indicating that almost all graft cells were cardiomyocytes. Importantly, these hESC-CM displayed increased maturation from 14 d to 84 d, evidenced by increased myofibril alignment, presence of sarcomeres and increase in cardiomyocyte diameter. This remarkable study demonstrates that hESCs can be grown, differentiated into cardiomyocytes and cryopreserved at a scale sufficient to treat a large animal model of MI. With further refinements in manufacturing, scale up to trials in human patients appears feasible. Despite previous studies of the same group in smaller animal models such as mice, rats, and guinea pigs gave no evidence of arrhythmias after hESC-CM engraftment [45], arrhythmias were consistently observed in the non-human primate study. The two most likely reasons appear to be differences in heart size and rate. Regarding size, the larger hearts of adult macaques (37–52 g) compared to hearts of mice (0.15 g), rats (1 g) and guinea pigs (3 g) allows for more hESC-CMs to be delivered, and resultant grafts are approximately ten-fold larger than the largest obtained in other species. Another important factor is the species-specific heart rates (macaques: 100–130 beats/min, vs. guinea-pigs: 230, rats: ~400, mice

~600). Faster spontaneous rates will favor ventricular capture from native conduction pathways. These factors are relevant to clinical translation given that the human heart is larger (300 g) with a slower basal rate (70 beats/min) than macaques used in this study. Since ventricular arrhythmias can be life-threatening, they need to be understood mechanistically and managed to guarantee safe clinical translation. Nevertheless, the extent of remuscularization and electromechanical coupling documented in this study with cryopreserved ESC-derived cardiomyocyes encourage further development of human cardiomyocyte transplantation as a clinical therapy for heart failure. Of note, this methodology does not require feeder cells or conditioned medium for the maintenance of undifferentiated hESCs, and cardiomyocyte differentiation is achieved by treatment of adherent cultures with only two growth factors, activin A and BMP-4, in a defined serum-free medium. The differentiated cells express appropriate cardiomyocyte markers and display electrophysiological phenotypes expected for cardiomyocytes. Furthermore, to our knowledge, this is the first time that efficient cryopreservation of hESC-derived cardiomyocytes is performed. The cryopreserved cardiomyocytes were shown to survive both *in vitro* and *in vivo* and have the ability to form viable grafts. The cryopreserved cells are capable of surviving even in the central regions of the infarct region for as long as 4 weeks after the transplantation. Together, this differentiation and cryopreservation technology will offer significant advantages to support a robust process for the production of cardiomyocytes on a scale suitable for cellular therapy. Mainly, highly specific and scalable cardiomyocyte differentiation of hESCs maintained in serum-free medium has been demonstrated, the resulting differentiated cells possess the characteristics of human cardiomyoyctes and have shown successful engraftment of the cryopreserved cardiomyocytes after transplantation. Further optimization of the process, such as the use of completely xeno-free reagents, will be key for safer clinical applications [65–69].

Umbilical Cord Blood (UCB) stored in biobanks has now been quite frequently used to treat hematopoietic malignancies, marrow failure, metabolic diseases, and immunodeficiency disorders. The plasticity of UCB cells and its readily availability as stored material have also encouraged its broader regenerative applications, such as the treatment of spinal cord injury and chronic wounds, including clinical trials on other non-hematopoietic diseases, such as epidermolysis bullosa, and neonatal hypoxic ischemic encephalopathy [70]. More recently, the potentiality of UCB has been explored in animal models of cardiac infarction, diabetes, and various neurological diseases. Specifically, in animal models of MI, direct injection of CD34+, CD133+, or mononuclear cells (MNCs) from UCB into the necrotic myocardium or into the border zone after coronary artery ligation, resulted in reduced infarction size and improved left ventricular (LV) function. In addition to the CD34+ or CD133+ populations and MNCs of UCB, UCB-derived Unrestricted Somatic Stem Cells (USSCs) have also been applied to treat MI in animal models. USSC is the first identified UCB stem cell population with intrinsic pluripotent differentiation potential. These $CD45^{neg}/CD34^{neg}$ cells were isolated from UCB based on their outgrowth in the presence of dexamethasone. USSCs could be differentiated *in vitro* into bone, cartilage, adipocytes, hematopoietic cells, and neural cells, and *in vivo* they might form myocardial cells. In a chronic porcine model of myocardial infarction (MI), direct injection of USSCs to the border of infarction 4 weeks post-MI resulted in reduction of infarct size, reduction in end-diastolic volume, and improvement in LV ejection fraction. The engrafted USSCs could be detected 4 weeks after transplantation. In contrast, in an acute porcine model of MI, injection of USSCs led to a surprising complete prevention of scar formation and functional improvement. However, long-term survival of the transplanted cells and cell-related regeneration of the myocardium were not observed. In contrast, in a similar porcine model of MI, Moelker and colleagues injected USSCs into the infarct-related coronary artery 1 week post-MI. Magnetic resonance imaging failed to document any improvement in regional or global LV function. The transplanted cells survived only in the infarct border zone at 5 weeks and did not express any cardiomyocyte or endothelial markers. Additionally, intracoronary administration of USSCs was accompanied by increases in inflammatory cell infiltration and calcification in the infarct zone, as compared to control animals [71–73]. Although encouraging results have been obtained in many preclinical studies of UCB cell therapy, challenges remain to be overcome. Of note, UCB stem cells offer many practical advantages, such as relative ease of collection, minimal risk to the donors, and mostly the opportunity to bank carefully screened samples for administration. Furthermore, there has been accumulating evidence for the existence in UCB of very primitive stem cells that share ESC properties but do not present a risk for teratoma formation upon transplantation. Therefore, these primitive UCB-derived stem cells could be a suitable alternative to ESCs for cardiac tissue regeneration.

A small pilot study was performed to test the safety and feasibility of freeze-controlled cryopreserved bone marrow-derived mesenchymal stromal cells (BM-MSC) to treat severe dilated ischemic cardiomyopathy [74]. MSCs were isolated from BM aspiration, expanded, harvested, and cryopreserved using 90 % autologous plasma and 10 % dimethyl sulfoxide (DMSO). The cells were frozen gradually using a rate-controlled freezer to –90 °C, and then transferred into vapor phase liquid nitrogen for storage. Three consecutive patients with end-stage ischemic heart failure were enrolled in the study. Patients underwent concurrent coronary artery bypass graft (CABG) with intramyocardial MSC injection. On the day of injection, MSCs were thawed in a water bath, washed once and then resuspended in sterile saline solution. MSC suspension was transferred into a 10 mL luer-lock syringe for injection. The whole duration of cryopreservation to the day of surgery was 5–16 days. The viability of the cells in the final suspension was higher than 90 % in all three cases. There was significant improvement in cardiac function, scar reduction and increased wall thickness for all patients on cardiac magnetic resonance imaging at 6 months compared with baseline. No arrhythmias were reported. The preliminary data of this small pilot study seem to suggest that intramyocardial injection of cryopreserved BM-MSCs is feasible and safe for ischemic cardiomyopathy. Cryopreserving autologous MSC in a rate-controlled manner does not lead to significant cell death and is an important practical opportunity to circumvent the timing of preparation of stem cells for injection.

On the overall, biobanking can be defined as a long-term storage and preservation activity that includes and standardizes collection, processing, and release of high-quality biological specimens needed for future scientific investigation or for therapeutic administration. Each sample deposited in a biobank possesses two main features: the biological material (collected, processed, and stored), and the database that contains information regarding demographic and clinical data for every sample in the inventory. Peripheral blood, plasma, serum, blood-derived cell types, solid tissues, urine, saliva, RNA, and DNA are among the most common biological samples being processed and stored in biobanks. The increasing relevance of extracellular vescicles (EVs), and in particular of EXOs in cancer, metabolic diseases, and other complex diseases strongly suggests the incorporation of EXOs in biobanks to provide a significant advance in the knowledge, diagnoses and treatment of these particular diseases in clinical research [58]. Most clinical studies on EXOs have a diagnostic purpose. The number and wide variety of clinical trials in this area indicate the high impact of EXOs mostly in clinical cancer research. In a clinical setting, and due to the plasticity of EXOs landscape to environmental and pathological conditions, investigators envision that their specific markers could be used to obtain a patient Exogram (particular combination of EXO-associated markers for a particular individual at a determined moment) with the purpose of precisely diagnosing or monitoring the response to treatments as part of precision medicine programs. In the context of associating patient clinical data with relevant Exogram markers, biobanks can provide the EXO-based research field an excellent opportunity to achieve its maximum development by providing the ideal platform connecting researchers and clinicians.

Biobanking of EXOs could have a relevant role in the growing field of clinical research [58, 75]. Currently, there are no strictly defined conditions for storing and also isolating EXOs. The only established guidelines are those identified for the isolation of platelets-derived EVs. There is no specific information regarding the collection and storage of EXOs, optimal time, temperature, storage period, freezing-thaw cycles, thawing conditions, or other variables. Preliminary studies included the evaluation of the stability of EXOs at different temperatures finding that there was an advantage in storing EXOs at – 80 °C. However, this may vary according to EXOs prepared from different sources and by different isolation procedures. In addition, the particular lipid composition of EXOs, which is also source-dependent, is expected to have an impact on optimal cryopreservation. Biobanking relies on standardization of conditions that result in sample quality homogeneity and allows proper sample characterization for release [58, 76]. The current difficulty to define EXOs and the hurdles to establish and implement standardized protocols for their isolation and storage, let us understand that there is an urgent need for biobanks to assume an active role also in this research area.

14.4 Conclusions

Many decades of remarkable research activity have greatly increased the understanding and the potentiality of the field of stem cells biology. The availability of different cardiac stem cell sources and stem cell types are giving us the opportunity of regenerating the diseased myocardium. First generation clinical trials on cardiovascular diseases have made an attempt to establish techniques for cell isolation, expansion, delivery, and for determining safety, feasibility, and preliminary efficacy in humans. We are also aware that in this exciting new era of regenerative medicine many challenges still have to be overcome. To name a few, adult progenitor cells still need to provide higher yields of functional cardiomyocytes, pluripotent stem cells have to demonstrate to be employed without the risk of tumorigenesis; additionally, the question of allogeneic vs. autologous use remains open. The need of expanding autologous stem cells to clinically relevant numbers at the time needed to be administered as cell therapy products precludes their use in any acute setting. Allogeneic cells will then provide the only "off the shelf" product, but we need to learn how to properly manage the immune response to prevent their immune rejection. All of these efforts will be advanced by implementation of biobanking activities which promote readily available and high quality controlled products. In this environment, biobanks represent a key resource both for research-based studies, and mostly for the manufacturing of readily available cell therapeutic products. Therefore, it is not surprising that clinics or academic and industrial counterparts have demonstrated an increasing interest in biobanking in the recent years. Concerning the field of cardiac regenerative medicine, the search for innovative treatment options to prevent adverse ventricular remodelling following AMI has been at the forefront of clinical research in cardiology. Evidence from the published clinical trials indicates that adult cell-based therapies seem to be safe. Hence the ability to bank autologous stem cells for later use has the potential to become a significant keystone in the development and implementation of regenerative and personalized medicine strategies. Regenerative medicine applications will most probably benefit from biobanking in order to allow this clinical effort to reach its fullest potential and serve those with the greatest and most immediate cardiovascular need.

References

1. Laflamme MA, Murry CE (2011) Heart regeneration. Nature 473:326–335
2. Poss KD, Wilson LG, Keating MT (2002) Heart regeneration in zebrafish. Science 298(5601):2188–2190
3. Jopling C, Sleep E, Raya M, Marti M, Raya A, Belmonte JC (2010) Zebrafish heart regeneration occurs by cardiomyocyte dedifferentiation and proliferation. Nature 464(7288):606–609

4. Carvalho AB, de Carvalho AC (2010) Heart regeneration: past, present and future. World J Cardiol 2(5):107–111
5. Laflamme MA, Zbinden S, Epstein SE, Murry CE (2007) Cell-based therapy for myocardial ischemia and infarction: pathophysiological mechanisms. Annu Rev Pathol 2:307–339
6. Povsic TJ, O'Connor CM (2010) Cell therapy for heart failure: the need for a new therapeutic strategy. Expert Rev Cardiovasc Ther 8(8):1107–1126
7. Whelan RS, Kaplinskiy V, Kitsis RN (2010) Cell death in the pathogenesis of heart disease: mechanisms and significance. Annu Rev Physiol 72:19–44
8. Segers VF, Lee RT (2008) Stem-cell therapy for cardiac disease. Nature 451(7181):937–942
9. Passier R, van Laake LW, Mummery CL (2008) Stem-cell-based therapy and lessons from the heart. Nature 453(7193):322–329
10. Orlic D, Kajstura J, Chimenti S, Jakoniuk I, Anderson SM, Li B et al (2001) Bone marrow cells regenerate infarcted myocardium. Nature 410(6829):701–705
11. Toma C, Pittenger MF, Cahill KS, Byrne BJ, Kessler PD (2002) Human mesenchymal stem cells differentiate to a cardiomyocyte phenotype in the adult murine heart. Circulation 105(1):93–98
12. Balsam LB, Wagers AJ, Christensen JL, Kofidis T, Weissman IL, Robbins RC (2004) Haematopoietic stem cells adopt mature haematopoietic fates in ischaemic myocardium. Nature 428(6983):668–673
13. Murry CE, Soonpaa MH, Reinecke H, Nakajima H, Nakajima HO, Rubart M et al (2004) Haematopoietic stem cells do not transdifferentiate into cardiac myocytes in myocardial infarcts. Nature 428(6983):664–668
14. Chen SL, Fang WW, Ye F, Liu YH, Qian J, Shan SJ et al (2004) Effect on left ventricular function of intracoronary transplantation of autologous bone marrow mesenchymal stem cell in patients with acute myocardial infarction. Am J Cardiol 94(1):92–95
15. Olivetti G, Melissari M, Capasso JM, Anversa P (1991) Cardiomyopathy of the aging human heart. Myocyte loss and reactive cellular hypertrophy. Circ Res 68(6):1560–1568
16. Beltrami AP, Urbanek K, Kajstura J, Yan SM, Finato N, Bussani R et al (2001) Evidence that human cardiac myocytes divide after myocardial infarction. N Engl J Med 344(23):1750–1757
17. Smith RR, Barile L, Cho HC, Leppo MK, Hare JM, Messina E et al (2007) Regenerative potential of cardiosphere-derived cells expanded from percutaneous endomyocardial biopsy specimens. Circulation 115(7):896–908
18. Behfar A, Terzic A (2006) Derivation of a cardiopoietic population from human mesenchymal stem cells yields cardiac progeny. Nat Clin Pract Cardiovasc Med 3(Suppl 1):S78–S82
19. Behfar A, Faustino RS, Arrell DK, Dzeja PP, Perez-Terzic C, Terzic A (2008) Guided stem cell cardiopoiesis: discovery and translation. J Mol Cell Cardiol 45(4):523–529
20. Behfar A, Yamada S, Crespo-Diaz R, Nesbitt JJ, Rowe LA, Perez-Terzic C et al (2010) Guided cardiopoiesis enhances therapeutic benefit of bone marrow human mesenchymal stem cells in chronic myocardial infarction. J Am Coll Cardiol 56(9):721–734
21. Bearzi C, Rota M, Hosoda T, Tillmanns J, Nascimbene A, De Angelis A et al (2007) Human cardiac stem cells. Proc Natl Acad Sci U S A 104(35):14068–14073
22. Hsieh PC, Segers VF, Davis ME, MacGillivray C, Gannon J, Molkentin JD et al (2007) Evidence from a genetic fate-mapping study that stem cells refresh adult mammalian cardiomyocytes after injury. Nat Med 13(8):970–974
23. Tallini YN, Greene KS, Craven M, Spealman A, Breitbach M, Smith J et al (2009) c-kit expression identifies cardiovascular precursors in the neonatal heart. Proc Natl Acad Sci U S A 106(6):1808–1813
24. Bergmann O, Bhardwaj RD, Bernard S, Zdunek S, Barnabe-Heider F, Walsh S et al (2009) Evidence for cardiomyocyte renewal in humans. Science 324(5923):98–102
25. Kajstura J, Urbanek K, Perl S, Hosoda T, Zheng H, Ogorek B et al (2010) Cardiomyogenesis in the adult human heart. Circ Res 107(2):305–315
26. Bolli R, Chugh AR, D'Amario D, Loughran JH, Stoddard MF, Ikram S et al (2011) Cardiac stem cells in patients with ischaemic cardiomyopathy (SCIPIO): initial results of a randomised phase 1 trial. Lancet 378(9806):1847–1857
27. Chugh AR, Beache GM, Loughran JH, Mewton N, Elmore JB, Kajstura J et al (2012) Administration of cardiac stem cells in patients with ischemic cardiomyopathy: the SCIPIO trial: surgical aspects and interim analysis of myocardial function and viability by magnetic resonance. Circulation 126(11 Suppl 1): S54–S64
28. Davis DR, Zhang Y, Smith RR, Cheng K, Terrovitis J, Malliaras K et al (2009) Validation of the cardiosphere method to culture cardiac progenitor cells from myocardial tissue. PLoS One 4(9):e7195
29. Chimenti I, Smith RR, Li TS, Gerstenblith G, Messina E, Giacomello A et al (2010) Relative roles of direct regeneration versus paracrine effects of human cardiosphere-derived cells transplanted into infarcted mice. Circ Res 106(5):971–980
30. Makkar RR, Smith RR, Cheng K, Malliaras K, Thomson LE, Berman D et al (2012) Intracoronary cardiosphere-derived cells for heart regeneration after myocardial infarction (CADUCEUS): a prospective, randomised phase 1 trial. Lancet 379(9819):895–904
31. Malliaras K, Makkar RR, Smith RR, Cheng K, Wu E, Bonow RO et al (2014) Intracoronary cardiosphere-derived cells after myocardial infarction: evidence of therapeutic regeneration in the final 1-year results of the CADUCEUS trial (CArdiosphere-Derived aUtol-

ogous stem CEIls to reverse ventricUlar dySfunction). J Am Coll Cardiol 63(2):110–122
32. Assmus B, Schächinger V, Teupe C, Britten M, Lehmann R, Döbert N et al (2002) Transplantation of progenitor cells and regeneration enhancement in acute myocardial infarction (TOPCARE-AMI). Circulation 106(24):3009–3017
33. Wollert KC, Meyer GP, Lotz J, Ringes-Lichtenberg S, Lippolt P, Breidenbach C et al (2004) Intracoronary autologous bone-marrow cell transfer after myocardial infarction: the BOOST randomised controlled clinical trial. Lancet 364(9429):141–148
34. Schächinger V, Erbs S, Elsässer A, Haberbosch W, Hambrecht R, Hölschermann H et al (2006) Intracoronary bone marrow-derived progenitor cells in acute myocardial infarction. N Engl J Med 355(12):1210–1221
35. Lunde K, Solheim S, Aakhus S, Arnesen H, Abdelnoor M, Egeland T et al (2006) Intracoronary injection of mononuclear bone marrow cells in acute myocardial infarction. N Engl J Med 355(12):1199–1209
36. Huikuri HV, Kervinen K, Niemelä M, Ylitalo K, Säily M, Koistinen P et al (2008) Effects of intracoronary injection of mononuclear bone marrow cells on left ventricular function, arrhythmia risk profile, and restenosis after thrombolytic therapy of acute myocardial infarction. Eur Heart J 29(22):2723–2732
37. Hirsch A, Nijveldt R, van der Vleuten PA, Tijssen JG, van der Giessen WJ, Tio RA et al (2011) Intracoronary infusion of mononuclear cells from bone marrow or peripheral blood compared with standard therapy in patients after acute myocardial infarction treated by primary percutaneous coronary intervention: results of the randomized controlled HEBE trial. Eur Heart J 32(14):1736–1747
38. Traverse JH, Henry TD, Pepine CJ, Willerson JT, Zhao DX, Ellis SG et al (2012) Effect of the use and timing of bone marrow mononuclear cell delivery on left ventricular function after acute myocardial infarction: the TIME randomized trial. JAMA 308(22):2380–2389
39. Traverse JH, Henry TD, Ellis SG, Pepine CJ, Willerson JT, Zhao DX et al (2011) Effect of intracoronary delivery of autologous bone marrow mononuclear cells 2 to 3 weeks following acute myocardial infarction on left ventricular function: the LateTIME randomized trial. JAMA 306(19):2110–2119
40. Karantalis V, DiFede DL, Gerstenblith G, Pham S, Symes J, Zambrano JP et al (2014) Autologous mesenchymal stem cells produce concordant improvements in regional function, tissue perfusion, and fibrotic burden when administered to patients undergoing coronary artery bypass grafting: the Prospective Randomized Study of Mesenchymal Stem Cell Therapy in Patients Undergoing Cardiac Surgery (PROMETHEUS) trial. Circ Res 114(8):1302–1310
41. Hare JM, Traverse JH, Henry TD, Dib N, Strumpf RK, Schulman SP et al (2009) A randomized double-blind placebo-controlled dose-escalation study of intravenous adult human mesenchymal stem cells (prochymal) after acute myocardial infarction. J Am Coll Cardiol 54(24):2277–2286
42. Bartunek J, Behfar A, Dolatabadi D, Vanderheyden M, Ostojic M, Dens J et al (2013) Cardiopoietic stem cell therapy in heart failure: the C CURE (Cardiopoietic stem Cell therapy in heart failURE) multicenter randomized trial with lineage-specified biologics. J Am Coll Cardiol 62(25):2457–2458
43. Kehat I, Kenyagin-Karsenti D, Snir M, Segev H, Amit M, Gepstein A et al (2001) Human embryonic stem cells can differentiate into myocytes with structural and functional properties of cardiomyocytes. J Clin Invest 108(3):407–414
44. Zhu WZ, Santana LF, Laflamme MA (2009) Local control of excitation-contraction coupling in human embryonic stem cell-derived cardiomyocytes. PLoS ONE 4(4):e5407
45. Shiba Y, Fernandes S, Zhu WZ, Filice D, Muskheli V, Kim J et al (2012) Human ES-cell-derived cardiomyocytes electrically couple and suppress arrhythmias in injured hearts. Nature 489(7415):322–325
46. Nussbaum J, Minami E, Laflamme MA, Virag JA, Ware CB, Masino A et al (2007) Transplantation of undifferentiated murine embryonic stem cells in the heart: teratoma formation and immune response. Faseb J 21(7):1345–1357
47. Takahashi K, Tanabe K, Ohnuki M, Narita M, Ichisaka T et al (2007) Induction of pluripotent stem cells from adult human fibroblasts by defined factors. Cell 131(5):861–872
48. Ieda M, Fu JD, Delgado-Olguin P, Vedantham V, Hayashi Y, Bruneau BG et al (2010) Direct reprogramming of fibroblasts into functional cardiomyocytes by defined factors. Cell 142(3):375–386
49. Efe JA, Hilcove S, Kim J, Zhou H, Ouyang K, Wang G et al (2011) Conversion of mouse fibroblasts into cardiomyocytes using a direct reprogramming strategy. Nat Cell Biol 13(3):215–222.50
50. Théry C, Amigorena S, Raposo G, Clayton A (2006) Isolation and characterization of exosomes from cell culture supernatants and biological fluids. Curr Protoc Cell Biol, Chapter 3:Unit 3.22
51. Corrado C, Raimondo S, Chiesi A, Ciccia F, de Leo G, Alessandro R (2013) Exosomes as intercellular signaling organelles involved in health and disease: basic science and clinical applications. Int J Mol Sci 14(3):5338–5366
52. Urbanelli L, Magini A, Buratta S, Brozzi A, Sagini K, Polchi A et al (2013) Signaling pathways in exosomes biogenesis, secretion and fate. Genes (Basel) 4(2):152–170
53. Yamashita T, Takahashi Y, Nishikawa M, Takakura Y (2016) Effect of Exosome isolation methods on physicochemical properties of exosomes and clearance of exosomes from the blood circulation. Eur J Pharm Biopharm 98:1–8
54. Günter J, Wolint P, Bopp A, Steiger J, Cambria E, Hoerstrup SP et al (2016) Microtissues in cardiovas-

cular medicine: regenerative potential based on a 3D microenvironment. Stem Cells Int. 2016:9098523 doi:10.1155/2016/9098523. Epub 2016 Mar 17
55. Zakharova L, Mastroeni D, Mutlu N, Molina M, Goldman S, Diethrich E et al (2010) Transplantation of cardiac progenitor cell sheet onto infarcted heart promotes cardiogenesis and improves function. Cardiovasc Res 87(1):40–49
56. Caspi O, Lesman A, Basevitch Y, Gepstein A, Arbel G, Habib IH et al (2007) Tissue engineering of vascularized cardiac muscle from human embryonic stem cells. Circ Res 100(2):263–272
57. Zimmermann WH, Melnychenko I, Wasmeier G, Didie M, Naito H, Nixdorff U et al (2006) Engineered heart tissue grafts improve systolic and diastolic function in infarcted rat hearts. Nat Med 12(4):452–458
58. Mora EM, Álvarez-Cubela S, Oltra E (2015) Biobanking of exosomes in the era of precision medicine: are we there yet? Int J Mol Sci 24(17(1)):13
59. Kinkorová J (2016) Biobanks in the era of personalized medicine: objectives, challenges, and innovation: overview. EPMA J 7:4
60. Olson JE, Bielinski SJ, Ryu E, Winkler EM, Takahashi PY, Pathak J et al (2014) Biobanks and personalized medicine. Clin Genet 86(1):50–55
61. Mirabet V, Carda C, Solves P, Novella-Maestre E, Carbonell-Uberos F, Caffarena JM (2008) Long-term storage in liquid nitrogen does not affect cell viability in cardiac valve allografts. Cryobiology 57(2):113–121
62. Neuenschwander S, Hoerstrup SP (2004) Heart valve tissue engineering. Transpl Immunol 12(3–4):359–365
63. Schmidt D, Mol A, Breymann C, Achermann J, Odermatt B, Gössi M et al (2006) Living autologous heart valves engineered from human prenatally harvested progenitors. Circulation 114(1 Suppl):I125–31
64. Chong JJ, Yang X, Don CW, Minami E, Liu YW, Weyers JJ et al (2014) Human embryonic-stem-cell-derived cardiomyocytes regenerate non-human primate hearts. Nature 510(7504):273–277
65. Bardelli S, Moccetti M (2016) Ex-vivo expansion of cardiac stem cell therapy products for clinical use: the importance of moving towards the optimization of process development. Int J Stem Cell Res Ther 3:1
66. Dolgos H, Trusheim M, Gross D, Halle JP, Ogden J, Osterwalder B et al (2016) Translational medicine guide transforms drug development processes: the recent Merck experience. Drug Discov Today 21(3):517–526
67. Li Y, Powell S, Brunette E, Lebkowski J, Mandalam R (2005) Expansion of human embryonic stem cells in defined serum-free medium devoid of animal-derived products. Biotechnol Bioeng 91:688–698
68. Xu XQ, Graichen R, Soo SY, Balakrishnan T, Rahmat SN, Sieh S et al (2008) Chemically defined medium supporting cardiomyocyte differentiation of human embryonic stem cells. Differentiation 76(9):958–970
69. Yao S, Chen S, Clark J, Hao E, Beattie GM, Hayek A et al (2006) Long-term self-renewal and directed differentiation of human embryonic stem cells in chemically defined conditions. Proc Natl Acad Sci U S A 103(18):6907–6912
70. Liao Y, Geyer MB, Yang AJ, Cairo MS (2011) Cord blood transplantation and stem cell regenerative potential. Exp Hematol 39(4):393–412
71. Möllmann H, Nef H, Elsässer A, Hamm C (2009) Stem cells in myocardial infarction: from bench to bedside. Heart 95(6):508–514
72. Ma N, Ladilov Y, Kaminski A, Piechaczek C, Choi YH, Li W et al (2006) Umbilical cord blood cell transplantation for myocardial regeneration. Transplant Proc 38(3):771–773
73. Henning RJ, Abu-Ali H, Balis JU, Morgan MB, Willing AE, Sanberg PR (2004) Human umbilical cord blood mononuclear cells for the treatment of acute myocardial infarction. Cell Transplant 13(7–8):729–739
74. Chin SP, Poey AC, Wong CY, Chang SK, Teh W, Mohr TJ et al (2010) Cryopreserved mesenchymal stromal cell treatment is safe and feasible for severe dilated ischemic cardiomyopathy. Cytotherapy 12(1):31–37
75. Vandergriff AC, de Andrade JB, Tang J, Hensley MT, Piedrahita JA, Caranasos TG et al (2015) Intravenous cardiac stem cell-derived exosomes ameliorate cardiac dysfunction in doxorubicin induced dilated cardiomyopathy. Stem Cells Int 2015:960926. doi:10.1155/2015/960926. Epub 2015 Aug 17
76. Xu C, Police S, Hassanipour M, Li Y, Chen Y, Priest C et al (2011) Efficient generation and cryopreservation of cardiomyocytes derived from human embryonic stem cells. Regen Med 6(1):53–66

Preservation of Ocular Epithelial Limbal Stem Cells: The New Frontier in Regenerative Medicine

15

Zala Lužnik, Marina Bertolin, Claudia Breda,
Barbara Ferrari, Vanessa Barbaro,
Petra Schollmayer, and Stefano Ferrari

Abstract

Significant advances have been made in the field of ocular regenerative medicine. Promising stem cell-based therapeutic strategies have been translated into the clinical practice over the last few decades. These new stem cell-based therapies offer the possibility of permanently restoring corneal epithelium in patients with severe disabling and blinding ocular surface disease. The European Union has already classified stem cell-based therapies as "medicinal products". Therefore, manipulation is strictly regulated according to the defined conditions of good manufacturing practice, with the production of stem cell therapeutics at only accredited production sites authorized by the national regulatory agencies. In this regard, as first medical products are licensed for commercial use in Europe enabling a more widespread access to a stem cell-based therapy, the need for safe, validated and reproducible techniques for ex vivo cultured tissue preservation and distribution are coming to the forefront of research. However, these provide various new challenges for biobanking industry such as the retention of viability, good functionality of stem cells and sterility issues. This chapter provides an overview of the current advances in the field of corneal/limbal epithelial stem cell culture preservation techniques using either hypothermic storage or cryopreservation methods, that were used in different culturing steps (from stem cell isolation to the

Z. Lužnik • P. Schollmayer
Eye Hospital, University Medical Centre,
Ljubljana, Slovenia
e-mail: zala_luznik@yahoo.com;
petra.schollmayer@gmail.com

M. Bertolin • C. Breda • B. Ferrari • V. Barbaro
S. Ferrari (✉)
Fondazione Banca degli Occhi del Veneto,
Venice, Italy
e-mail: marina.bertolin@fbov.it; claudia.breda@fbov.it;
barbara.ferrari@fbov.it; vanessa.barbaro@fbov.it;
stefano.ferrari@fbov.it

F. Karimi-Busheri, M. Weinfeld (eds.), *Biobanking and Cryopreservation of Stem Cells*,
Advances in Experimental Medicine and Biology 951, DOI 10.1007/978-3-319-45457-3_15

ex vivo epithelial graft preparation), with the reported impact on the post-thawing product recovery.

Keywords
Corneal epithelium • Ocular surface • Corneal disease • Stem cells • Regenerative medicine • Culture methods • Cryopreservation • Vitrification • Hypothermic storage • Cell-based therapy

Abbreviations

ΔNp63α	DeltaNp63alpha
ABCG2	ATP-binding cassette sub-family G member 2
SSEA-4	Stage-specific embryonic antigen-4
CD	Cluster of differentiation
CK19	Cytokeratin 19
CLET	Cultured limbal epithelium transplantation
CPAs	Cryoprotective agents
DMSO	Dimethyl sulfoxide
DMEM	Dulbecco modified Eagle medium
DNA	Deoxyribonucleic acid
EG	Ethylen glycol
EU	European Union
FBS	Fetal bovine serum
GMP	Good manufacturing practice
HLA-DR	Human leukocyte antigen – antigen D related
LSCD	Limbal stem cell deficiency
LESCs	Limbal epithelial stem cells
NF-kB	Nuclear factor kappa-light-chain-enhancer of activated B cells
SCs	Stem cells
TACs	Transient amplifying cells

15.1 Introduction

Significant advances have been made in the field of regenerative medicine with promising stem cell-based therapeutic strategies being translated into the clinical practice over the last few decades. These new stem cell-based therapies offer the possibility of permanently restoring or replacing previously irreparable tissue and organ damage with cultured cells [1]. Clinically verified examples include the use of ex vivo cultured epithelial stem cells (SCs) to regenerate many types of squamous epithelia [2]. Cultured corneal/limbal SC transplantation revolutionized the current treatment options for patients with advanced painful and vision threatening ocular surface disease, such as total limbal stem cell deficiency (LSCD) [1].

The human ocular surface is covered with two phenotypically distinct squamous non-keratinized epithelial cell types [1], the corneal and conjunctival squamous epithelia (Fig. 15.1). The conjunctival stratified epithelium is well vascularized and contains mucin secreting goblet cells which contribute to the tear film stability and consequently to the optical clarity of the corneal surface [3]. In contrary, the flat corneal epithelium is devoid of goblet cells and is lying directly on a transparent sheet of tissue known as Bowman's layer, which separates the corneal epithelium with its basement membrane from the beneath lying corneal stroma. It is renewed by a small population of SCs found in the limbal palisades of Vogt, a narrow zone between the bulbar conjunctiva and the peripheral cornea [4] (Fig. 15.1). Complete loss of corneal/limbal SCs due to various ocular diseases or injuries leads to corneal re-epithelialization by bulbar conjunctival cells [5], followed by corneal vascularization, recurrent corneal epithelial defects, chronic inflammation, and corneal stromal scarring, causing painful visual loss in affected patients [6].

Total and severe LSCD is difficult to manage [6]. Autologous or allogeneic (in bilateral disease) transplantation of limbal epithelial SCs is necessary to restore vision and the corneal sur-

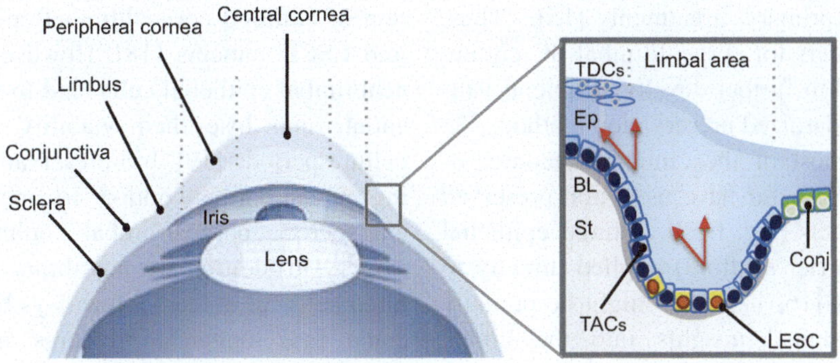

Fig. 15.1 The ocular surface and the limbal niche in cross-section. The corneal epithelium is renewed by limbal epithelial stem cells (LESCs). LESCs are located in the basal layers of the limbus (the limbal stem cell niche). Stem cells auto-regenerate and give origin to fast dividing daughter cells – the transient amplifying cells (TACs), which are migrating centripetally into the cornea and to the surface layers of the epithelium (*red arrows*). TACs give rise to more differentiated postmitotic cells (terminally differentiated cells TDCs). *Ep* corneal epithelium, *BL* Bowman's layer, *St* corneal stroma, *Con* conjunctival epithelium

face [6]. Since 1997, when Pellegrini and colleagues first demonstrated transplantation of *in vitro* expanded limbal epithelium for treatment of LSCD from a small limbal biopsy taken from the patient's contralateral healthy eye [7], transplantation of cultured corneal/limbal epithelium is currently the most successful alternative to corneal surface reconstruction in patients with unilateral LSCD and provides a therapeutic chance to patients with severe bilateral LSCD [8]. To prevent or minimize the theoretical risk of iatrogenic damage to the healthy or less injured patient's eye, and reducing the need for long-term immunosuppression, autologous oral mucosa SCs [9] and cultured conjunctival SCs [10] have been used in the treatment of total bilateral LSCD in humans as well [2]. To date, almost 1,000 cultured limbal epithelium transplantation (CLET) cases have been reported worldwide with an approximately 75 % overall success rate [11].

However, although a variety of expansion protocols for limbal SCs culturing have shown good clinical outcomes [12], including limbal explant or limbal cell suspension culture, culture on intact or epithelially denuded human amniotic membrane or other cell culture scaffolds, co-cultivation with mitotically inactivated murine feeder layers (e.g. 3T3 fibroblasts), and air-lifting, limbal epithelial SC-based therapy still faces challenges regarding tissue safety and sterility, tissue transportation, surgery logistics and availability of cultured tissue [12]. As the European Union (EU) has classified SC-based therapies as "medicinal products" [13], manipulation is strictly regulated according to the manufacture of biological medicinal products for human use under defined conditions of good manufacturing practice (GMP), with the production of SC therapeutics at only accredited production sites authorized by the national regulatory agencies [13, 14]. Importantly, these conditions also represent a barrier to a more widespread use of SC-based therapies, while demand is anticipated to increase due to successful clinical outcomes.

In this regard, as first medical products are licensed for commercial use in Europe (e.g. Holoclar) enabling wider distribution of SC-based therapies, the need for safe, validated and reproducible techniques for ex vivo cultured tissue preservation are coming to the forefront [14]. Appropriate tissue storage (short and long-term) methods allow cultured tissue transportation from centralized laboratories to the operating theater and between eye banks, thus offering the logistic flexibility in scheduling transplantation surgery [12]. Moreover, tissue storage increases the opportunity for quality controls (microbiological testing, SC identification) [2]. In addition, long-term cryopreservation of surplus cultured SCs could enable consecutive surgeries in case of unsuccessful primary treatment [15]. Thus, accredited centers for corneal/limbal SC culture are challenged to further develop efficient, validated and standardized preservation methods.

However, most of the clinical outcomes of CLET reported so far have used non-preserved terminal products (e.g. fresh cultured epithelial grafts on a carrier scaffold) applied directly to LSCD patients [16, 17]. This might be partially due to the limited insights into the limbal SC-based product's safety and stability after preservation. It is mandatory that the final products recovered from hypothermic or cryopreservation storage are physiologically and biochemically identical to its pre-preservation state, retaining high viability, good functionality and tissue morphology [14]. Given this, the current preservation methods still raise fundamental questions regarding the optimal storage temperature and storage media formulations (short or long term) not only for single limbal SC suspensions but also for the final products (e.g. the cultured epithelial sheets on different scaffolds). Therefore, storage methods and cryopreservation techniques for corneal SC delivery have recently been under investigation at an accelerated rate.

In this chapter we provide an overview of the current advances in the field of corneal/limbal epithelial SC culture preservation techniques using either hypothermic storage or cryopreservation methods, that were used in different culturing stages (from SC isolation to ex vivo epithelial graft preparation). As the preservation protocols applied in the regenerative medicine must deliver functional products at clinical scale, which also comply with the GMP regulations, we additionally focused on the post-thawing product biochemical and physiological recovery.

15.2 Bio-preservation of Limbal Epithelial Stem Cells

Developments in SC-based technology and tissue engineering have already enabled proper limbal SC isolation, *in vitro* expansion and amplification of cells on different scaffolds and enabled successful clinical application with the transplantation of cultured corneal/limbal epithelial sheets into LSCD patients [18]. However, as the corneal/limbal epithelial cells tend to further differentiate and lose their viability in prolonged culture periods [19], bio-preservation is becoming an important option to preserve tissue-engineered corneal/limbal epithelium until surgery. In addition, the possibility of storing cultured cells for at least a few days before surgery could ease logistical problems and would be especially mandatory for proper transportation of cultured tissue [3].

To date, there are only limited studies reported on the bio-preservation of cultured ocular epithelia. The most common bio-preservation methods under investigation for increasing the storage intervals by limiting the negative effects associated with prolonged tissue culture are the **hypothermic and cryopreservation techniques**, both having their advantages and disadvantages. Cryopreservation is the standard method for long-term storage for cells in suspension. However, as cryopreservation requires expensive facilities and expertise, when only short-term storage is required, hypothermic storage methods (or above-zero degree storage) are reported to be more advantageous than cryopreservation methods [3].

Although all above strategies have generally reported sufficient survival of preserved cells measured immediately after thawing, surviving cells may lose key functional SC characteristics that may not be recovered. It is known that the long-term corneal epithelium renewal after ex vivo cultured limbal epithelial transplantation depends mainly on the sufficient number of limbal epithelial SCs (holoclones) transferred in the cultured epithelial sheets [20]. Therefore, optimal clinical outcomes after transplantation primarily depend on the quality of the limbal cultures used to prepare the corneal sheets, with the best quality control being the accurate determination of limbal epithelial SCs in the cultured epithelial sheets [13]. To date, transcriptional factor $\Delta Np63\alpha$ is still thought to be the best positive marker to determine the percentage of SCs in cultured corneal/limbal sheets, as a positive clinical correlation was already reported [13, 21].

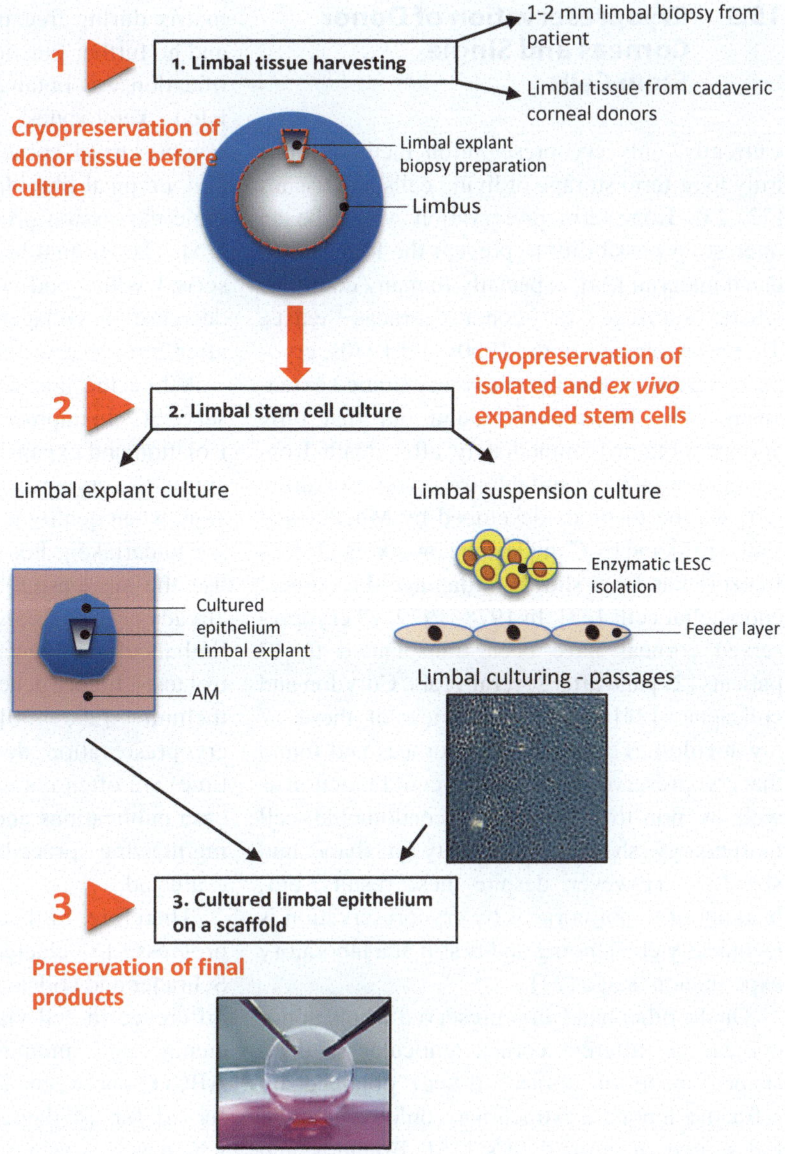

Fig. 15.2 Preservation strategies in different stages of the production process of bio-engineered corneal/limbal epithelial sheets. The cryopreservation methods for limbal stem cell preservation fall into two main approaches: cryopreservation of limbal stem cells before *in vitro* corneal/limbal tissue generation (*1*) e.g. cryopreservation of whole donor corneas, corneoscleral rims, corneal stromal lenticules, etc.); cryopreservation of suspended amplified stem cells (*2*) e.g. after several passages) or cryopreservation of corneal/limbal bioengineered sheets on different scaffolds (*3*). Preservation of final products includes: long-term preservation: **cryopreservation** of cultured adherent ocular epithelia and short-term preservation: **hypothermic storage** of cultured adherent ocular epithelia

Thus, as the limbal SCs are the key component of tissue-engineered corneal/limbal sheets that provide tissue specificity and the bioactivity required for achieving a therapeutic effect, the impact of bio-preservation methods on the limbal SC survival and function is essential in determining the post-thaw tissue quality.

As a result, different studies have focused on SC bio-preservation in different stages of the production process of bio-engineered corneal/limbal epithelial sheets (Fig. 15.2). Currently, the cryopreservation methods for limbal SC preservation have fallen into two main approaches: **cryopreservation of limbal SCs before *in vitro* limbal/corneal tissue generation** (e.g. cryopreservation of donor corneas, cryopreservation of suspended amplified SCs – after several passages) or **cryopreservation of corneal/limbal bioengineered sheets on different scaffolds**. In addition, **hypothermic storage methods** are investigated for short-term storage of corneal/limbal sheets.

15.3 Cryopreservation of Donor Corneas and Single Stem Cells

Currently, only cryopreservation methods offer truly long-term storage of living cells and tissues [22, 23]. Long-term preservation would be an interesting possibility to prevent the loss of outdated material [24], especially in many countries where shortage of donor corneas exists. Therefore, already in the 1960s and 1970s, cryopreservation of donor corneas was studied extensively [24]. Researchers found out that only corneas obtained immediately after death from young donors were suitable for cryopreservation [24] via the methods developed by Mueller and Smith [25] and by Capella and coworkers [26], as freezing has been shown to damage the corneal endothelial cells [23]. In 1978–1979, 17 cryopreserved corneas have been transplanted to 17 patients [27] and after several years, Corydon and colleagues [24] reported a review of these 17 organ-cultured cryopreserved corneas and found that cryopreserved endothelium could function as well as non-frozen, although endothelial cell morphology showed irregularity in shape and size [24]. However, despite these results, biobanking of donor corneas by cryopreservation is technically challenging and is still at a laboratory experimental stage [24].

On the other hand, cryopreserved whole donor corneas or different cornea lenticules (a disc-shaped piece of corneal tissue) obtained by refractive lenticule extractions, could be a potential source of corneal SCs [15]. Bratanov and coworkers [28] successfully cultured limbal explants obtained from cryopreserved corneas. Although the cryopreserved limbal tissue was subjected to long lasting storage (up to 12 months) in liquid nitrogen, it enabled proliferation of epithelial cells and phenotypically identified limbal SCs (p63 positive cells), which was found to be as efficient as the limbal explant cultures obtained from non-cryopreserved corneas [28]. However, the researchers reported a different growing pattern of expanded cells from the cryopreserved limbal explants, which might be due to the destabilization of the extracellular matrix during freezing-thawing procedures [28] and no further functional test for limbal SC identification was obtained. In addition, *in vitro* isolated keratocytes derived from fresh and cryopreserved corneal stroma lenticules exhibited a typical fibroblastic phenotype and comparable expression profiles for the tested markers [15]. The stromal lenticule architecture was preserved with good cellular viability, although a decrease in collagen fibril density was shown after cryopreservation [15].

While the above strategies tried to cryopreserve SCs in unprocessed corneal tissues for later isolation and expansion, another promising strategy is SC cryopreservation after *in vitro* expansion, when quality tests and purification methods are undertaken. For example, some centers that use the suspension limbal cell culture method already need to freeze (cryopreserve) the primary limbal culture to await the results of quality control tests, before generating the final cultured epithelium [29, 30]. However, the exact cryopreservation details (e.g. cryopreservation time) are often not explicitly reported in clinical case publications and the exact impact of freezing/thawing procedures is often not further evaluated.

Therefore, further experimental studies are in progress to elucidate this important issue. Schrader and colleagues [31] already reported no difference in cell viability, colony-forming efficiency and immunoreactivity to p63α and ABCG2 between conjunctival cells cryopreserved for 14 days, and more than 6 months, which were frozen from primary cultures isolated from bulbar conjunctiva biopsies and expanded on arrested 3T3 feeder layer cells [31]. The researchers concluded that conjunctival epithelial cells could be efficiently cryopreserved with maintenance of the progenitor cell-like characteristics and function *in vitro* over several culture passages [31].

Furthermore, cryopreservation of suspended limbal SCs after *in vitro* expansion of up to passage level one showed a non-immunogenic nature of de-frosted limbal cells [32]. Human limbal epithelial SC cultures retained the expression of major limbal epithelial SC markers such as p63,

SSEA-4, ABCG2, cytokeratin 19 (CK19), integrin β1 and vimentin, however no HLA-DR gene expression was observed with the expanded and cryopreserved samples [32]. Cultured limbal SCs were additionally unable to stimulate allogeneic T cell proliferation *in vitro* even in the presence of pro-inflammatory cytokines [32]. Therefore, cryopreserved limbal SC cultures could express negative immunoregulatory molecules, which may be critical for their survival in an allogeneic environment [32] and would enable better allograft survival.

Although some clinically used techniques already required a short cryopreservation step [29, 30] before the routine final graft production, to date, only one experimental study explicitly reported successful reconstruction of corneal epithelium in a goat model from cultured epithelial grafts generated from long-term cryopreserved limbal SCs [33]. From the primary culture ABCG2-positive and CD34-negative cells were collected and amplified for additional two to four passages, when the cells were cryopreserved in liquid nitrogen in storage medium containing 10 % dimethyl sulfoxide (DMSO) and 25 % fetal bovine serum (FBS) [33]. After 1 year of cryopreservation and the thawing process, the SC viability was 85 % [33]. The cryopreserved limbal SCs were then seeded onto a human amniotic membrane to construct corneal/limbal epithelial sheets and transplanted into a damaged corneal surface in a goat model [33]. Although it is not yet known how stable these corneal/limbal sheets will be over time, the study clearly demonstrated that transplantation of long-term cryopreserved limbal SCs could successfully reconstruct damaged corneal surface in a goat model [33].

15.4 Cryopreservation of Bioengineered Sheets on Different Scaffolds

In recent years, much attention has been focused on cryopreservation techniques for bioengineered tissue storage, as only effective preservation procedures of these terminal (final) products would enable a stable supply without delay and efficient transportation over longer distances [19, 34, 35]. However, tissue cryopreservation faces biobanks with new challenges of adapting strategies and protocols for post-freezing product integrity. When cell suspensions are frozen, there are two principal mechanisms of cell damage: intracellular freezing at high cooling rates, and solution effect injury at low cooling rates [23]. In case of tissue cryopreservation, there are additional factors that render cells more susceptible to intracellular freezing and cultured tissue damage [23].

Therefore, various permeating cryoprotective agents (CPAs) such as DMSO, glycerol and ethylene glycol (EG) have been studied in freezing media to protect the cells from the extreme temperature excursions used in cryopreservation processing and enable better cellular viability after cryopreservation [36]. CPAs are thought to protect cellular membranes and proteins, although the cryoprotective mechanisms are largely unknown [36]. The most commonly used CPAs for the cryopreservation of cultured epithelial and mucosal cell sheets are glycerol and DMSO, where glycerol has been reported to be superior to DMSO by some researchers [19]. However, it is known that the optimal conditions for cryopreservation depend on the cell type and species of origin [22].

Kito and colleagues [19] investigated rabbit limbal epithelial cell survival obtained after freezing storage in either 10 % glycerol or DMSO at −80 °C or −196 °C for 4 or 12 weeks and reported that the samples stored at a lower storage temperature (−196 °C) showed an improved survival compared with those at a higher temperature (−80 °C) [19]. Importantly, the structural integrity of the cultured corneal/limbal sheets was destroyed in all tested cryopreservation protocols, although the structural damage was more clearly observed in the corneal epithelial sheets cryopreserved with DMSO than in those with glycerol at −80 °C [19]. However, cell viability after thawing was reported to be up to 70 %, and the remained cells were able to regenerate a new cell sheet [19].

On the other hand, Yeh and colleagues [35] reported successful 8-week cryopreservation of limbal explant cultures expanded on human amniotic membrane [35]. In their study they reported that the optimal formula of cryoprotectants was 60 % Dulbecco modified Eagle medium (DMEM), 30 % FBS, and 10 % DMSO, which was superior to 10 % glycerol [35]. Under this conditions they yielded thawed cell sheets with high viability (around 50 %) and good growth ability [35]. In comparison, Oh and colleagues [34] cryopreserved human limbal and rabbit conjunctival cultured epithelial sheets for 1 week at −196 °C by using various cryopreservative conditions (they varied the FBS concentrations and used DMSO and glycerol as CPAs). They reported the best result being obtained by 50 % FBS in limbal SC cultures for cell viability and 10 % glycerol seemed to be superior to 10 % DMSO in cell viability of the rabbit conjunctival epithelium after cryopreservation [34].

15.5 Vitrification of Corneal Limbal Stem Cells

To date, only one study compared cultivated fresh rabbit limbal and oral mucosal explant cultures on human amniotic membrane, co-cultured with 3T3 fibroblasts and using the air-lifting method, that had similar morphology to cultures obtained from 2 months cryopreserved limbal and oral mucosal tissue with vitrification method [37]. As vitrification technique avoids intracellular and extracellular ice formation with step-wise addition of high molar concentrations of CPAs during the cooling process compared to the conventional controlled rate cooling cryopreservation techniques [38], it might offer a better option for cryopreservation of more complex tissues such as cultured epithelial sheets.

As discussed above, this initial experiment showed that cryopreservation of adherent stratified ocular epithelia has been more challenging and less successful compared to single cell suspension cryopreservation techniques, therefore further research is needed. These disadvantages, in addition to the needed complicated and expensive freeze/thaw schedules and specialized equipment, makes reliable storage of ocular epithelial sheets at above-freezing temperatures a promising alternative [2], as will be discussed in the next section.

15.6 Hypothermic Storage of Final Products

Refrigeration or above-freezing temperature storage might be a promising short-term resolution for cultured corneal epithelial sheet preservation [2]. However, there are several issues to be solved regarding the optimal preservation temperature and storage medium composition before widespread clinical use. Studies on cornea preservation methods already showed morphological corneal epithelial changes such as epithelial detaching and intracellular vacuoles after 1 week of organ-culture preservation at 31 °C [12]. On the other side, hypothermic storage of corneas at 5 °C in Optisol-GS also showed separation of cells below the superficial epithelial layer and pronounced intracellular edema [12]. Therefore, recent studies are focusing on the optimal temperature for cultured epithelial corneal/limbal sheets preservation. Utheim and colleagues [39] first reported a method for short-term eye bank storage of cultured corneal sheets [12, 39]. They reported maintenance of the original multilayered structure and undifferentiated phenotype after 1 week organ culture storage at 23 °C of corneal/limbal sheets obtained from 3-week limbal explant culture on intact amniotic membrane [12, 39]. In addition, their data indicated that storage of cultured corneal/limbal sheets at ambient temperature is superior to organ culture storage at 31 °C and Optisol-GS storage at 5 °C, with minimal apoptosis after eye bank storage of cultured epithelial sheets [12, 39]. Furthermore, in a later study they reported successful storage of cultured human conjunctival epithelial sheets on amniotic membrane for at least 4 days in serum-free conditions at 23 °C [40].

In contrast, the results of a recent study, which evaluated the effect of different temperatures spanning from 4 to 37 °C, on the cell morphology, viability, proliferation and metabolic status of cultured epidermal sheets stored over a 2 week period in a xenobiotic–free system, showed some differences [2]. Reduced cell viability was reported at several temperatures, with the best viability being obtained at 24 °C, however, preservation of morphology and maintenance of proliferation function was best at 12 °C and 20 °C [2] compared to non-stored control. Similarly, Eidet et al. [3] reported results on human conjunctival epithelial culture storage in a xenogenic-free medium condition at nine different temperatures between 4 and 37 °C for 4 and 7 days and only storage at 12 °C appeared optimal for preserving viability, morphology and total cell numbers, which may be related to temperature-associated effects on cell metabolism [3].

Importantly, hypothermic storage might not only suppress cellular metabolism, it could also have an important effect on reducing immunoreactions after transplantation [12]. Interesting alterations have been observed in multigene profiling in cultured human limbal cells after eye bank storage under 23 °C and 5 °C, where a reduction in nuclear factor (NF)-kB activity and several apoptosis-regulating genes were reduced in their expression [12]. The research group has just recently published a study where they further investigated the molecular mechanisms underlying activation of cell death pathways using genome-wide transcriptional analysis in human limbal epithelial cultures following conventional hypothermic storage in Optisol-GS for 2, 4 and 7 days at 4 °C [41]. With increasing storage time larger differences in absolute gene expression values were observed with up-regulated expression of histone-coding genes together with down-regulation of genes affecting cell differentiation and DNA repair mechanisms, which might be responsible for increased cell death after storage [41].

Given this, while researchers continue the search for the optimum physiological approach to preserve cultured corneal/limbal sheets at the genomic, proteomic, structural and functional levels, there still remains the initial variability of different culture techniques and cell phenotypes used, making direct comparison and conclusions on the optimal preservation method difficult.

15.7 Conclusion

Development of proper standardized and safe short and long-term preservation methods for cultured corneal epithelial SCs will provide tremendous advances in regenerative medicine. While today's well-established laboratory and clinical standards enable transplantation of cultured limbal epithelial sheets only in specialized centers worldwide, cryopreservation methods might enable a more widespread use. Short and long-term storage will not only contribute to the improved transportation possibilities, it will give more opportunity for quality controls and increase the flexibility in scheduling transplantation surgeries. Furthermore, stored reserve tissue for repeated operations might increase safety and be even less immunogenic, which might improve the long-term rehabilitation of patients with debilitating severe bilateral ocular surface disease.

However, to date, there are still several issues that need to be overcome before cryopreservation methods can be used routinely in a day-to-day clinical practice. As such, reproducible and standardized freezing and thawing protocols need to be further developed in regard to preserve a sufficient SC content and their recovery, with the whole structural integrity of cultured epithelial sheets. Due to various culturing techniques used for corneal SC culture, including cell suspension or explant culture, different media and scaffold carrier usage, the results of cryopreservation outcomes between these techniques are difficult to assess and compare. Therefore, the selection of optimal freezing temperature and storage medium for cryopreservation currently depend mainly on the few reported cases of different cell type used (e.g. conjunctival, corneal, oral mucosal) and the complexity of the cultured tissue, which can differ significantly.

Without doubt cryopreservation of suspended cells yielded better results than whole epithelial sheet storage, as was discussed in this chapter. Studies showed that the first step in cultured epithelial sheet generation could be successfully postponed for longer time periods. However, as we strive to storage more complex cultured tissues, the morphological integrity after thawing processes comes to the front. Cryopreservation of adherent fully stratified ocular epithelia has been shown more challenging. Currently, there is no consensus regarding the final functional and morphological outcomes after cryopreservation of cultured corneal epithelial sheets based on previous *in vitro* studies. Therefore, further studies testing different biological scaffolds, freezing media compositions and cryopreservation techniques need to be prepared to further elucidate the biological mechanisms underlying these changes. In addition, studies on hypothermic preservation methods for cultured epithelial sheets are implicating more convenient and lower-cost strategies for short tissue storage.

In this regard, biobanking industry is facing new challenges at providing GMP compliant, accessible, reliable, validated, pretested, and robust storage protocols for cryopreservation and hypothermic preservation of desired *in vitro* cultured ocular surface epithelial sheets, which will enable a new era of industrialized medical SC-based treatment strategies.

References

1. De Luca M, Pellegrini G, Green H (2006) Regeneration of squamous epithelia from stem cells of cultured grafts. Regen Med 1:45–57
2. Jackson C, Aabel P, Eidet JR, Messelt EB, Lyberg T, von Unge M et al (2014) Effect of storage temperature on cultured epidermal cell sheets stored in xenobiotic-free medium. PLoS One 9(8), e105808
3. Eidet JR, Utheim ØA, Islam R, Lyberg T, Messelt EB, Dartt DA et al (2015) The impact of storage temperature on the morphology, viability, cell number and metabolism of cultured human conjunctival epithelium. Curr Eye Res 40(1):30–39
4. Dua HS, Shanmuganathan VA, Powell-Richards AO, Tighe PJ, Joseph A (2005) Limbal epithelial crypts: a novel anatomical structure and a putative limbal stem cell niche. Br J Ophthalmol 89:529–532
5. Dua HS, Forrester JV (1990) The corneoscleral limbus in human corneal epithelial wound healing. Am J Ophthalmol 110(6):646–656
6. Baylis O, Figueiredo F, Henein C, Lako M, Ahmad S (2011) 13 years of cultured limbal epithelial cell therapy: a review of the outcomes. J Cell Biochem 112:993–1002
7. Pellegrini G, Traverso CE, Franzi AT, Zingirian M, Cancedda R, De Luca M (1997) Long-term restoration of damaged corneal surface with autologous cultivated corneal epithelium. Lancet 349:990–993
8. Burman S, Sangwan V (2008) Cultivated limbal stem cell transplantation for ocular surface reconstruction. Clin Ophthalmol 2(3):489–502
9. Nishida K, Yamato M, Hayashida Y, Watanabe K, Yamamoto K, Adachi E et al (2004) Corneal reconstruction with tissue-engineered cell sheets composed of autologous oral mucosal epithelium. N Engl J Med 351:1187–1196
10. Ang LP, Tanioka H, Kawasaki S, Ang LP, Yamasaki K, Do TP et al (2010) Cultivated human conjunctival epithelial transplantation for total limbal stem cell deficiency. Invest Ophthalmol Vis Sci 51(2):758–764
11. Utheim TP (2013) Limbal epithelial cell therapy: past, present, and future. Methods Mol Biol 1014:3–43
12. Raeder S, Utheim TP, Utheim OA, Nicolaissen B, Roald B, Cai Y et al (2007) Effects of organ culture and Optisol-GS storage on structural integrity, phenotypes, and apoptosis in cultured corneal epithelium. Invest Ophthalmol Vis Sci 48(12):5484–5493
13. Di Iorio E, Ferrari S, Fasolo A, Böhm E, Ponzin D, Barbaro V (2010) Techniques for culture and assessment of limbal stem cell grafts. Ocul Surf 8(3):146–153
14. Massie I, Selden C, Hodgson H, Fuller B, Gibbons S, Morris GJ (2014) GMP cryopreservation of large volumes of cells for regenerative medicine: active control of the freezing process. Tissue Eng Part C Methods 20(9):693–702
15. Mohamed-Noriega K, Toh KP, Poh R, Balehosur D, Riau A, Htoon HM et al (2011) Cornea lenticule viability and structural integrity after refractive lenticule extraction (ReLEx) and cryopreservation. Mol Vis 17:3437–3449
16. Ti SE, Grueterich M, Espana EM, Touhami A, Anderson DF, Tseng SC (2004) Correlation of long term phenotypic and clinical outcomes following limbal epithelial transplantation cultivated on amniotic membrane in rabbits. Br J Ophthalmol 88(3):422–427
17. Qu L, Yang X, Wang X, Zhao M, Mi S, Dou Z et al (2009) Reconstruction of corneal epithelium with cryopreserved corneal limbal stem cells in a rabbit model. Vet J 179(3):392–400
18. Grueterich M, Espana EM, Tseng SC (2003) Ex vivo expansion of limbal epithelial stem cells: amniotic membrane serving as a stem cell niche. Surv Ophthalmol 48(6):631–646
19. Kito K, Kagami H, Kobayashi C, Ueda M, Terasaki H (2005) Effects of cryopreservation on histology and viability of cultured corneal epithelial cell sheets in rabbit. Cornea 24(6):735–741

20. Rama P, Bonini S, Lambiase A, Golisano O, Paterna P, De Luca M et al (2001) Autologous fibrin-cultured limbal stem cells permanently restore the corneal surface of patients with total limbal stem cell deficiency. Transplantation 72:1478–1485
21. Pellegrini G, Golisano O, Paterna P, Lambiase A, Bonini S, Rama P et al (1999) Location and clonal analysis of stem cells and their differentiated progeny in the human ocular surface. J Cell Biol 145(4):769–782
22. Woods EJ, Benson JD, Agca Y, Critser JK (2004) Fundamental cryobiology of reproductive cells and tissues. Cryobiology 48(2):146–156
23. Armitage WJ (2009) Cryopreservation for corneal storage. Dev Ophthalmol 43:63–69
24. Corydon C, Hjortdal J, Ehlers N (2009) Re-examination of organ-cultured, cryopreserved human corneal grafts after 27 years. Acta Ophthalmol 87(2):173–175
25. Mueller FO, Smith AU (1963) Some experiments on grafting frozen corneal tissue in rabbits. Exp Eye Res 44:237–246
26. Capella JA, Kaufman HE, Robbins JE (1965) Preservation of viable corneal tissue. Arch Ophthalmol 74:669–673
27. Ehlers N, Sperling S (1983) Ultrastructure of cryopreserved, functioning human corneal endothelium. Acta Ophthalmol 61:245–253
28. Bratanov M, Neronov A, Nikolova E (2009) Limbal explants from cryopreserved cadaver human corneas. Immunofluorescence and light microscopy of epithelial cells growing in culture. Cryo Letters 30(3):183–189
29. Pellegrini G, Rama P, Di Rocco A, Panaras A, De Luca M (2014) Concise review: hurdles in a successful example of limbal stem cell-based regenerative medicine. Stem Cells 32(1):26–34
30. Corradini F, Venturi B, Pellegrini G, De Luca M (2012) Methods for characterization/manipulation of human corneal stem cells and their applications in regenerative medicine. Methods Mol Biol 916:357–372
31. Schrader S, Notara M, Beaconsfield M, Tuft S, Geerling G, Daniels JT (2009) Conjunctival epithelial cells maintain stem cell properties after long-term culture and cryopreservation. Regen Med 4(5):677–687
32. Vasania VS, Prasad P, Gill RK, Mehta A, Viswanathan C, Sarang S et al (2011) Molecular and cellular characterization of expanded and cryopreserved human limbal epithelial stem cells reveal unique immunological properties. Exp Eye Res 92(1):47–56
33. Mi S, Yang X, Zhao Q, Qu L, Chen S, Meek MK et al (2008) Reconstruction of corneal epithelium with cryopreserved corneal limbal stem cells in a goat model. Mol Reprod Dev 75(11):1607–1616
34. Oh JY, Kim MK, Shin KS, Shin MS, Wee WR, Lee JH et al (2007) Efficient cryopreservative conditions for cultivated limbal and conjunctival epithelial cells. Cornea 26(7):840–846
35. Yeh HJ, Yao CL, Chen HI, Cheng HC, Hwang SM (2008) Cryopreservation of human limbal stem cells ex vivo expanded on amniotic membrane. Cornea 27(3):327–333
36. Spindler R, Wolkers WF, Glasmacher B (2011) Dimethyl sulfoxide and ethylene glycol promote membrane phase change during cryopreservation. Cryo Letters 32(2):148–157
37. Promprasit D, Bumroongkit K, Tocharus C, Mevatee U, Tananuvat N (2015) Cultivation and phenotypic characterization of rabbit epithelial cells expanded ex vivo from fresh and cryopreserved limbal and oral mucosal explants. Curr Eye Res 40(3):274–281
38. Fan WX, Ma XH, Ge D, Liu TQ, Cui ZF (2009) Cryoprotectants for the vitrification of corneal endothelial cells. Cryobiology 58(1):28–36
39. Utheim TP, Raeder S, Utheim ØA, Cai Y, Roald B, Drolsum L et al (2007) A novel method for preserving cultured limbal epithelial cells. Br J Ophthalmol 91(6):797–800
40. Eidet JR, Utheim OA, Raeder S, Dartt DA, Lyberg T, Carreras E et al (2012) Effects of serum-free storage on morphology, phenotype, and viability of ex vivo cultured human conjunctival epithelium. Exp Eye Res 94(1):109–116
41. Utheim TP, Salvanos P, Aass Utheim Ø, Ræder S, Pasovic L, Olstad OK et al (2016) Transcriptome analysis of cultured limbal epithelial cells on an intact amniotic membrane following hypothermic storage in optisol-GS. J Funct Biomater 18;7(1). pii: E4

Cryopreservation of Hair-Follicle Associated Pluripotent (HAP) Stem Cells Maintains Differentiation and Hair-Growth Potential

16

Robert M. Hoffman, Satoshi Kajiura, Wenluo Cao, Fang Liu, and Yasuyuki Amoh

Abstract

Hair follicles contain nestin-expressing pluripotent stem cells which originate above the bulge area of the follicle, below the sebaceous gland. We have termed these cells hair follicle-associated pluripotent (HAP) stem cells. We have established efficient cryopreservation methods of the hair follicle that maintain the pluripotency of HAP stem cells as well as hair growth. We cryopreserved the whole hair follicle by slow-rate cooling in TC-Protector medium or in DMSO-containing medium and storage in liquid nitrogen or at −80 °C. After thawing and culture of the cryopreserved whisker follicles, growing HAP stem cells formed hair spheres. The hair spheres contained cells that differentiated to neurons, glial cells, and other cell types. The hair spheres derived from slow-cooling cryopreserved hair follicles were as pluripotent as hair spheres from fresh hair follicles. We have also previously demonstrated that cryopreserved mouse whisker hair follicles maintain their hair-growth potential. DMSO better cryopreserved mouse whisker follicles compared to glycerol. DMSO-cryopreserved hair follicles also maintained the HAP stem cells, evidenced by $P75_{ntr}$ expression. Subcutaneous transplantation of DMSO-cryopreserved hair follicles in nude mice resulted in extensive hair fiber growth over 8 weeks, indicat-

R.M. Hoffman (✉)
AntiCancer, Inc, 7917 Ostrow St, San Diego, CA 92111, USA

Department of Surgery, University of California, San Diego, CA, USA
e-mail: all@anticancer.com; http://www.anticancer.com

S. Kajiura • Y. Amoh
Department of Dermatology, Kitasato University, Sagamihara, Japan

W. Cao
AntiCancer, Inc, 7917 Ostrow St, San Diego, CA 92111, USA

Department of Surgery, University of California, San Diego, CA, USA

Department of Anatomy, Second Military Medical University, Shanghai, China

F. Liu
Department of Anatomy, Second Military Medical University, Shanghai, China

F. Karimi-Busheri, M. Weinfeld (eds.), *Biobanking and Cryopreservation of Stem Cells*, Advances in Experimental Medicine and Biology 951, DOI 10.1007/978-3-319-45457-3_16

ing the functional recovery of hair-shaft growth of cryopreserved hair follicles. HAP stem cells can be used for nerve and spinal-cord repair. This biobanking of hair follicles can allow each patient the potential for their own stem cell use for regenerative medicine or hair transplantation.

Keywords
Hair follicle • Stem cells • Pluripotent • Hair growth • Bio-banking

Abbreviations

HAP	Hair follicle-associated pluripotent
ES	Embryonic stem
GFP	Green fluorescent protein
ND-GFP	Nestin-driven-GFP
SMA	Smooth muscle actin
GFAP	Glial fibrillary acidic protein
RT-PCR	Real-time polymerase chain reaction

16.1 Introduction

Nestin-expressing stem cells of the hair follicle, discovered by our laboratory [1], have been shown to be able to form outer root sheaths of the follicle as well as neurons and many other non-follicle cell types [2]. We have termed the nestin-expressing stem cells of the hair follicle hair follicle-associated pluripotent (HAP) stem cells. We have shown that the HAP stem cells from the hair follicle can effect the repair of peripheral-nerve and spinal-cord injury [2–5]. The hair-follicle stem cells differentiate into neuronal and glial cells after transplantation to the injured peripheral nerve and spinal cord, and enhance injury repair and locomotor recovery [2–5].

When the excised hair follicles with their nerve stump were placed in Gelfoam ® 3D histoculture, HAP stem cells grew and extended the hair follicle nerve which consisted of βIII-tubulin-positive fibers with F-actin expression at the tip. These findings indicate that βIII-tubulin-positive fibers elongating from the whisker follicle sensory-nerve stump were growing axons. The growing whisker sensory nerve was highly enriched in HAP stem cells, which appeared to play a major role in its elongation and interaction with other nerves in 3D Gelfoam® histoculture, including the sciatic nerve, the trigeminal nerve, and the trigeminal nerve ganglion. These results suggest that a major function of the HAP stem cells in the hair follicle is for growth of the follicle sensory nerve [6].

Recently, we have shown that HAP stem cells can differentiate into beating cardiac muscle cells [7, 8]. HAP stem cells have critical advantages for regenerative medicine over embryonic stem (ES) cells and induced pluripotent stem (iPS) cells in that they are highly accessible from each patient, thereby eliminating immunological issues since they are autologous, require no genetic manipulation, are non-tumorigenic, and do not present ethical issues for regenerative medicine [9].

Efficient cryopreservation methods of the hair follicle which maintained the pluripotency of HAP stem cells and hair shaft growth are presented here [8].

16.2 Materials and Methods

16.2.1 Mice

Transgenic C57/B6-green fluorescent protein (GFP) mice [10] and transgenic mice with nestin-driven (ND)-GFP and non-transgenic (*nu/nu*) nude mice (AntiCancer, Inc., San Diego, CA) were used in these studies [11, 12]. The C57/B6-GFP mice expressed the *Aequorea victoria* GFP under the control of the chicken β-actin promoter and cytomegalovirus enhancer. ND-GFP mice express GFP under control of the nestin regulatory element. All of the tissues from this transgenic line, with the exception of erythrocytes and hair, were fluorescent green under excitation light. All animal experiments were conducted according to the *Guidelines for Animal Experimentation* at the Kitasato University [13].

16.2.2 Isolation of Vibrissa Hair Follicles

To isolate the vibrissa follicles from mice, the upper lip containing the vibrissa pad was cut under anesthesia and the inner surface was exposed. Entire vibrissa hair follicles were dissected under a binocular microscope and plucked from the pad by pulling them gently by the neck with fine forceps. The isolated vibrissae were washed in Dulbecco's modified Eagle's medium (DMEM)/F12 (GIBCO-BRL) with 2 % B-27 (GIBCO-BRL) and 50 μg/mL gentamicin (Gibco-BRL). The follicles were divided into three parts using a surgical knife and fine forceps under a binocular microscope as previously described [14]. All surgical procedures were done under a sterile environment.

16.2.3 Hair-Follicle and Hair-Sphere Culture

The upper part of the vibrissa hair follicle was isolated and cultured in DMEM with 10 % fetal bovine serum (FBS). After 4 weeks of culture, cells growing out from the upper follicle were treated enzymatically with Accumax (Innovative Cell Technologies, Inc.) to detach them. The detached cells were then transferred to non-adhesive culture dishes with DMEM/F12 containing 2 % B-27. After 1 week of culture, the growing cells formed hair spheres containing nestin-expressing HAP stem cells. After the change of medium to DMEM containing 10 % FBS and 2 days of additional culture, the HAP stem cells differentiated to β-III tubulin-positive neurons, glial fibrillary acidic protein (GFAP)-positive glial cells, K15-positive keratinocytes, and smooth muscle actin (SMA)-positive smooth muscle cells [13, 14].

16.2.4 Cryopreservation of the Whole Hair Follicle

Five whole vibrissa follicles were transferred to cryovials, and TC-Protector medium (DS Pharma Biomedical Co.) was added (500 μL). Eighteen cryovials containing the vibrissa follicles were stored in a −80 °C freezer overnight and then transferred to a liquid nitrogen tank. Three mice were used for this method for three independent experiments involving one mouse each.

The cryopreserved vibrissa follicles were thawed at 37 °C in a water bath for 60–90 s (slow recovery) with gentle shaking and separated in three parts (upper, middle, and lower). The upper part of hair follicle was isolated and cultured in DMEM with 10 % FBS [13].

Whisker hair follicles were also cryopreserved in either 10 % glycerol or DMSO with 90 % FBS added (1 ml) as the freezing medium. For short time periods, cryovials containing hair follicles were stored in an −80 °C freezer [12].

16.2.5 Cryopreservation of Hair Spheres

The upper part of hair follicle was isolated and cultured in DMEM with 10 % FBS. After 4 weeks of culture, cells that grew out from the follicles were transferred to DMEM/F12 with 2 % B-27 as described above. After 1 week of culture, hair spheres formed. Fifty spheres were transferred to cryovials and TC-Protector medium was added (500 μL). Nine cryovials with spheres were stored in a −80 °C freezer overnight and then transferred to a liquid-nitrogen tank. The stored spheres were thawed at 37 °C in a water bath with gentle shaking. The spheres were cultured in DMEM with 10 % FBS [13].

16.2.6 Immunohistochemistry of Hair Follicles

Hair follicles were fixed in 4 % paraformaldehyde overnight at 4 °C, after dehydration in 15 % and then 30 % sucrose overnight at 4 °C. Tissues were then embedded in the OTC freezing compound (Thermo Scientific, Kalamazoo, MI) and frozen sections of 10 μm were prepared with a CM1850 cryostat (Leica, Buffalo Grove, IL). The frozen sections were washed with PBS three times. After incubation with 0.4 % Triton X-100 for 10 min, the sections were incubated with 20 % BSA in PBS containing 0.4 % Triton X-100 for 1 h at room temperature (RT). Primary antibody, anti- $p75^{NTR}$ (rabbit, 1:400, Cell Signaling), was applied at 4 °C overnight. After three washes in PBS, secondary antibody, Alexa Fluor® 555-conjugated anti-rabbit (1:100, Jackson Laboratories), was applied in PBS at RT for 1 h. After three washes in PBS, the nuclei were stained with DAPI (1:2000, Invitrogen) for 5 min. Images were obtained with a confocal FV1000 laser-scanning microscope [12].

16.2.7 Immunohistochemistry of Hair Spheres

After a 1-week culture of hair spheres in DMEM with 10 % FBS, the hair spheres were stained in individual wells with the following antibodies: anti-III-β-tubulin mAb (1:500, Tuj1 clone, Covance Research Products); anti-GFAP chicken polyclonal Ab (1:200, Abcam); anti-SMA mAb (1:400, Lab Vision); and anti-K15 mAb (1:100, Lab Vision). Secondary antibodies used were as follows: Alexa Fluor 568-labeled goat anti-mouse IgG (1:400, Molecular Probes) was used for anti-III-β-tubulin, anti-SMA, and anti-K15 and Alexa Fluor 568-labeled goat anti-chicken IgG (1:1000, Molecular Probes) was used for anti-GFAP. For quantification of the percentage of cells producing a given marker protein, at least four fields were photographed in any given experiment. The number of positive cells was determined relative to the total number of cells with 4′,6-diamidino-2-phenylindole dihydrochloride (Molecular Probes) nuclear staining [12].

16.2.8 Expression of Stem-Cell Marker Genes

The mRNA levels of stem-cell marker genes (nestin, Sox2, SSEA-1) were examined using real-time polymerase chain reaction (RT-PCR). After 1 week of culture in DMEM/F12 containing 2 % B-27, the growing cells formed hair spheres. Total RNA from 100 hair spheres was extracted using the RNeasy Plus Mini kit (Qiagen). c-DNA was synthesized by high-capacity RNA with a cDNA kit (ABI). Real-time PCR on CFX96 (Bio-Rad) with TaqMan Gene Expression Assays (ABI) and TaqMan Probes were used as follows: r18s:Hs99999901_s1; Nestin: Mm00450205_m1; Sox2: Mm03053810_s1; and SSEA-1: Mm00487448_s1. The mRNA levels were normalized by comparison with r18s.

Western blot analysis was also used to detect the expression of SSEA-1. After 1 week of culture in DMEM/F12 containing 2 % B-27, the growing cells formed hair spheres. Total proteins (30 μg/well) from hair spheres were subjected to 4–20 % SDS-PAGE (sodium dodecyl sulfate-polyacrylamide gel electrophoresis) and then transferred to Immobilon-P membranes (Millipore Corporation). SSEA-1 was detected by an anti-SSEA-1 primary antibody (1:250, BioLegend), followed by a mixture of peroxidase-conjugated anti-mouse IgA, IgG and IgM (Chemicon) with enhanced chemiluminescence plus a Western blotting detection system (Amersham Biosciences). Results are expressed as mean±standard deviation (SD) for three samples each.

16.2.9 Gelfoam® Histoculture of Whisker Hair Follicles

After 1 week, the cryopreserved hair follicles were thawed at 37 °C in a water bath with gentle shaking. The hair follicles were washed at

DMEM medium for in vitro culture and in vivo transplantation. The cryopreserved hair follicles or fresh-isolated hair follicles were carefully placed on sterile Gelfoam® (Pharmacia and Upjohn Co., Kalamazoo, MI), hydrated with DMEM-F12 (GIBCO, Life Technologies, Grand Island, NY) containing B-27 (2.5 %) (GIBCO), N2 (1 %) (GIBCO) and 1 % penicillin and streptomycin (GIBCO). The Gelfoam® follicles cultures [6, 11, 12, 15] were incubated at 37 °C, 5 % CO_2. The medium was changed every other day. High-resolution images of hair follicle growth on Gelfoam® were obtained with a confocal laser-scanning microscope (Fluoview FV1000, Olympus, Tokyo, Japan) [6, 11, 12, 15].

16.2.10 Subcutaneous Transplantation of Hair Follicles

Non-transgenic nude mice were anesthetized with a ketamine mixture (intramuscular injection of a 0.02 ml solution of 20 mg/kg ketamine, 15.2 mg/kg xylazine, and 0.48 mg/kg acepromazine maleate). Thawed hair follicles were transplanted into the subcutis on the right flank of nude mice and fresh isolated hair follicles were transplanted to the left flank as controls. A skin flap was made to expose the growing hair follicle every 2 weeks (2, 4, 6, and 8 weeks) in order to observe hair growth. Images were obtained with a Dino-Lite microscope portable imager (AM4113TGFBW Dino-Lite Premier; AnMo Electronics Corporation, Taiwan) [12, 16].

16.2.10 Measurement of Hair Shaft Length

The in vivo hair follicle images obtained with the Dino-Lite were used to determine the length of each hair shaft using Image Pro Plus 6.0 software. The average length of the ten longest hair follicles in each group are presented as mean±SEM [12].

16.2.10 Statistical Analysis

All experiments were conducted in three independent cultures. Results are expressed as mean±SD. *P*-values were calculated with the paired Student's *t*-test. Group differences were obtained using the ANOVA test. The significant level for all tests was $P < 0.05$ [12].

16.3 Results and Discussion

16.3.1 Effects of Cryopreservation of the Whisker Hair Follicle and HAP Stem Cells

After thawing and 4 weeks culture in DMEM with 10 % FBS, the upper parts of hair follicles were placed in dishes. The attachment rate of the hair follicles and number of cells growing out from the follicle were quantified as follows: 95.2 % ± 7.7 % of fresh hair follicles (nonfrozen control) attached and $5.9 \pm 0.9 \times 10^4$ cells per hair follicle grew out. 90.8 % ± 13.3 % of hair follicles cryopreserved by the slow-rate cooling method attached and $3.7 \pm 1.3 \times 10^4$cells per hair follicle grew out [13].

The upper part of the fresh vibrissa had 38.0 ± 9.1 hair spheres growing, with each hair sphere containing $\sim 1 \times 10^2$ nestin-expressing HAP stem cells. Slow-rate cooling-cryopreserved hair follicles formed 41.6 ± 5.9 hair spheres per upper part of hair follicle, with each hair sphere containing $\sim 1 \times 10^2$ nestin-expressing HAP stem cells [13].

With hair spheres derived from fresh hair follicles, ~4.3 % ± 3.6 % cells differentiated to anti-III-β-tubulin-positive neurons; 30.2 % ± 8.1 % of the cells differentiated to GFAP-positive glial cells; 40.2 % ± 8.0 % of the cells differentiated to SMA-positive smooth muscle cells; and 7.2 % ± 5.7 % of the cells differentiated to K15-positive keratinocytes. With hair spheres derived from slow-rate cooling-cryopreserved hair follicles, ~12.6 % ± 3.2 % of the cells differentiated to anti-III-β-tubulin-positive neurons; 34.0 % ± 6.6 % of the cells differentiated to GFAP-positive glial cells; 63.7 % ± 10.0 % of the cells differentiated to SMA-positive smooth muscle cells; and 11.0 % ± 2.1 % of the cells differentiated to K15-positive keratinocytes [13].

Nestin RNA levels, relative to r18s, were 1.00 ± 0.11 in fresh hair follicles and 1.01 ± 0.02 in slow-rate cooled hair follicles. Sox2 mRNA levels, relative to r18s, were 1.00 ± 0.02 in fresh hair follicles and 0.81 ± 0.07 in slow-rate cooled hair follicles. SSEA-1 mRNA levels, relative to r18s, were 1.00 ± 0.02 in fresh hair follicles and 0.90 ± 0.03 in slow-rate cooled hair follicles. Thus, the mRNA levels of stem-cell marker genes were maintained in the slow-rate cooling method [13].

Western blot analysis was performed to detect the expression of SSEA-1. The protein levels of SSEA-1 were maintained in the slow-rate cooling method [13].

16.3.2 Cryopreservation of Hair Spheres

Fresh hair spheres completely attached when placed in plastic dishes. In contrast, after slow-rate cooling cryopreservation, 5.4 % ± 3.2 % of the hair spheres survived. With vitrification rapid-cooled cryopreservation, ~12.5 % + 10.8 % of the hair spheres survived. However, after thawing, slow rate cooled and vitrification rapid-cooled hair spheres did not differentiate. Our results demonstrate that slow-rate cooling cryopreservation of hair follicles preserves the same attachment rate and production of pluripotent hair spheres as fresh hair follicles [13].

16.3.3 Whisker Hair Follicle and HAP Stem Cell Behavior in Gelfoam® Histoculture After Cryopreservation

After thawing and culture on Gelfoam® for 4 weeks, the ND-GFP fluorescence of the HAP stem cells was measured. Confocal real-time 3D images of ND-GFP HAP hair follicle on Gelfoam® showed that ND-GFP HAP stem-cell-fluorescence in fresh hair follicle had dramatic increases at 10, 16 and 26 days. The increases in HAP stem cell ND-GFP fluorescence after DMSO cryopreservation of hair follicles was slower than in fresh follicles. Glycerol-cryopreserved follicles had less increases in ND-GFP fluorescence than the DMSO-preserved follicles (Fig. 16.1). At day 26, 2/6 hair follicles cryopreserved in glycerol recovered to grow ND-GFP HAP stem cells, 2/6 had poor growth, and 2/6 were consider dead. In contrast, 6/6 hair follicles frozen in DMSO recovered [12].

After 26 days in Gelfoam® histoculture, 3D confocal images showed that ND-GFP HAP stem cell outgrowth occurred both at the upper and lower part of fresh hair follicles. DAPI nuclear staining showed that the fresh hair follicles maintained normal structures in the hair shaft, inner root sheath, and outer root sheath, and a large amount of ND-GFP HAP stem cells grew around the bulb area. In DMSO-cryopreserved follicles, ND-GFP 3D images indicated similar HAP-stem cell outgrowth both in the upper and lower part of hair follicle, but with less cell outgrowth than in fresh hair follicles. With glycerol-cryopreservation, the recovered hair follicles had less ND-GFP HAP stem cells. The cultured hair follicles had ND-GFP HAP stem cells mostly in the outer root sheath and the bulb area after both DMSO- and glycerol-cryopreservation. In contrast, the inner structure of cultured hair follicles had little DAPI nuclear staining remained after either method of cryo-

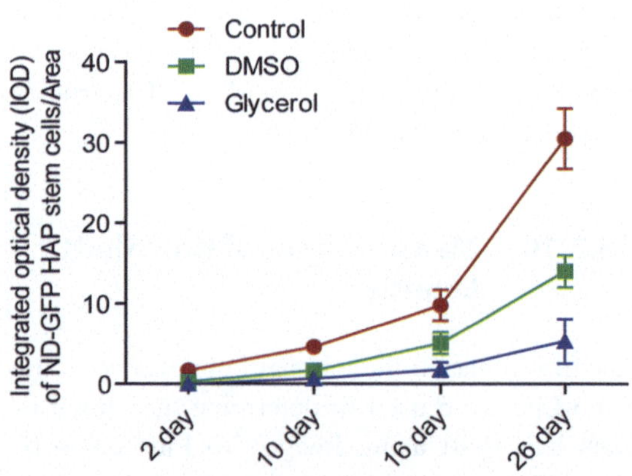

Fig. 16.1 Quantitative analysis of ND-GFP HAP stem cells fluorescence in cultured whisker follicles. DMSO-cryopreserved follicles had more extensive growth of HAP stem cells than in glycerol-cryopreserved follicles, but less than in fresh follicles

preservation. These results indicate cell death of the inner part and structural damage of the frozen hair follicle in both cryopreservation media. The bulb area of the hair follicle had ND-GFP-nestin HAP stem cells growing out, and the dermal papilla area maintained DAPI nuclear staining after both cryopreservation methods. These results suggest the survival of bulb area and dermal papilla cells in both DMSO and glycerol-cryopreserved hair follicles. p75^{NTR} immunostaining also indicated ND-GFP HAP stem cells were maintained in both conditions of cryopreservation [12].

16.3.4 Functional Recovery of Cryopreserved Hair Follicles After Subcutaneous Transplantation

After the hair follicles were cryopreserved in DMSO for 1 week at −80 °C, the hair follicles were thawed and transplanted subcutaneously in nude mice. Each flank was transplanted with five to six hair follicles and a total of three mice were used. Eight of 15 transplanted cryopreserved hair follicles grew out extensive long hair shafts (Fig. 16.2) [12].

In fresh hair follicles, the hair follicle established blood vessel connection with the host nude mice and grew rapidly by week 2 (1.26 mm ± 0.15 mm). At week 4, the average of the ten longest hair-shaft length was 3.82 mm ± 0.31 mm. At week 6, the average of the ten longest hair-shaft length was 5.73 mm ± 0.31 mm), and at week 8, the hair shafts length was 7.66 mm ± 0.63 mm. The DSMO-cryopreserved hair follicles also established blood vessel connections with the host nude mice as indicated by red blood in the lower part and upper part of the hair follicle. At week 4, there were more blood vessels surrounding the hair follicles. There were obvious hair roots at the bulb area as well as longer hair shafts shaft at week 4, with the longest length of 1.88 mm ± 0.55 mm, compared to week 2, when the longest hair shaft length was 0.85 mm ± 0.07 mm. These results indicate that the DSMO-cryopreserved hair follicle started to rapidly regrow hair. At week 6, the longest hair shafts grew up to 2.85 mm ± 0.66 mm and more hair roots in the bulb area were observed, as well as long hair shafts growing out. Eventually, at week 8, there were a large amount of long hair shafts with the longest with a length of 4.92 mm ± 0.97 mm. Our results showed that DSMO-cryopreserved hair follicles maintained the ability to produce hair shafts after subcutaneous transplantation [12].

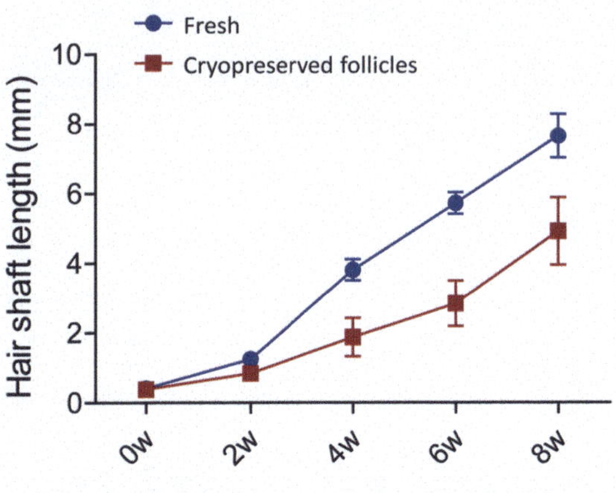

Fig. 16.2 Quantitative data of hair-shaft growth in fresh and DMSO-cryopreserved hair follicles after hair-follicle subcutaneous transplantation in nude mice. Each data point represents the ten longest hair follicles for each group

16.4 Conclusions

Hair follicles have important potential including hair transplantation and use of HAP stem cells for regenerative medicine for nerves, spinal cord, heart and other applications.

The present chapter suggests that all these potential applications are maintained after hair follicle cryopreservation. Each potential patient can, therefore, have a bio-bank of their own hair follicles.

References

1. Li L, Mignone J, Yang M, Matic M, Penman S, Enikolopov G, Hoffman RM (2003) Nestin expression in hair follicle sheath progenitor cells. Proc Natl Acad Sci U S A 100:9958–9961
2. Amoh Y, Li L, Katsuoka K, Penman S, Hoffman RM (2005) Multipotent nestin-positive, keratin-negative hair-follicle-bulge stem cells can form neurons. Proc Natl Acad Sci U S A 102:5530–5534

3. Amoh Y, Li L, Katsuoka K, Hoffman RM (2008) Multipotent hair follicle stem cells promote repair of spinal cord injury and recovery of walking function. Cell Cycle 7:1865–1869
4. Liu F, Uchugonova A, Kimura H, Zhang C, Zhao M, Zhang L, Koenig K, Duong J, Aki R, Saito N, Mii S, Amoh Y, Katsuoka K, Hoffman RM (2011) The bulge area is the major hair follicle source of nestin-expressing pluripotent stem cells which can repair the spinal cord compared to the dermal papilla. Cell Cycle 10:830–839
5. Amoh Y, Kanoh M, Niiyama S, Hamada Y, Kawahara K, Sato Y, Hoffman RM, Katsuoka K (2009) Human hair follicle pluripotent stem (hfPS) cells promote regeneration of peripheral-nerve injury: an advantageous alternative to ES and iPS cells. J Cell Biochem 107:1016–1020
6. Mii S, Duong J, Tome Y, Uchugonova A, Liu F, Amoh Y, Saito N, Katsuoka K, Hoffman RM (2013) The role of hair follicle nestin-expressing stem cells during whisker sensory-nerve growth in long-term 3D culture. J Cell Biochem 114:1674–1684
7. Yashiro M, Mii S, Aki R, Hamada Y, Arakawa N, Kawahara K, Hoffman RM, Amoh Y (2015) From hair to heart: nestin-expressing hair-follicle-associated pluripotent (HAP) stem cells differentiate to beating cardiac muscle cells. Cell Cycle 14:2362–2366
8. Yamazaki A, Yashiro M, Mii S, Aki R, Hamada Y, Arakawa N, Kawahara K, Hoffman RM, Amoh Y (2016) Isoproterenol directs hair follicle-associated pluripotent (HAP) stem cells to differentiate *in vitro* to cardiac muscle cells which can be induced to form beating heart muscle tissue sheets. Cell Cycle 15:760–765
9. Hoffman RM (2016) Chap. 1: Introduction to hair-follicle-associated pluripotent (HAP) stem cells. In: Hoffman RM (ed) Multipotent stem cells of the hair follicle: methods and protocols, (Walker JM, series ed.). Methods Mol Biol 1453:0–5. Humana Press (Springer Science+Business Media), New York. doi:10.1007/978-1-4939-3786-8_1
10. Okabe M, Ikawa M, Kominami K, Nakanishi T, Nishimune Y (1997) 'Green mice' as a source of ubiquitous green cells. FEBS Lett 407:313–319
11. Duong J, Mii S, Uchugonova A, Liu F, Moossa AR, Hoffman RM (2012) Real-time confocal imaging of trafficking of nestin-expressing multipotent stem cells in mouse whiskers in long-term 3-D histoculture. In Vitro Cell Dev Biol Anim 48:301–305
12. Cao W, Li L, Tran B, Kajiura S, Amoh Y, Liu F, Hoffman RM (2015) Extensive hair shaft growth after mouse whisker follicle isolation, cryopreservation and transplantation in nude mice. PLoS ONE 10:e0145997
13. Kajiura S, Mii S, Aki R, Hamada Y, Arakawa N, Kawahara K, Li L, Katsuoka K, Hoffman RM, Amoh Y (2015) Cryopreservation of the hair follicle maintains pluripotency of nestin-expressing hair follicle-associated pluripotent stem cells. Tissue Eng Part C 21:825–831
14. Amoh Y, Mii S, Aki R, Hamada Y, Kawahara K, Hoffman RM, Katsuoka K (2012) Multipotent nestin-expressing stem cells capable of forming neurons are located in the upper, middle and lower part of the vibrissa hair follicle. Cell Cycle 11:3513–3517
15. Mii S, Uehara F, Yano S, Tran B, Miwa S, Hiroshima Y, Amoh Y, Katsuoka K, Hoffman RM (2013) Nestin-expressing stem cells promote nerve growth in long-term 3-dimensional Gelfoam®-supported histoculture. PLoS ONE 8:e67153
16. Hiroshima Y, Maawy A, Sato S, Murakami T, Uehara F, Miwa S, Yano S, Momiyama M, Chishima T, Tanaka K, Bouvet M, Endo I, Hoffman RM (2014) Hand-held high-resolution fluorescence imaging system for fluorescence-guided surgery of patient and cell-line pancreatic tumors growing orthotopically in nude mice. J Surg Res 187:510–517

Cryopreservation and Banking of Dental Stem Cells

17

Petra Hilkens, Ronald B. Driesen, Esther Wolfs, Pascal Gervois, Tim Vangansewinkel, Jessica Ratajczak, Yörg Dillen, Annelies Bronckaers, and Ivo Lambrichts

Abstract

Over the past decade, dental tissues have become an attractive source of mesenchymal stem cells (MSCs). Dental stem cells (DSCs) are not only able to differentiate into adipogenic, chondrogenic and osteogenic lineanges, but an increasing amount of research also pointed out their potential applicability in numerous clinical disorders, such as myocardial infarction, neurodegenerative diseases and diabetes. Together with their multilineage differentiation capacity, their easy availability from extracted third molars makes these stem cells a suitable alternative for bone marrow-derived MSCs. More importantly, DSCs appear to retain their stem cell properties following cryopreservation, a key aspect in their long-term preservation and upscale production. However, the vast number of different cryopreservation protocols makes it difficult to draw definite conclusions regarding the behavior of these stem cells. The routine application and banking of DSCs is also associated with some other pitfalls, such as interdonor variability, cell culture-induced changes and the use of animal-derived culture medium additives. Only thorough assessment of these challenges and the implementation of standardized, GMP procedures will successfully lead to better treatment options for patients who no longer benefit from current stem cell therapies.

P. Hilkens (✉) • R.B. Driesen • E. Wolfs • P. Gervois
T. Vangansewinkel • J. Ratajczak • Y. Dillen
A. Bronckaers • I. Lambrichts
Laboratory of Morphology, Biomedical Research Institute (BIOMED), Hasselt University, Agoralaan, Building C, 3590 Diepenbeek, Belgium
e-mail: petra.hilkens@uhasselt.be; ronald.driesen@uhasselt.be; esther.wolfs@uhasselt.be; pascal.gervois@uhasselt.be; tim.vangansewinkel@uhasselt.be; jessica.ratajczak@uhasselt.be; yorg.dillen@uhasselt.be; annelies.bronckaers@uhasselt.be; ivo.lambrichts@uhasselt.be

F. Karimi-Busheri, M. Weinfeld (eds.), *Biobanking and Cryopreservation of Stem Cells*, Advances in Experimental Medicine and Biology 951, DOI 10.1007/978-3-319-45457-3_17

Keywords

Dental stem cells • Mesenchymal stem cells • Paracrine effects • Multilineage differentiation • Cryopreservation • Dental stem cell banking • Good manufacturing practice

Abbreviations

MSCs	Mesenchymal stem cells
DSCs	Dental stem cells
DPSCs	Dental pulp stem cells
BM-MSCs	Bone marrow-derived MSCs
ASCs	Adipose tissue-derived stem cells
FSCs	Follicle precursor cells
PDL	Periodontal ligament
PDLSCs	Periodontal ligament stem cells
SCAPs	Stem cells from the apical papilla
UMSCs	Umbilical cord MSCs
UCBC	Umbilical cord blood cells
ERM	Epithelial cell rests of Malassez
DePDL	Deciduous periodontal ligament
SHEDs	Stem cells from human exfoliated deciduous teeth
GMP	Good Manufacturing Practice
HLA-D	Human leukocyte antigen
vWF	von Willebrand factor
DFO	Deferoxamine
ISCBI	International Stem Cell Banking Initiative

17.1 Introduction

Mesenchymal stem cells (MSCs) are derived from adult tissues and have been extensively studied as therapeutic application for tissue regeneration in a wide range of diseases. This is due to their multipotent ability to transdifferentiate into other cell types such as chondrocytes, osteocytes and cardiac myocytes, which have high potential to replace damaged tissue. Sources for MSC isolation include the bone marrow (BM-MSCs) [1], adipose tissue (ASCs) [2], Wharton's Jelly in the umbilical cord (UMSCs) [3], umbilical cord blood (UCBC) [4] and as discussed in this chapter, dental tissues.

In 2000, Gronthos et al. discovered an alternative source of MSCs, i.e. odontogenic progenitor cells, which were retrieved from the dental pulp. This finding counted as a starting point for the discovery of a heterogeneous population of dental stem cells (DSCs) which are defined by their localization in the developing tooth and its associated tissues (Fig. 17.1). These populations comprise dental pulp stem cells (DPSCs) [5], stem cells from the apical papilla (SCAPs) [6], periodontal ligament stem cells (PDLSCs) [7] and dental follicle precursor cells (FSCs) [8]. The periodontal ligament (PDL) is believed to contain another source of DSCs residing in the epithelial cell rests of Malassez (ERM) which are present in the periodontal ligament matrix [9]. Full-grown or developing teeth are therefore a valuable source of DSCs. In addition, several studies reported that deciduous teeth can be utilized to isolate stem cells from the pulp of human exfoliated deciduous teeth (SHEDs) [10] and deciduous periodontal ligament (DePDL) [11]. DSCs originate from the neural crest [12, 13] however, a large fraction of these cells are of glial origin [14].

The application of DSCs offers several advantages over more conventionally used MSC populations, such as BM-MSCs. Not only are these cells easily obtained with minimally invasive surgery but once in culture they display a high proliferative capacity, making them a suitable candidate for cell banking. The heterogeneity of DSCs encouraged numerous researchers to explore their multilineage differentiation potential, which resulted in a broad range of differentiation protocols for various cell types. Emerging therapeutic strategies are focusing on the

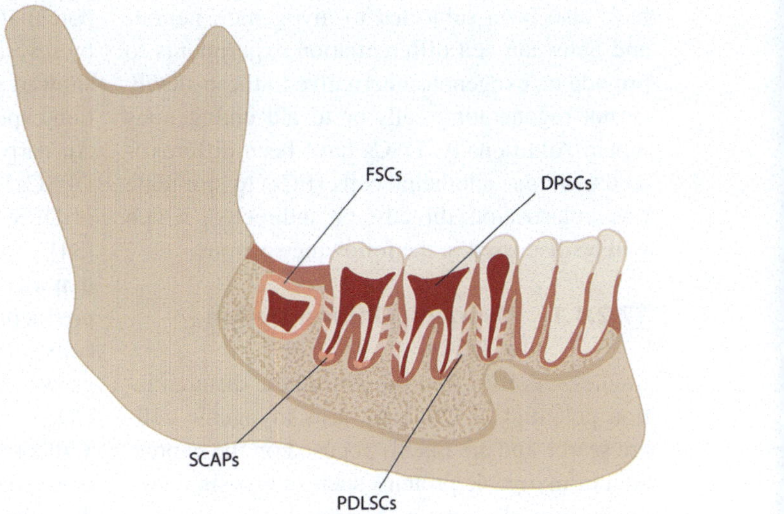

Fig. 17.1 Schematic representation of the different sources of DSCs. *FSCs* dental follicle precursor cells, *DPSCs* dental pulp stem cells, *SCAPs* stem cells from the apical papilla, *PDLSCs* periodontal ligament stem cells (Figure was adapted with permission from [15])

integration of pre-differentiated and preconditioned DSCs in the diseased host tissue exerting a beneficial effect through either intercellular contacts or via paracrine mediated interactions. Autologous as well as allogeneic transplantation of DSCs in clinical studies thus requires the need for a quality control system which provides a detailed assessment of the specific DSC phenotype and its matching paracrine profile. Upscaling of DSCs and the establishment of a dental stem cell banking system should follow standard Good Manufacturing Practice (GMP) regulations. In this chapter we will provide in-depth knowledge on the heterogeneity of DSCs and their concomitant differentiation potential *in vitro* and *in vivo*. We will focus specifically on the current strategies of cryopreservation to maintain phenotypic stability and to prevent risks of cell death and contamination. An overview of the pitfalls which could arise in creating a dental stem cell bank is also given as well as practical guidelines and prospects for application in clinical studies.

17.2 Dental Stem Cells

In order to be classified as MSCs, several requirements need to be fulfilled. DSCs are characterized by the expression of the surface markers CD73, CD90 and CD105 but lack the expression of CD34, CD14, CD45 and human leukocyte antigen (HLA)-D. Moreover, they adhere to plastic and are capable of multilineage differentiation into bone-, cartilage- and fat producing cells [5–8, 10, 16] which classifies them as MSCs as proposed by the International Society for Cellular Therapy [17].

17.2.1 Dental Stem Cells: A Heterogenous Pool of Stem Cells with Multilineage (trans) Differentiation Potential

DSCs form a very heterogenous cell population in culture. High interdonor variability exists in terms of protein marker expression including CD31, CD117 (c-kit), Stro-1 and the low affinity growth factor receptor p75 (LANGFRp75) [18–21]. Remarkably, DSC variants have different stem cell characteristics as shown by studies that used subsets of DSCs for differentiation experiments. For example, Stro-1$^+$/CD146$^+$ SCAPs show an enhanced colony forming unit capacity and osteo/odontogenic mineralization potential [22], a feature which was previously shown to enhance the cementoblastic differentiation of PDLSCs [7]. Stro-1$^+$ DPSCs, but also CD34$^+$/CD117$^+$ DPSCs have been shown to enhance bone formation [23].

In addition to multilineage differentiation towards classical mesodermal lineages, DSCs have also been subjected to myogenic-, neural- and Schwann cell differentiation experiments to provide an exogenous alternative to these slowly or not regenerating cells or to aid endogenous repair. Additionally, DSCs have been differentiated towards endothelial cells (ECs) to stimulate revascularization directly or indirectly, which will be discussed in the following sections.

17.2.1.1 Myogenic Differentiation of DSCs

Studies reporting and supporting the differentiation potential of DSCs towards myogenic cells are scarce and are mainly focused on the expression of myogenic proteins such as myosin heavy chain, titin, desmin and α-smooth muscle actin. Zhang et al. were one of the first to investigate the differentiation potential of DPSCs towards myogenic lineages *in vitro* and *in vivo* [24, 25]. To induce myogenic differentiation, DPSCs were kept in culture with 0.1 μM dexamethasone, 50 μM hydrocortisone, 50 μg/ml gentamycin and with both fetal calf and horse serum. The differentiated cells showed immunoreactivity for myosin heavy chain *in vitro* as well as in DPSC/collagen scaffolds after transplantation into subcutaneous tissue of immunocompromised mice. Additional research by Nakatsuka et al. showed that administration of 0.5 mM 5-Aza-20-deoxycytidine induced myotube formation and myosin heavy chain expression in mouse DPSCs [26]. A similar protocol promoted myogenic differentiation in PDLSCs, which expressed desmin [27]. While DPSCs were investigated first, the myogenic differentiation potential of SHEDs was studied into more detail. Kerkis et al. reported the expression of myogenic proteins in SHEDs such as titin and actin but they also observed the formation of Z-discs [28]. Unfortunately, none of these studies performed a functional assessment of the myogenic differentiated cells and experiments were solely restricted to evaluate expression levels of myogenic protein markers.

17.2.1.2 Endothelial Differentiation of DSCs

Due to the need for new clinically applicable revascularization strategies, the differentiation potential of DSCs towards ECs was investigated by several research groups as other MSC sources showed great promise in endothelial differentiation experiments (i.e. [29–32], reviewed in [33]). An early report described the differentiation of DPSCs towards EC-like cells through incubation of DPSCs for 40 days in high serum conditions [34]. The outcome of the study was that the majority of DPSCs differentiated towards bone-producing cells. Surprisingly, a fraction of the cells differentiated towards vascular endothelial growth factor receptor 1 (VEGFR1)$^+$/CD44$^+$/CD54$^+$ EC-like cells which also expressed von Willebrand factor (vWF), CD31 and angiotensin-converting enzyme. By following the protocol described in [29], Marchionni et al. were able to obtain EC-like cells from DPSCs which expressed CD54, CD34 and vWF. Furthermore, these cells were able to form tubes *in vitro* [35]. Remarkably, a CD31$^-$/CD146$^-$ subset of DPSCs highly expressed CD34 and VEGFR2 and was able to form extensive networks and had a high proliferative and migratory capacity. Moreover, these cells displayed functional endothelial activity as they gained capacity to take up acetylated low density lipoprotein (Ac-LDL) and release vWF upon histamine stimulation after being cultured in 10 ng/ml vascular endothelial growth factor (VEGF) and basic fibroblast growth factor (bFGF) for 14 days [36]. Similar to DPSCs, SHEDs formed tubes *in vitro* and expressed VEGFR2, CD31 and VE-cadherin after exposure to 50 ng/ml VEGF in endothelial growth medium [37]. Amorim et al. recently showed that differentiation of PDLSCs in endothelial cells resulted in expression of VEGFR2 and lectin and the formation of tube-like structures [38]. A report by Bakopoulou et al. demonstrated the endothelial-like differentiation of SCAPs under serum and oxygen/glucose deprivation as evidenced by their capillary forming capacity [39]. To date, only limited evidence for the *in vivo* endothelial differentiation potential of DSCs is available. A recent study of Zhang et al., for example, reported the presence of human CD31$^+$ blood vessels following the transplantation of DPSCs and SHEDs in a tooth slice model [40]. However, transplanted DSCs mainly appear to act as pericytes as they are often found in juxtaposition to host endothelium [36, 41, 42].

17.2.1.3 Neural Differentiation Potential of DSCs

The neuroectodermal and/or glial origin of DSCs makes these cells interesting candidates to study their neurogenic differentiation potential. Although DPSCs are the most meticulously studied subtype of DSCs, the neurogenic differentiation potential of SCAPs, SHEDs, FSCs and PDLSCs has also been a focus of investigation. Remarkably, DSCs are able to differentiate both towards neuronal-like cells and peripheral glial cells with Schwann cell-like properties. Multiple approaches have been envisaged to differentiate DSCs towards neuronal cell lineages. The two most often used methods *in vitro* are either the application of neuro-inductive medium which contains specific neuronal inducing chemicals and/or cytokines or through neurosphere formation. Arthur et al. were one of the first to demonstrate that exposure of DPSCs to epidermal growth factor (EGF) and bFGF, increased the expression of neuronal-related markers such as neural cell adhesion molecule (NCAM), neurofilament-M (NF-M) and NF-H. Contrarily, the expression of early neuronal markers i.e. nestin and beta-III tubulin were markedly decreased. Moreover, patch-clamp recordings revealed that differentiated DPSCs acquired functional neuronal characteristics such as inward sodium currents but lacked action potential firing [43]. Follow-up research aimed to optimize the differentiation protocol of Arthur et al. to evoke generation of action potentials. Kiràly et al. enhanced neurogenic induction via epigenetic reprogramming whereas Gervois et al. used neurosphere formation in the presence of EGF and bFGF signaling. Both procedures were followed by a maturation period based on cyclic AMP and growth factor signaling [44, 45]. The outcome of these studies was similar to the study of Arthur et al., showing decreased expression of nestin and a marked increase in expression of neuronal markers. In addition, these studies revealed the existence of tetrodotoxin- and tetraethylammonium- sensitive sodium and potassium currents. The study of Gervois et al. was also able to demonstrate the generation of a single action potential. Other studies which focused on the neurogenic differentiation potential of DPSCs used neurosphere formation [46] or commercially available differentiation media [47].

Neurogenic differentiation of SHEDs is based on neuronal induction with EGF and bFGF [48], sonic hedgehog, bFGF and FGF-8 [49, 50] and/or neurosphere formation [50, 51]. Jarmalaviciute et al. were able to differentiate SHEDs towards peripheral sensory neurons after neuronal maturation by exposing the neuronally induced SHEDs to elevated cyclic AMP levels, glial cell derived neurotrophic factor (GDNF), brain-derived neurotrophic factor (BDNF) and nerve growth factor (NGF). PDLSCs were also differentiated towards neuronal cells with commercially available medium [47], EGF and bFGF [52] supplemented with glucose, progesteron, insulin, transferrin and selenite [53] or neurosphere formation [54]. SCAPs were differentiated towards neuronal cells by Lee et al. [47] with commercially available medium while both SCAPs and FSCs could be differentiated by neurosphere formation [51, 55]. In addition to directed differentiation towards neuronal cells, DPSCs were successfully differentiated towards Schwann cells with a multistep protocol. The differentiated cells expressed the glial markers CD104, glial fibrillary acidic protein (GFAP), p75 and laminin but lacked nestin expression. Moreover, these cells were able to myelinate and guide neurites of dorsal root ganglia *in vitro* [56]. PDLSCs from dogs were differentiated towards Schwann cell-like cells by Li et al. using different protocols [57]. These differentiated cells expressed GFAP and S100, but also nestin. SHEDs were differentiated towards Schwann cells after exposure to a mixture of growth factors such as bFGF, EGF, BDNF, GDNF, NGF and cyclic AMP.

17.2.2 DSCs as an Ambulant Growth Factor Delivery System: Paracrine Mediated Actions of DSCs

In addition to direct cell replacement strategies, DSCs have the potential to stimulate endogenous repair mechanisms through paracrine interactions. Current research is focusing on the pro-angiogenic, neuroregenerative and neuroprotective properties of DSCs which are all attributed to their rich secretome. To induce angiogenesis, pro-angiogenic growth factors promote ECs to migrate, proliferate and form hollow tubes. Each of these prerequisites are evaluated *in vitro* on commercially available EC-lines such as human umbilical cord vein endothelial cells (HUVECs) or human microvascular endothelial cells (HMECs). The secretome of DSCs contains a significant amount of pro-angiogenic growth factors including platelet derived growth factor, angiopoietin, bFGF, colony stimulating factors, VEGF and endothelin-1 [58–61]. On the other hand, DSCs also secrete anti-angiogenic proteins such as plasminogen activator inhibitor-1, endostatin, thrombospondin-1, tissue inhibitor of matrix metalloproteinase -1/4 and pentraxin-3 [58, 59]. Iohara et al. used the secretome of $CD31^-/CD146^-$ DPSCs to investigate the proliferative effect on HUVECs and they showed a significant increase in HUVEC proliferation [36]. However, when HMECs were used, the same effect could not be repeated with the secretome of either DPSCs, SCAPs or FSCs [58]. The second most commonly investigated paracrine effect of DSCs on EC function is the migration of ECs along a gradient of chemotactic proteins. Endothelial migration, evaluated by means of a transwell assay, was found to be significantly stimulated by DPSCs and SCAPs, whereas FSCs had no apparent effect [39, 58]. The tube formation assay is a third assay which evaluates the influence of angiogenesis-promoting substances. Tube-promoting effects have been observed for all DSCs, excluding FSCs, when they were in direct co-culture with ECs [41, 60, 62]. Although these results support the pericyte-like role for DSCs in angiogenesis, paracrine mediated tube formation has also been observed. An indirect co-culture of DPSCs and HUVECs demonstrated tubular network formation [63]. Similar results were achieved when HMECs were incubated with the secretome of DPSCs [58] or HUVECs with the secretome of SCAPs [39] or SHEDS [64].

Preconditioning of DSCs with hypoxia-mimicking agents such as deferoxamine (DFO) or oxygen/glucose deprivation are current strategies to enhance the *in vitro* angiogenic properties of DSCs. Hypoxic preconditioning enhanced VEGF expression but did not affect DPSC proliferation. By applying the secretome of hypoxic preconditioned DPSCs to HMEC cultures, proliferation and sprouting of HMECs was evident [65]. Bakopoulou et al. demonstrated that deprivation of SCAPs from nutrients and oxygen induced the secretion of pro-angiogenic growth factors, which had an additional positive influence on EC migration and tube formation [39]. Moreover, oxygen and nutrient deprivation also seemed to enhance SCAP differentiation towards EC [39, 66]. Hypoxic preconditioning of PDLSCs increased VEGF and IL-6 secretion. Hypoxia-mimetic agents have also been used to precondition DSCs. prolyl hydroxylase inhibitors (PHD) such as cobalt chloride ($CoCl_2$), DFO, L-mimosine and dimethyloxalglycine promoted the secretion of VEGF and increased the expression of HIF-1α in DPSCs [67], PDLSCs [68] and SCAPs [62]. When SCAPs and HUVECs were treated with $CoCl_2$, increased tube formation was observed [62]. Similar to PHD inhibitors, the iron chelator hinokitiol upregulated HIF-1α expression and VEGF secretion in DPSCs. Moreover, the conditioned medium of these hinokitiol-exposed DPSCs enhanced angiogenesis *in vitro* and *in vivo* [69]. Together these data indicate a promising future for the use of hypoxia and hypoxia-mimicking agents to increase the angiogenic properties of DSCs.

Similar to the paracrine mediated effect of DSCs on angiogenesis, the paracrine effects of DSCs on neuroprotection, neurogenesis and neuritogenesis have been evaluated. Human DSCs display a rich secretome including BDNF, NGF, neurotrophin-3 (NT-3) and GDNF, which are considered hallmark factors involved in neurotrophic and neuroregulatory signaling pathways [45, 56, 61, 64, 70]. Moreover, DPSCs and SCAPs secrete a significantly higher amount of these factors *in vitro* compared to bone marrow-derived and adipose-derived MSCs [61, 71]. The DPSC secretome has been shown to enhance the survival of axotomized retinal ganglion cells and stimulate their neurite outgrowth [71]. DPSCs themselves and their secretome were able to protect astrocytes from ischemia [72]. Apel et al. observed a neuroprotective effect of DPSCs when co-cultured with amyloid beta peptide 1–42 and 6-hydroxy-dopamine (6-OHDA) treated neurons which are *in vitro* models for Alzheimer's and Parkinson's disease respectively [73]. Furthermore, neuroprotection by DPSCs was observed in cultures of dopaminergic neurons treated with either 6-OHDA [74] or a combination of MPP+ and rotenone [75]. Finally, the secretome of DPSCs and SHEDs enhanced Schwann cell survival, proliferation and migration *in vitro* [64, 76].

17.2.3 Immunomodulatory Properties of DSCs

DPSCs have the ability to modulate the immune response [77, 78]. Demircan et al. showed that the suppressive actions of DPSCs on T-cells were mediated via paracrine effects by means of a transwell and mixed lymphocyte reaction (MLR) assay. Increased levels of hepatocyte growth factor (HGF), HLA-G, transforming growth factor beta (TGF-β), CD54, IL-6, IL-10, CD106, and VEGF were found in DPSC/T-cell co-cultures. Moreover, in the transwell system, the expression of pro-inflammatory cytokines by T-cells such as interferon-gamma (IFN-γ), IL-2, IL-6 receptor, IL-12, IL-17A and tumor necrosis factor- alpha (TNF-α) were decreased whereas the expression of the anti-inflammatory cytokine inducible protein-10 was upregulated. Interestingly, DPSCs induced the expression of the regulatory T-cell markers CD4, CD25 and Foxp3. Apoptosis of T-cells was increased after 24 h incubation with DPSCs [79]. A similar paracrine mediated immunosuppression was exerted by the secretome of porcine and human DPSCs [21, 80, 81]. Additional insight into the immunomodulatory properties of DSCs has arisen from a study by Wada et al. who showed that PDLSCs suppressed peripheral blood mononuclear cell (PBMCs) proliferation after stimulation with a mitogen or in a MLR. Moreover, they were able to demonstrate that IFN-γ, produced by activated PBMCs, was partially responsible for the immunomodulatory effects of PDLSCs. PDLSCs cultured in the presence of activated PBMCs increased the expression of HGF, TGF-β and indoleamine 2,3-dioxygenase (IDO) [82]. In addition to these factors, Fas ligand (FasL) also appears to play a role in DSC-mediated immunosuppressive effects, as knockdown of FasL reduced the capacity of DPSCs to induce T-cell apoptosis [83]. Similarly, Toll-like receptor 3 (TLR3) agonists improved the immunosuppressive effect of DPSCs and FSCs, by upregulating TGF-β and IL-6 production. Interestingly, TLR4 agonists only improved the immunosuppressive effect of FSCs and completely abolished the effect of DPSCs by decreasing TG-β and IDO production [84]. Ding et al. confirmed that TGF-β is a key regulatory molecule in the immunomodulatory properties of DPSCs [85]. SCAPs were also found to secrete immunomodulatory proteins although in lower levels compared to BM-MSCs [61].

17.3 Applications in Regenerative Medicine

Increasing evidence indicates that stem cells represent a potential therapeutic tool for tissue engineering and regeneration in a wide variety of diseases and disorders [86, 87]. The main goal of stem cell-based therapies in regenerative medicine is to replace, repair or to enhance endogenous repair processes of injured tissues and organs [88, 89]. Until now, only limited stem cell-based therapies have been established as a clinical standard, including hematopoietic stem cell transplantation for leukaemia and epithelial stem cell-based interventions for burn wounds and corneal disorders [90, 91]. DSCs have gained great interest during the last decade because of their accessibility, (trans)differentiation potential and high proliferative capacity. Elaborate preclinical research has highlighted the underlying biological and cellular properties of different DSCs, as well as their potential application in the treatment of medical conditions [92–97], such as myocardial infarction [98], ischemic disease [99], nerve injury and other neurodegenerative disorders [47, 50, 70, 100–105], inflammatory diseases [106, 107], diabetes [108–112], muscular dystrophy [26, 113, 114], bone/cartilage defects [115–119], hair follicle loss [120], skin injuries [121], salivary gland defects [122], corneal epithelial defects [123–126], and the regeneration of dental tissues [15, 127–132].

17.3.1 Dentin, Pulp and Periodontal Tissue Regeneration

The main applications for DSCs are in the regeneration of damaged tooth-related structures such as dentin, pulp and periodontal ligament [15, 128–135]. A significant number of studies in animal models have accomplished regeneration of dentin/pulp tissue and of the cementum/periodontal complex by applying DSCs in combination with an appropriate scaffold or supporting matrix [5, 7, 134, 136, 137]. DPSCs have a potent dentinogenic potential and they promote repair and reconstruction of a dentin-pulp-like complex when they are transplanted alone or in combination with growth factors (e.g. bone morphogenetic protein-2 (BMP-2)) in the pulp cavity [138, 139]. Several growth factors such as TGF-β, bFGF and dentin matrix protein 1 (DMP1) can induce odontoblast differentiation of DPSCs [140], and Almushayt et al. demonstrated that these differentiated DPSCs form reparative dentin that covers pulp tissue [141]. Prescott et al. transplanted dentin slices containing DPSCs, collagen and DMP1 subcutaneously in nude mice, which resulted in the formation of a well-organized matrix similar to pulp tissue [142]. Tran and Doan recently achieved regeneration of dentin-like tissue *in vivo* by culturing human DPSCs onto dentin-derived scaffolds [143]. In line with these findings, Huang et al. reported that a vascularized dentin-pulp-like complex can be regenerated de novo in an emptied root canal space by DPSCs and SCAPs after subcutaneous transplantation in mice [132]. Additionally, also SHEDs combined with a scaffold formed dental pulp-like structures within human tooth slices [144]. Furthermore, an ongoing phase 1 clinical trial aims to repair immature permanent teeth with necrotic pulps by using SHEDs (ClinicalTrials.gov NCT01814436).

New therapeutic strategies for periodontal tissue diseases (e.g. periodontitis) focus on the addition of exogenous growth factors (e.g. bone morphogenetic proteins, platelet derived growth factor) and stem cell therapy [128, 145]. Extensive research studies have explored the use of PDLSCs to improve periodontal tissue regeneration. PDLSCs showed the best capacity to regenerate periodontium compared to other DSC populations [146, 147]. Transplantation of autologous PDLSCs combined with tricalcium phosphate/hydroxyapatite (TCP-HA) into periodontal defects improved bone formation and regeneration of cementum and periodontal ligament [148, 149]. Seo et al. also transplanted PDLSCs loaded onto TCP-HA into immunocompromised rodents, which resulted in the formation of a cementum/periodontal ligament-like structure and contributed to periodontal tissue repair [7]. Moreover, PDLSCs applied together with titanium dental implants improved periodontal ligament regeneration in a rat molar implant model [150], but also in humans [151]. In addition to the abovementioned animal models, a retrospective pilot study performed in humans provided evidence that autologous transplantation of periodontal ligament stem/progenitor cells might provide therapeutic improvement for periodontal defects without adverse effects [152]. Apart from PDLSCs, also implantation of DPSCs into periodontal defects of dogs results in bone formation with neovascularization [153, 154]. A clinical study is currently ongoing in chronic periodontal disease patients which receive a local injection of allogeneic human DPSCs in order to improve periodontal tissue regeneration (ClinicalTrials.gov NCT02523651).

The ultimate goal in dentistry is the regeneration of a functional and living tooth. Recent advances in DSC biotechnology explored the use of stem cells and scaffolds to engineer teeth with the appropriate functional properties. This approach represents a promising therapeutic strategy for the replacement of a diseased or damaged tooth [133, 155, 156]. Sonoyama et al. transplanted both human SCAPs and PDLSCs with TCP-HA as a carrier in a minipig model, resulting in the formation of a viable root/periodontal complex formation that is capable of supporting a porcelain crown [155]. Moreover, Khorsand et al. succeeded to regenerate peridontium using autologous DPSCs loaded onto Bio-Oss scaffolds in dogs that received an experimental defect [157]. Furthermore, Nakashima and coworkers were able to induce whole pulp regeneration after pulpectomy in a dog model using a combination of autologous DPSCs and stromal cell derived factor-1 loaded onto a 3D collagen scaffold [130, 158]. Implantation of FSCs combined with a treated dentin matrix in the alveolar fossa of rats also generated root-like tissues with a pulp–dentin complex and a periodontal ligament that connected the cementum-like layer to host alveolar bone [159, 160].

17.3.2 Osseous Regeneration

Tooth-derived stem cells have also been extensively used in several preclinical studies of osseous regeneration, such as craniofacial and alveolar bone healing [92, 161]. DSCs represent a promising cell-based therapy for repair of bone-related diseases and orthopedic surgeries due to their osteoregenerative capacities. Especially DPSCs are considered to be a potent cell source to enhance bone regeneration [115]. Because of their ectomesenchymal origin, DPSCs express bone-specific markers and they exhibit an osteogenic differentiation profile [34, 162]. Moreover, DPSCs can differentiate into (pre)osteoblasts and they deposit extracellular matrix that forms mineralized woven bone [163–166]. Graziano et al. showed that $CD34^+$ DPSCs transplanted subcutaneously in rats forms several bone nodules [167]. Another study in rats demonstrated that human DPSCs promote repair of a large-scale cranial bone defects [115–117]. Clinical studies performed by d'Aquino et al. demonstrated that a construct composed of DPSCs and a collagen sponge scaffold promotes bone repair following oro-maxillofacial defects [168]. Several studies have indicated that DPSCs gain osteogenic differentiation potential when combined with different biomaterials, such as collagen and titanium [169–173]. Another important factor that stimulates the osteogenic properties of DSCs is BMP-2 [23, 174, 175]. Liu et al. showed that DPSCs expressing BMP-2 have better mineralization capacities, and they generate more bone after alveolar bone defects in a rabbit model [176]. Other DSC populations, such as PDLSCs and SHEDs also represent a promising tool for bone regeneration [177–180]. As reported by Seo et al., SHEDs combined with TCP-HA can repair critical-sized calvarial defects in mice with substantial new bone formation [180]. In line with these findings, Zheng et al. demonstrated that stem cells from miniature pig deciduous teeth were able to engraft and induce bone to repair critical size mandibular defects [181]. A clinical trial is currently ongoing that uses SHEDs for alveolar bone tissue engineering for cleft lip and palate patients (ClinicalTrials.gov – NCT01932164). Another study showed that human PDLSCs expanded *ex vivo* and seeded in three-dimensional scaffolds (e.g. fibrin sponge) also generated new bone [147]. When SCAPs are combined with HA scaffolds and implanted subcutaneously in immunocompromised rats, both bone- and dentin-like mineralized tissues are formed [182].

17.3.3 Neural Tissue Regeneration

Several DSC populations such as DPSCs and PDLSCs originate from ectomesenchyme, indicating its interaction with the neural crest during embryonic development [183–186]. This implicates that DSCs may possess distinct functions compared to other MSC populations (e.g.

BM-MSCs) [183, 187]. As already mentioned, DSCs also have potent neural characteristics besides their classical MSC-properties. This indicates that DSCs may have a therapeutic benefit for the treatment of neurological diseases and disorders. Studies describing the effect of DSCs in animal models of central- or peripheral nervous system pathology are scarce and mainly focused on DPSCs. Arthur et al., pioneers in the use of DPSCs for neuronal regeneration, showed that DPSCs acquired a neuronal morphology, expressed neuronal markers and were able to attract trigeminal neurons after transplantation into chicken embryos [43, 188]. Moreover, neurogenic pre-differentiated DPSCs were able to migrate to the host brain after injection in the cerebrospinal fluid where they integrated in the host circuitry [189]. The therapeutic effect of DPSCs is also studied in various animal models of neurological dysfunction. These animal models include hypoxic ischemic encephalopathy [100], spinal cord injury [101] and ischemic stroke [190], in which DPSCs ameliorated the outcome of the disease. Preclinical evidence in a rodent model of middle cerebral artery ischemic stroke indicates improvement in neurobehavioral function with adult human DSC therapy [190–192]. Recently, the "first-in-human" autologous DPSC therapy clinical trial, TOOTH – The Open study Of dental pulp stem cell Therapy in Humans, was initiated and it evaluates the safety and feasibility of autologous human adult DPSC therapy in patients with chronic disability after stroke [193]. SHEDs have also been successfully applied in heatstroke animals [194], cerebral ischemia [102], perinatal hypoxia/ischemia [103], spinal cord injury [104, 105] and Parkinson's disease [50] with an improved disease outcome after transplantation. Research of PDLSCs in *in vivo* models is lacking, but Bueno et al. were able to observe engraftment of PDLSCs in the host brain after transplantation [53]. FSCs were used in a spinal cord injury model after being seeded on aligned poly(ε caprolactone)/ poly-DL-lactide-co-glycolide (PCL/PLGA) fibers which supported nerve fiber outgrowth. Unfortunately, no functional improvement was observed after transplantation of these constructs [195]. In addition, SCAPs were transplanted in a hydrogel or in the apical pad of a rat spinal cord injury model [196]. Remarkably, functional improvement was only observed in the experimental group which received the entire tissue graft [197]. The proposed mechanism of disease amelioration by the transplanted cells in these studies was mainly based on paracrine effects of the transplanted cells which stimulated proliferation and differentiation of endogenous neural stem cells [102, 198] or stimulated host repair mechanisms [100–102, 190].

In vivo studies that apply DSCs for peripheral nerve injury were previously limited to reports that used artificial tubes seeded with DSCs in animal models of facial nerve injury. Sasaki et al. indicated that rat DPSCs promoted remyelination, blood vessel formation and normal nerve regeneration when applied in combination with silicon or PLGA tubes as nerve conduits. Paracrine factors secreted by the DPSCs were believed to be responsible for this effect [199, 200]. Later, Sasaki et al. showed that a silicon tube containing a DPSCs/type I collagen gel mixture improved the electrophysiological and functional characteristics of the facial nerve, with an outcome similar to nerve autografts [201]. More recently, the paracrine effects of DPSCs and SHEDs on Schwann cells and peripheral nerve function were evaluated in a sciatic nerve model [64, 76]. The study by Yamamato et al. showed that transplanted DPSCs promoted regeneration of myelinated fibers with a larger axon-to fiber ratio (G-ratio) and revascularization. Direct differentiation into Schwann cells was not observed, but endogenous Schwann cells showed reduced apoptosis and increased proliferation [76]. Similarly, Sugimura-Wakayama demonstrated that silicon conduits containing the SHED secretome significantly increased the number of myelinated axons and G-ratio. Moreover, the sciatic functional index and gastrocnemius wet weight ratio suggested reinnervation of the target muscle with functional recovery [64].

17.3.4 Diabetes

Another important research area in which DSCs may be successful is diabetes, a chronic endocrinal disease that is associated with pancreatic islet dysfunction. Transplantation of insulin-producing pancreatic islet cells represents an important therapeutic strategy for this condition. Several preclinical studies indicate that DSCs can differentiate towards a pancreatic cell lineage [108–112]. Govindasamy et al. showed that differentiated DPSCs resemble islet-like cell aggregates, which release insulin and C-peptide in a glucose-dependent manner [108]. Other studies confirmed these findings, and transplantation of islet-like cell clusters derived from human DPSCs and SHEDs restored normoglycemia in diabetic mice [109, 111, 112]. As shown by Lee et al., PDLSCs can also transdifferentiate into functional pancreatic islet-like cells and therefore represent an alternative stem cell population for pancreatic repair [110]. These preclinical studies implicate that DSCs offer a source of human tissue that can be used as an autologous stem cell therapy for diabetes in order to improve islet regeneration.

17.3.5 Angiogenesis: A Role in Myocardial Infarction and Ischemic Disease

As DSCs display a pronounced angiogenic potential, they are also utilized in clinically relevant disease models, such as myocardial infarction (MI), hindlimb ischemia and a pulpectomized tooth model. Intracardial injection of GFP-positive DPSCs in a rat model of MI resulted in increased capillary density, a reduction in infarct size and a reduced thickening of the anterior ventricular wall. Interestingly, these effects were thought to be caused by paracrine effects of the transplanted DPSCs, as no GFP-positive cells were observed in the tissue [98]. The study by Iohara et al. in which a sub-fraction of $CD31^{-}$/$CD146^{-}$ porcine DPSCs was injected in a mouse model of hindlimb ischemia, demonstrated a higher capillary density. The transplanted cells were found in proximity to the newly formed vessels, supporting the pericyte and paracrine function of DPSCs [36]. Interestingly, the same subpopulation of DPSCs was capable of increasing vasculogenesis and neurogenesis after transplantation in a rat model of focal cerebral ischemia [21, 192]. Finally, vasculogenesis was enhanced in a pulpectomized tooth model after activation of chemotaxis in DPSCs by granulocyte-colony stimulating factor [80, 202].

17.4 Cryopreservation of Dental Stem Cells

As stated above, DSCs are an attractive source of mesenchymal stem cells for multilineage differentiation and regenerative medicine. However, long-term culture of these stem cells may be accompanied with deleterious effects such as phenotypic instability, cell death, senescence or contamination [203]. Therefore, adequate long term storage procedures are required that not only keep the cells viable, but also safeguard the phenotypic stability and differentiation capacity. The advances made in rapidly growing fields such as (stem) cell therapy, personalized medicine and cancer research, further drive the need to overcome this hurdle in order to move forward towards clinical applications [204, 205]. For allogeneic patients, cryopreservation allows the transport of cellular products and provides a time window to screen the cells prior to transplantation. For autologous patients, cells can simply be stored for later clinical use [206]. Currently a vast number of protocols are being tested for the cryopreservation of DSCs, each with their own criteria to evaluate the potential effect on the intrinsic behavior of these stem cells (Table 17.1).

During cryopreservation, the ultra-low temperatures bring an indefinite halt to the cell's metabolism. However, cryopreservation represents a physical insult on the cells and is currently only effective for single cell suspensions and a few simple tissues [204, 205]. Since cryopreservation aims at preserving living cells, it is essential to design protocols that minimize the injuries that are associated with the freeze-thaw process

Table 17.1 Overview of crypreservations protocols for dental stem cells and dental tissues

Author	Method	Criteria tested	Comparison between control and cryopreserved cells/tissues
DPSCs			
Davies et al. [1]	90 % FBS + 10 % DMSO	Viability	Slightly reduced
	4 °C for 1 h, then at −20 °C for 2 h and −80 °C overnight	CD29/CD90 ratio	Upregulated after cryopreservation
		CD73, CD44	Upregulated after cryopreservation
		CD105 marker expression	No effect
		Pluripotency markers cMyc, KLF-4,Nanog, Lin28	Upregulated after cryopreservation
Ducret et al. [2]	Serum-free; medicinal manufacturing approach	Cell proliferation	No effect
	10 % DMSO + 90 % serum-free medium (Cryo-3)	Karyotyping	No effect
	Isopropanol filled cryobox	Osteogenic differentiation	No effect
Hata et al. [3]	CELLBANKER®	MSC marker expression	No effect
		Cell proliferation	No effect
		Osteogenic differentiation	No effect
		Adipogenic differentiation	No effect
		In vivo diabetic polyneuropathy rat model	No effect
Kumar et al. [4]	7 different protocols	MSC marker expression	No effect
		Cell proliferation	No effect
		Osteogenic differentiation	No effect
		Adipogenic differentiation	No effect
		Neural differentiation	No effect
Papaccio et al. [5]	10 % DMSO in 90 % FBS	Osteogenic differentiation	No effect
		In vivo bone formation	No effect
Lee et al. [6]	Magnetic cryopreservation in serum-free culture medium + 3 % DMSO	Cell proliferation	No effect
		MSC marker expression	Reduced after cryopreservation
		Osteogenic differentiation	No effect
		Adipogenic differentiation	No effect
		DNA damage	No effect
Perry et al. [7]	10 % FBS in Mesencult™ direct at −1 °C/min to −85 °C	Proliferation	No effect
		MSC marker expression	No effect
		Osteogenic differentiation	No effect
		Adipogenic differentiation	No effect
		Chondrogenic differentiation	No effect
Umemura et al. [8]	DMEM + 10 % ethylene glycol, 1.0 M sucrose and 0.00075 M polyvinylpyrrolidone	Proliferation	No effect
		Ultrastructure	No effect
		MSC markers	No effect
Woods et al. [9]	Various protocols	MSC marker expression	No effect

(continued)

Table 17.1 (continued)

Author	Method	Criteria tested	Comparison between control and cryopreserved cells/tissues
Zhang et al. [10]	Medium not defined in full text	Proliferation	No effect
		Expression of STRO-1	No effect
		Osteogenic differentiation	No effect
		Adipogenic differentiation	No effect
		Chondrogenic differentiation	No effect
		Myogenic differentiation	No effect
		Neurogenic differentiation	No effect
DPSCs from cryopreserved pulp			
Chen et al. [11]	Culture medium + 10 % DMSO	Hepatic-like differentiation	No effect
Chen et al. [12]	Culture medium + 10 % DMSO	Viability	No effect
	2 h at 4 °C, then 1 °C/min to −80 °C	MSC markers	No effect
Malekfar et al. [13]	90 % FBS + 10 % DMSO	MSC marker expression	No effect
		Osteogenic differentiation	No effect
		Adipogenic differentiation	No effect
Temmerman et al. [14]	90 % FBS + 10 % DMSO	Cell proliferation	No effect
	1 °C/min to −80 °C		
Woods et al. [9]	(0.5 M- 1 M–1.5 M) of ethylene glycol, propylene glycol, or dimethyl sulfoxide	Osteogenic differentiation	No effect
		Adipogenic differentiation	No effect
		Chondrogenic differentiation	No effect
DPSCs from cryopreserved teeth			
Abedini et al. [15]	Programmed freezer combined with a magnetic field, known as Cells Alive System (CAS)	VEGF and NGF secretion	Decreased after cryopreservation
Chen et al. [12]	Culture medium + 10 % DMSO	Viability	No effect
	2 h at 4 °C	MSC markers	No effect
	1 °C/min to −80 °C		
Perry et al. [7]	10 % FBS in Mesencult™	Proliferation	No effect
	1 h at 4 °C	MSC marker expression	No effect
	1 °C/min to −80 °C	Osteogenic differentiation	No effect
		Adipogenic differentiation	No effect
		Chondrogenic differentiation	No effect
SHEDs from intact cryopreserved teeth			
Ji et al. [16]	90 % FBS + 10 % DMSO	Viability	Decreased after cryopreservation
	1 h at 4 °C		
	1 °C/min to −80 °C	Cell outgrowth capacity	Decreased after cryopreservation
Gioventu et al. [17]	RPMI 1,640 + 10 % DMSO + 10 % human albumin	Viability	No effect
		Proliferation	No effect
		MSC marker expression	No effect

(continued)

Table 17.1 (continued)

Author	Method	Criteria tested	Comparison between control and cryopreserved cells/tissues
Lindemann et al. [18]	90 % FCS + 10 % DMSO	MSC marker expression	No effect
	1 h at 4 °C	Cell proliferation	No effect
	1 °C/min to −80 °C	Osteogenic differentiation	No effect
		Adipogenic differentiation	No effect
		Chondrogenic differentiation	No effect
Lee [19]	90 % FBS + 10 % DMSO	Cell proliferation	No effect
	1 h at 4 °C	CFU-F	No effect
	1 °C/min to −80 °C	MSC marker expression	No effect
		Adipogenic differentiation	No effect
		Osteogenic differentiation	No effect
		Ectopic bone formation in mice	No effect
SHEDs from intact cryopreserved pulp			
Ma et al. [20]	90 % FBS + 10 % DMSO in -80 °C	Cell proliferation	No effect
	4 °C for 1 h, then −80 °C overnight.	CFU-F	No effect
		MSC marker expression	No effect
		CD105 marker expression	Slightly reduced after cryopreservation
		Osteogenic differentiation	No effect
		Adipogenic differentiation	No effect
		Chondrogenic differentiation	No effect
		Hepatic-like differentiation	No effect
		Endothelial differentiation	No effect
		Neural differentiation	No effect
PDLSCs			
Vasconcelos et al. [21]	90 % FBS + 10 % DMSO	Cell proliferation	No effect
	2 h at 4 °C, 18 h in −25 °C, next −80 °C	Cell adhesion	No effect
Kamada et al. [22]	Programmed freezer combined with a magnetic field, known as Cells Alive System (CAS)	mRNA expression of collagen I	No effect
		mRNA expression of alkaline phosphastase (ALP)	Slightly reduced after cryopreservation
Kaku et al. [23]	Programmed freezer combined with a magnetic field, known as Cells Alive System (CAS)	Cell survival	Increased in comparison to a conventional freezing method
PDLSCs from cryopreserved periodontal ligament tissue			
Seo et al. [24]	90 % FBS + 10 % DMSO	Stro-1 expression	No effect
		MSC differentiation	No effect
		Cementum/periodontal ligament-like tissue generation *in vivo*	No effect
Abedini et al. [15]	Programmed freezer combined with a magnetic field, known as Cells Alive System (CAS)	Expression of collagen I, ALP and VEGF	No effect

(continued)

Table 17.1 (continued)

Author	Method	Criteria tested	Comparison between control and cryopreserved cells/tissues
PDLSCs from cryopreserved teeth			
Min et al. [25]	DMEM + 10 % FBS + 10 %DMSO	FGFR2 expression	Dramatically decreased after cryopreservation
FSCs from cryopreserved dental follicle			
Park et al. [26]	0.05M glucose, 0.05M sucrose and 1.5M ethylene glycol	MSC marker expression	No effect
		Osteogenic differentiation,	No effect
	In PBS	Adipogenic differentiation	No effect
		Chondrogenic differentiation	No effect
		p53 expression	Increased after cryopreservation
Kang et al. [27]	0.05M glucose, 0.05M sucrose and 1.5M ethylene glycol in PBS	Expression of immunological markers	No effect
		In vivo bone formation	No effect
SCAPs			
Ding et al. [28]	Three different methods:	MSC marker expression	No effect
	90 % FCS + 10 % DMSO	Osteogenic differentiation	No effect
	90 % FCS + 10 % glycerol	Adipogenic differentiation	No effect
	90 % FCS + 10 % ethyleen glycol	Cell proliferation	No effect
		CFU-F	No effect
		In vitro immunomodulatory potential	No effect
		Bioroot formation in minipig	No effect

Articles were retrieved by using the search terms 'cryopreservation' and 'dental stem cells' or 'dental tissues' in the Pubmed website. Case reports or articles without quantitative data were not included in the table

References

1. Davies OG, Smith AJ, Cooper PR, Shelton RM, Scheven BA (2014) The effects of cryopreservation on cells isolated from adipose, bone marrow and dental pulp tissues. Cryobiology 69 (2):342–347. doi:10.1016/j.cryobiol.2014.08.003
2. Ducret M, Fabre H, Farges JC, Degoul O, Atzeni G, McGuckin C, Forraz N, Mallein-Gerin F, Perrier-Groult E (2015) Production of Human Dental Pulp Cells with a Medicinal Manufacturing Approach. Journal of endodontics 41 (9):1492–1499. doi:10.1016/j.joen.2015.05.017
3. Hata M, Omi M, Kobayashi Y, Nakamura N, Tosaki T, Miyabe M, Kojima N, Kubo K, Ozawa S, Maeda H, Tanaka Y, Matsubara T, Naruse K (2015) Transplantation of cultured dental pulp stem cells into the skeletal muscles ameliorated diabetic polyneuropathy: therapeutic plausibility of freshly isolated and cryopreserved dental pulp stem cells. Stem cell research & therapy 6:162. doi:10.1186/s13287-015-0156-4
4. Kumar A, Bhattacharyya S, Rattan V (2015) Effect of uncontrolled freezing on biological characteristics of human dental pulp stem cells. Cell Tissue Bank 16 (4):513–522. doi:10.1007/s10561-015-9498-5
5. Papaccio G, Graziano A, d'Aquino R, Graziano MF, Pirozzi G, Menditti D, De Rosa A, Carinci F, Laino G (2006) Long-term cryopreservation of dental pulp stem cells (SBP-DPSCs) and their differentiated osteoblasts: a cell source for tissue repair. Journal of cellular physiology 208 (2):319–325. doi:10.1002/jcp.20667
6. Lee SY, Huang GW, Shiung JN, Huang YH, Jeng JH, Kuo TF, Yang JC, Yang WC (2012) Magnetic cryopreservation for dental pulp stem cells. Cells, tissues, organs 196 (1):23–33. doi:10.1159/000331247
7. Perry BC, Zhou D, Wu X, Yang FC, Byers MA, Chu TM, Hockema JJ, Woods EJ, Goebel WS (2008) Collection, cryopreservation, and characterization of human dental pulp-derived mesenchymal stem cells for banking and clinical use. Tissue Eng Part C Methods 14 (2):149–156. doi:10.1089/ten.tec.2008.0031
8. Umemura E, Yamada Y, Nakamura S, Ito K, Hara K, Ueda M (2011) Viable cryopreserving tissue-engineered cell-biomaterial for cell banking therapy in an effective cryoprotectant. Tissue Eng Part C Methods 17 (8):799–807. doi:10.1089/ten.tec.2011.0003

(continued)

Table 17.1 (continued)

9. Woods EJ, Perry BC, Hockema JJ, Larson L, Zhou D, Goebel WS (2009) Optimized cryopreservation method for human dental pulp-derived stem cells and their tissues of origin for banking and clinical use. Cryobiology 59 (2):150–157. doi:10.1016/j.cryobiol.2009.06.005
10. Zhang W, Walboomers XF, Shi S, Fan M, Jansen JA (2006) Multilineage differentiation potential of stem cells derived from human dental pulp after cryopreservation. Tissue engineering 12 (10):2813–2823. doi:10.1089/ten.2006.12.2813
11. Chen YK, Huang AH, Chan AW, Lin LM (2013) Human dental pulp stem cells derived from cryopreserved dental pulp tissues of vital extracted teeth with disease demonstrate hepatic-like differentiation. Journal of tissue engineering and regenerative medicine. doi:10.1002/term.1763
12. Chen YK, Huang AH, Chan AW, Shieh TY, Lin LM (2011) Human dental pulp stem cells derived from different cryopreservation methods of human dental pulp tissues of diseased teeth. Journal of oral pathology & medicine : official publication of the International Association of Oral Pathologists and the American Academy of Oral Pathology 40 (10):793–800. doi:10.1111/j.1600-0714.2011.01040.x
13. Malekfar A, Valli KS, Kanafi MM, Bhonde RR (2016) Isolation and Characterization of Human Dental Pulp Stem Cells from Cryopreserved Pulp Tissues Obtained from Teeth with Irreversible Pulpitis. Journal of endodontics 42 (1):76–81. doi:10.1016/j.joen.2015.10.001
14. Temmerman L, Beele H, Dermaut LR, Van Maele G, De Pauw GA (2010) Influence of cryopreservation on the pulpal tissue of immature third molars in vitro. Cell Tissue Bank 11 (3):281–289. doi:10.1007/s10561-009-9148-x
15. Abedini S, Kaku M, Kawata T, Koseki H, Kojima S, Sumi H, Motokawa M, Fujita T, Ohtani J, Ohwada N, Tanne K (2011) Effects of cryopreservation with a newly-developed magnetic field programmed freezer on periodontal ligament cells and pulp tissues. Cryobiology 62 (3):181–187. doi:10.1016/j.cryobiol.2011.03.001
16. Ji EH, Song JS, Kim SO, Jeon M, Choi BJ, Lee JH (2014) Viability of pulp stromal cells in cryopreserved deciduous teeth. Cell Tissue Bank 15 (1):67–74. doi:10.1007/s10561-013-9375-z
17. Gioventu S, Andriolo G, Bonino F, Frasca S, Lazzari L, Montelatici E, Santoro F, Rebulla P (2012) A novel method for banking dental pulp stem cells. Transfus Apher Sci 47 (2):199–206. doi:10.1016/j.transci.2012.06.005
18. Lindemann D, Werle SB, Steffens D, Garcia-Godoy F, Pranke P, Casagrande L (2014) Effects of cryopreservation on the characteristics of dental pulp stem cells of intact deciduous teeth. Archives of oral biology 59 (9):970–976. doi:10.1016/j.archoralbio.2014.04.008
19. Lee HS, Jeon MJ, Kim SO, Kim SH, Lee JH, Ahn SJ, Shin Y, Song JS (2015) Characteristics of stem cells from human exfoliated deciduous teeth (SHED) from intact cryopreserved deciduous teeth. Cryobiology 71 (3):374–383. doi:10.1016/j.cryobiol.2015.10.146
20. Ma L, Makino Y, Yamaza H, Akiyama K, Hoshino Y, Song G, Kukita T, Nonaka K, Shi S, Yamaza T (2012) Cryopreserved dental pulp tissues of exfoliated deciduous teeth is a feasible stem cell resource for regenerative medicine. PloS one 7 (12):e51777. doi:10.1371/journal.pone.0051777
21. Vasconcelos RG, Ribeiro RA, Vasconcelos MG, Lima KC, Barboza CA (2012) In vitro comparative analysis of cryopreservation of undifferentiated mesenchymal cells derived from human periodontal ligament. Cell Tissue Bank 13 (3):461–469. doi:10.1007/s10561-011-9271-3
22. Kamada H, Kaku M, Kawata T, Koseki H, Abedini S, Kojima S, Sumi A, Motokawa M, Fujita T, Ohtani J, Ohwada N, Tanne K (2011) In-vitro and in-vivo study of periodontal ligament cryopreserved with a magnetic field. American journal of orthodontics and dentofacial orthopedics : official publication of the American Association of Orthodontists, its constituent societies, and the American Board of Orthodontics 140 (6):799–805. doi:10.1016/j.ajodo.2011.04.024
23. Kaku M, Kamada H, Kawata T, Koseki H, Abedini S, Kojima S, Motokawa M, Fujita T, Ohtani J, Tsuka N, Matsuda Y, Sunagawa H, Hernandes RA, Ohwada N, Tanne K (2010) Cryopreservation of periodontal ligament cells with magnetic field for tooth banking. Cryobiology 61 (1):73–78. doi:10.1016/j.cryobiol.2010.05.003
24. Seo BM, Miura M, Sonoyama W, Coppe C, Stanyon R, Shi S (2005) Recovery of stem cells from cryopreserved periodontal ligament. Journal of dental research 84 (10):907–912
25. Min KS, Lee HW, Lee HS, Lee JH, Park SH (2010) Comparison of gene expression in human periodontal ligament cells cultured from teeth immediately after extraction and from teeth cryopreserved for 1 week. Cryobiology 60 (3):326–330. doi:10.1016/j.cryobiol.2010.02.008
26. Park BW, Jang SJ, Byun JH, Kang YH, Choi MJ, Park WU, Lee WJ, Rho GJ (2014) Cryopreservation of human dental follicle tissue for use as a resource of autologous mesenchymal stem cells. Journal of tissue engineering and regenerative medicine. doi:10.1002/term.1945
27. Kang YH, Lee HJ, Jang SJ, Byun JH, Lee JS, Lee HC, Park WU, Lee JH, Rho GJ, Park BW (2015) Immunomodulatory properties and in vivo osteogenesis of human dental stem cells from fresh and cryopreserved dental follicles. Differentiation 90 (1–3):48–58. doi:10.1016/j.diff.2015.10.001
28. Ding G, Wang W, Liu Y, An Y, Zhang C, Shi S, Wang S (2010) Effect of cryopreservation on biological and immunological properties of stem cells from apical papilla. Journal of cellular physiology 223 (2):415–422. doi:10.1002/jcp.22050

in order to ensure maximum recovery of viable and functional cells [206]. One of the most important aspects of freezing cells is the rate of cooling. When cells are cooled too slowly they will dehydrate and shrink due to osmotic stress, whereas rapid cooling results in intracellular ice formation [206, 207]. However, while intracellular ice formation is lethal to cells in suspension, it could be innocuous to cell monolayers due to its propagation via gap junctions. When exposed to 100 % intracellular ice formation, DPSC monolayers have been shown to retain membrane integrity, however they lost the ability to proliferate [208]. Despite the fact that different cell types have a different membrane permeability, cooling rates of 1 °C/min are often applied for several mammalian cell types [205, 206]. Besides carefully controlling cooling rates, media can also be supplemented with cryoprotectant agents in order to minimize cryoinjury. Cryopreservation solutions for MSCs are most commonly supplemented with 10 % dimethyl sulfoxide (DMSO) [205, 209]. DMSO is a permeating cryoprotectant as it is able to penetrate the cell membrane, thereby preventing cell rupture. Several DMSO-based cryopreservation protocols have been tested for the preservation of DPSCs [203, 204, 210–212], SCAPs [213] and PDLSCs [214] (Table 17.1). Woods et al. reported that 7.8–11.6 % DMSO was optimal for DPSCs cooled at −1 °C/min in an isopropanol bath to −85 °C for 24 h followed by storage in liquid nitrogen [212]. DPSCs have been shown to retain their stem cell markers [203, 204, 210, 212] as well as their multilineage differentiation potential post thawing [25, 203]. Ding et al. reported no difference in cell viability, colony forming efficiency, proliferation rate, multilineage differentiation potential, MSC marker expression, karyotype anomalies and immunomodulatory capacities between freshly isolated SCAPs and cryopreserved (10 % DMSO + 90 % fetal bovine serum (FBS)) SCAPs [213]. Cryopreservation of PDLSCs using 10 % DMSO did not affect their proliferative capacity [214].

Despite the widespread use of DMSO as a cryoprotectant, it is known to be potentially cytotoxic. DMSO-preserved bone marrow cells have been shown to cause adverse effects post transplantation [215–217]. Since MSCs have shown great resilience during cryopreservation many other strategies have been explored to reduce cryoprotectant toxicity, minimize cryoinjuries and to develop xeno-free freezing solutions [206]. Hoping to lower or even completely remove DMSO from the freezing solutions, several other compounds have been investigated. Boron is a micro-mineral that is involved in membrane integrity and is important for membrane structure and function and could therefore minimize/inhibit intracellular ice formation. Demirci and colleagues reported an increased viability when tooth germ- derived stem cells from a single donor were preserved in a solution containing 20 μg/mL borate and 5 % DMSO. Furthermore, this solution did not affect surface antigen expression nor did it alter the osteogenic and chondrogenic differentiation potential of these cells [218]. Umemura et al. developed a DMSO-, serum- and xeno-free cryopreservation method by encapsulating DPSCs in an alginate gel. Cryopreserved DPSCs displayed a normal morphology, showed high proliferative capacity and maintained MSC marker expression [219]. Furthermore, Ducret and colleagues recently described a medicinal manufacturing approach for producing DPSCs. In this study, DPSCs were cultured under xeno-free conditions and were cryopreserved for 510 days. Cryopreserved DPSCs were reported to be free of karyotype abnormalities and to have stable doubling times, comparable to those of fresh DPSCs [220].

Using a freezing protocol involving a magnetic field, DPSCs can be frozen in a serum-free freezing solution containing only 3 % DMSO [221]. Magnetic cryopreservation (Cells Alive System, CAS) was originally designed by the ABI Corporation for the preservation of food, but has now found its way into the scientific world as well. By using magnetically induced non-thermal vibrations this freezing method prevents the formation of ice crystals [222] (Table 17.1). Within the field of dentistry, tooth auto-transplantation is a useful technique for replacing missing teeth with several advantages such as maintaining bone volume and PDL regeneration capacity as well as allowing dentofacial development [223]. It occurs that patients no longer possess a suitable donor tooth, due to prior extractions. In these cases tooth banking could greatly expand the usage of auto-transplantation [224] (Table 17.1). The survival and regeneration of the PDL is crucial for the success of the transplantation and the prognosis. Kaku et al. reported an increased survival rate of PDL cells after they were frozen with a magnetic field compared to those without a magnetic field. A magnetic field of 0.01 mT, hold time of 15 min and plunging temperature of −30 °C were determined as the optimal cryopreservation parameters for PDL cells [224]. Furthermore, no difference in mRNA expression or collagen type I was demonstrated and only a minor decrease in alkaline phosphatase was reported between cryopreserved PDL cells and control groups [222]. More importantly, there was no progressive root resorption after reimplantation of cryopreserved teeth into Wistar rats, in contrast to the widespread root resorption and ankyloses in dried teeth [222]. When intact teeth were frozen for 5 years using the CAS freezer, Abedini et al. showed no difference in collagen type I, alkaline phosphatase and VEGF both at mRNA level as well as at protein level [225]. Moreover, a clinical case illustrated PDL regeneration 1 month after the transplantation of a third molar which was cryopreserved for 3 months. Sixteen months post-surgery there was sufficient bone regeneration and complete apexogenesis was achieved [225]. Abedini and colleagues also investigated dental pulp cells isolated from either mature or immature fresh third molars or mature or immature 3-month-cryopreserved molars. Cryopreservation resulted in a delay in cellular outgrowth and pulp cells from cryopreserved molars with complete root formation could not survive properly until confluence was reached. These effects could probably be ascribed to the difficulty of the cryoprotectant to reach the pulp chamber through the narrow tooth apex. Dental pulp cells from cryopreserved immature teeth showed a decreased protein expression of VEGF and NGF compared to their fresh immature counterparts [225].

As demonstrated by the case study of Abedini et al., the field of dentistry would greatly benefit from the ability to cryopreserve intact teeth or at least dental tissues prior to stem cell isolation (Table 17.1). Studies investigating cryopreservation of teeth have mainly focused on the transplantation of permanent teeth and therefore questioned the survival of the periodontal ligament and pulp tissue of the cryopreserved teeth [225–228]. Oh and colleagues reported no significant difference in the viability and osteogenic differentiation capacity of PDL cells from 1-week-cryopreserved teeth [226]. Min et al. found a decrease in bFGF receptor expression in PDL cells isolated from 1-week-cryopreserved teeth [227]. However, after cryopreservation of intact teeth, Woods et al. could only establish DPSC cultures from 20 % of the frozen teeth [212]. The low yield could be correlated to the dimensions of the apical foramen, as Temmerman et al. obtained maximal pulp viability after cryopreservation of teeth with an apical opening of at least 9.42 mm^2 [229]. For this reason, cryopreservation of pulp tissue could offer a suitable alternative. DPSCs isolated from cryopreserved pulp tissues showed no difference in CD marker expression and no loss in osteogenic and adipogenic differentiation potential compared to their fresh counterparts [230]. Gioventu and colleagues tried to circumvent the issue of a limited apical foramen by laser piercing deciduous teeth prior to cryopreservation and found that DPSCs from cryopreserved teeth showed similar growth rates and marker expression than DPSCs isolated from fresh tissue [228]. Since deciduous teeth cannot be replanted, the ability to cryopreserve them for future stem cell isolation would greatly decrease costs and efforts associated with stem cell banking from deciduous teeth [231]. Lindemann et al. only reported a 30 % isolation success rate of 1-week-cryopreserved teeth. Despite the lower culture rates and morphological aberrations, no changes in immunophenotype and differentiation potential were seen [232]. Increasing serum levels could possibly improve the culture rate after cryopreservation as Bressan and colleagues found that increasing serum concentrations from 10 % to 20 % resulted in a more than 50 % increase in culture rate [233]. The abovementioned studies only cryopreserved teeth for a short period of time and according to Ji et al. the period of cryopreservation also plays a crucial role in the viability of the resulting pulp cells. They reported a reduction in proliferation rate and increased apoptosis when teeth were cryopreserved for over 3 months [231]. Perry et al. reported a 70 % success rate of isolation from 1-month cryopreserved whole teeth [211]. Dental follicle tissue has also been shown to yield viable DSCs after 3 months of tissue storage. Tissue samples were stored in phosphate-buffered saline (PBS) supplemented with glucose, sucrose and ethylene glycol for cryoprotection. DSCs isolated from cryopreserved dental follicle showed no difference in the expression of stem cell markers or the ability for multilineage differentiation. However, cells from cryopreserved tissues did show a higher intensity of p53 protein, indicating damage from the freeze-thaw process [234]. Furthermore, upon transplantation into mandibular defects in miniature pigs, both FSCs from both fresh and cryopreserved dental follicle significantly enhanced new bone formation after 8 weeks [235]. Freezing of the intact PDL, only yielded 40 % of single-colony derived PDLSCs compared to the number of PDLSC colonies recovered from fresh periodontal ligament tissue. These cells showed a high proliferation rate and maintained their multilineage differentiation potential. Moreover, upon transplantation, PDLSCs derived from cryopreserved PDL generated a typical cementum/periodontal ligament-like structure [236]. The possibility of storing tissue samples for longer periods of time without significant damage to the inherent stem cell population offers great perspectives for regenerative medicine and stem cell banking.

17.5 From Bench to Bedside

Stem cell research holds great promise for the development of treatment strategies for various injuries and diseases. The scope of potential stem cell therapies is expanding due to advances in stem cell research. The emerging demand of stem cell therapeutics necessitates the establishment and collaboration of centralized biobanks at an international level. Initiatives like the International Stem Cell Banking Initiative (ISCBI) and the International Society for Stem Cell Research (ISSCR) strive for the creation of such centralized biobanks to provide high-quality stem cells for therapeutic purposes in a standardized approach [237, 238]. Though, this seems challenging given the heterogeneous ethical, regulatory and legal setting. In order to provide a standardized stem cell-based therapy, strict rules have to be followed regarding the different processes that are performed from isolation to transplantation [239, 240]. These processes vary highly depending on the stem cell source, type and clinical application.

Standardized procedures are essential to ensure optimal reproducibility, safety and efficacy when DSCs are used for clinical applications (Fig. 17.2). However, current manufacturing procedures for DSC-based products are not fully in line with the international guidelines for GMP for stem cell therapies. GMP requires quality control regarding donor eligibility, isolation procedure, labeling, transportation, processing, storage, lab equipment, reagents, distribution to the patient and documentation [241–243]. Briefly, the manufacturing steps of DSC-based products can be generally defined as (1) tooth extraction and pulp tissue collection, (2) DSC isolation, (3) cell culture and expansion, (4) cryobanking (5) safety control and quality testing, (6) transplantation [244]. Extraction of third molars for the isolation of DSCs is being performed in donors of different ages and with third molar in different stages of development [72, 190, 211, 245, 246]. This impairs the comparison of the experimental results and makes it difficult to provide a standardized therapy. There are no guidelines that specify the developmental stage of the tooth for the isolation of DSCs and DPSCs in particular. However, it has been suggested to use impacted third molars between Nolla developmental stage 5 (almost completed dental crown) and stage 7 (root completed for one third) in young donors, since the use of impacted molars in this developmental stage reduces the risk for contamination

Fig. 17.2 From bench to beside. Standardized procedure for the cryopreservation and manufacturing of dental stem cell-based products

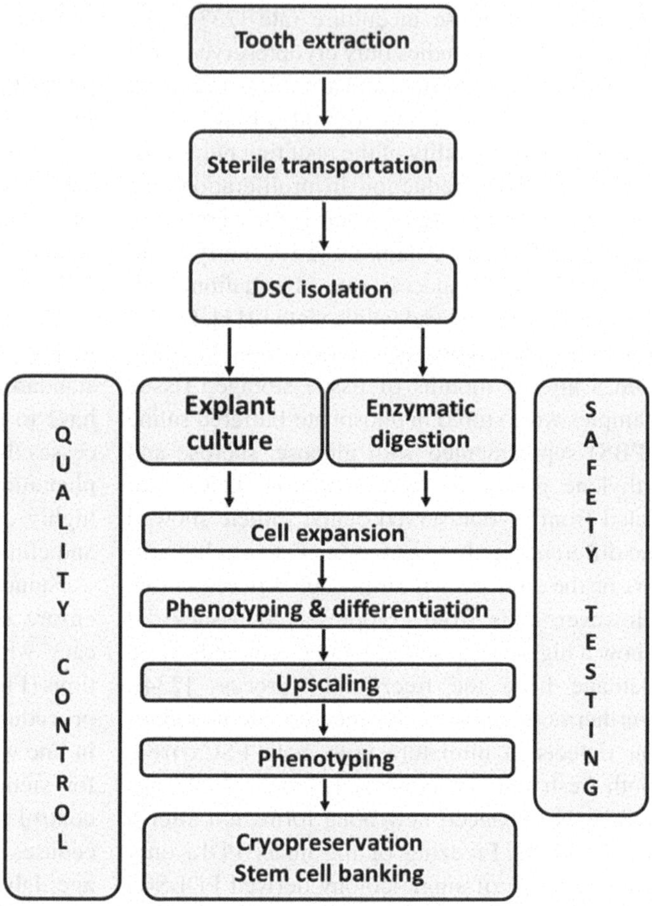

and avoids the use of mechanical separation [244]. Dental pulp tissue, for example, can be collected in PBS for transportation to the lab, as DPSCs remain viable for up to 5 days in this medium [212]. For the isolation of DSCs two widely used methods are available, namely enzymatic digestion and explant culture. Considering translation to clinical applications the explant method seems to be the preferred approach. Explant culture is easier, safer, less expensive and more compliant with the GMP guidelines [246]. Cell culture and expansion is arguably the most important process in providing a cell-based therapy, as it is essential for the stem cell products to be free of any microbiological contamination. A major pitfall of standard DSC culture is the continuing use of animal culture medium additives, which will be elaborately discussed in the following section. Therefore, the development of xeno-free, serum-free media is a critical objective for standardized manufacturing of DSC-based products. Additionally, the use of xeno-free dissociation buffers such as TrypLe® or Accutase® is highly recommended for passaging the cells. Nevertheless, further research is necessary to optimize the culture properties and to ensure the phenotype of the cells is not negatively altered.

To provide an adequate amount of cells for clinical trials and commercialization of the cell-based product, efficient expansion or 'upscaling' is crucial for the establishment of a novel successful therapy. For autologous applications, the goal is to produce an appropriate number of cells for the treatment of one patient, mostly in the range of one to five billion cells per manufacturing process [239]. In this context, the use of conventional cell culture using culture flasks or small bioreactors is feasible. However, in the case of allogenic applications where multiple patients are targeted, the upscaling has to be performed to a much higher extent. Here, the use of cell factories with large scale bioreactors becomes inevitable. Providing a cell-based therapy often involves the storage of the cell product or an intermediate form of the cell product for future use. Cryopreservation and long-term storage of cells permits completion of safety control, quality testing and stable transportation to the clinical site [216]. Optimal cryopreservation conditions are necessary to maximize cell stability, recovery and function, as extensively discussed in the previous section. Furthermore, it is required to phenotypically characterize the cells and perform additional tests to ensure the safety of the cell-based product. Flow cytometry is an ideal technique to analyze various phenotypic markers simultaneously on the cellular level. It is advised to use a panel of markers that provides a high level of specificity for DSCs. However, this seems to be challenging since there is no unique marker for DSCs and many markers that were proposed by the ISCT to identify MSCs are also expressed in other cell types, including mature fibroblasts [247, 248]. Supplementary testing of the trilineage differentiation capacity should be performed to guarantee the stem cell properties of the isolated cells.

Despite the current lack of standardization in the upscale production and long-term storage of DSCs, there are already several commercial tooth banking services available. In Japan, for example, both the Hiroshima University and Nagoya University have founded their own tooth bank. In 2008, a collaboration between the Norwegian Institute of Public Health and the University of Bergen led to the first European tooth bank [249]. Commercial tooth banking companies can also be found in the United States, such as BioEden (Austin, Texas, USA), Store-A-Tooth (Provia Laboratories, Littleton, Massachusetts, USA) and StemSave (Stemsave Inc, New York, USA).

It has to be emphasized that all variables in the manufacturing process of DSC products have to be optimized for the specific application before continuing into clinical trials, since these can affect the properties of the cell-based product. Consequently, if the product successfully passes clinical trials, it can be produced without altering the manufacturing process.

17.6 Pitfalls Associated with Dental Stem Cell Banking

When contemplating the long-term cryopreservation as well as the upscale production of (dental) stem cells, one certainly needs to take into account certain pitfalls associated with these procedures. Although considered to be common nutritional components, the use of animal-derived culture medium additives, for example, has been the subject of debate for quite some time. Next to animal welfare concerns and the high costs associated with the use of FBS and other animal-derived proteins, the addition of these cell nutrients to culture media has some other drawbacks as well as they are considered to be potentially hazardous [250–253]. More specifically, not only are these additives a potential source of endotoxins and other infectious agents such as bacteria, viruses, fungi and prions, but the xenogeneic antigens could also evoke a severe immunological response within the host [250–252]. Together with the high batch-to-batch variety in protein content, these disadvantages urged the need for alternative, serum-free cell culture systems [250, 252, 254].

With regard to the culturing of DSCs in serum-free conditions, Tarle et al. developed a chemically defined serum-free culture medium for SHEDs and PDLSCs. There were no significant differences in comparison to the stem cells cultured in FBS-containing medium; both stem cell populations were able to expand and maintained their multipotent capacity [255]. Comparable observations were made by Hirata et al. and others, showing both (sub-optimal) proliferation and expression of stem cell markers by DPSCs, PDLSCs and FSCs when cultured in serum-free media containing a range of different growth factors [256–259]. Depending on the specific growth factor supplement, however, stem cell differentiation can also be induced. Bonnamain et al., for example, indicated the expression of neuronal and oligodendrocyte markers when incubating DPSCs with EGF and bFGF [260]. Neurogenic differentiation of DPSCs was also reported by Xiao and Tsutsui, showing the expression of neural markers after using a commercially available serum replacement [261]. In addition, a $CD117^+$ subset of DPSCs differentiated into pancreatic cells following incubation with a specific combination of growth factors and chemical substances [111]. An increase in the expression of endothelial markers was also mentioned by Karbanova et al., after incubating a serum-free DPSC culture with VEGF and insulin-transferrin-sodium selenite (ITS) [262]. In the search for appropriate alternatives for animal-derived culture medium additives, numerous publications have reported on the use of human serum or other blood-derived products in stem cell cultures [252, 254, 257, 263–271]. With respect to DSCs in particular, a recent study of Pisciolaro et al. indicated a significantly improved proliferation rate and mineralization potential when incubating DPSCs with human autologous serum [268]. A more prominent osteogenic and adipogenic differentiation of DPSCs following incubation with allogenic human serum was also found by Khanna-Jain et al., while Govindasamy et al. demonstrated the pronounced expansion of DPSCs in medium containing human platelet lysate [257, 266]. However, the use of human blood derivatives still holds several disadvantages. As with any donor-derived product, inherent differences between samples and batch-to-batch variability cannot be excluded. Besides the extensive amount that would be needed for the upscale production of (dental) stem cells, the current lack of consistent study set-ups demands for further analysis of the precise composition as well as the potential impact of these human-derived blood products on the intrinsic properties and behavior of the stem cells [252, 264, 265]. Further research is thus required before any decision can be taken by scientists as well as regulatory agencies regarding the future application of alternative culture medium additives.

Another important aspect one definitely has to keep in mind when considering the application of autologous and/or allogenic DSCs is potential donor-related variability, as the inherent behavior of these cells can be affected by a range of different factors. The pressure and tension generated during the application of orthodontic force, for example, causes the release of various growth factors and cytokines and, subsequently, the creation of a supportive microenvironment for bone remodeling, root resorption and differentiation of residing (stem) cells [272–275]. With regard to the effect of the age of the donor at the time of stem cell isolation, contrasting results are described in literature. Iohora et al., for instance, reported an age-related decline of stem cell-mediated dental pulp regeneration in dogs [276]. Similar observations were made by Feng et al. and others, showing an age-dependent decrease in proliferation, migration and osteogenic differentiation potential of DPSCs and PDLSCs [277–280]. Atari et al., on the other hand, successfully isolated DPSCs from 14 to 60-year-old donors, with no significant differences in gene expression [281]. These results were also confirmed by Kellner et al. and others, stating that age is not a determining factor for the stem cell properties and regenerative potential of DPSCs [127, 282]. Next to orthodontic tooth movement and age, the (oral) health of the patient also seems to significantly influence the biological properties of DSCs. Nicotine, for example, does not only influence their osteogenic differentiation potential, but also alters the viability, growth and migration rate of PDLSCs in particular [283–286]. In 2010, Cooper et al., already described the complex interaction between inflammation and the natural regenerative potential of dental pulp, as inflammatory mediators can modulate the repair processes within the tissue [287]. Since then, a number of papers have been published reporting on the effects of caries and/or inflammatory processes on DSCs. Alongi et al., for example, showed a decreased expression of MSC markers as well as a diminished proliferation rate and osteogenic differentiation potential in vitro by DPSCs originating from inflamed dental pulps. However, no notable differences were observed when comparing the formation of pulp/dentin complexes in immunocompromised mice [288].

Similar results were found by Liu et al. and others, indicating an altered proliferation and migration rate as well as a differential protein expression profile and immunomodulatory properties for carious and/or inflamed DPSCs and MSCs from periapical lesions [289–293]. In contrast, several papers mentioned no significantly different behavior of DPSCs isolated from carious deciduous and/or permanent teeth, thereby suggesting their potential use in cell-based therapies [294–296]. When considering the therapeutic application of autologous stem cells in cancer patients, the potential detrimental effects caused by previous treatment with radiotherapy certainly need to be taken into account. With regard to the effects of ionizing radiation (IR) on the behavior of DSCs, a couple of studies reported the induction of cell cycle arrest, premature senescence and differentiation of DPSCs following exposure to different dosages of IR [297, 298]. In contrast, Abe et al., demonstrated a radio-resistant phenotype for SCAPs, albeit with a lower hard tissue forming capacity *in vivo* [299]. Before any therapeutic application is possible, the implementation of a dental stem cell banking system not only demands extensive evaluation at scientific as well as regulatory level but also requires additional characterization of its associated pitfalls such as culture medium additives and donor-related confounding factors.

17.7 Conclusion and Future Perspectives

Taken together, DSCs seem to hold great promise for various applications. One of the major advances made in the field of regenerative medicine is 3D-printing. This technique evolved from a simple 3D- printed construct towards the fabrication of a complex structure containing multiple cell types and biological constitutes/components [300–302]. These constructs can ultimately serve as a 3D-scaffold in which DSCs can be studied for multilineage differentiation or more specifically for dental pulp regeneration. However, to ensure future therapeutic applications, one should strive to maintain a cost-effective DSC cell banking system which guarantees phenotypic stability of the heterogeneous populations of DSCs. Some hurdles have to be overcome to establish clinical translation of DSC-based products, since current manufacturing methods are not in line with the GMP guidelines for stem cell therapy. Developing a more standardized and compliant manufacturing protocol will result in an increased efficacy, reproducibility and safety.

References

1. Friedenstein AJ, Chailakhjan RK, Lalykina KS (1970) The development of fibroblast colonies in monolayer cultures of guinea-pig bone marrow and spleen cells. Cell Tissue Kinet 3(4):393–403
2. Zuk PA, Zhu M, Ashjian P, De Ugarte DA, Huang JI, Mizuno H, Alfonso ZC, Fraser JK, Benhaim P, Hedrick MH (2002) Human adipose tissue is a source of multipotent stem cells. Mol Biol Cell 13(12):4279–4295. doi:10.1091/mbc.E02-02-0105
3. Weiss ML, Medicetty S, Bledsoe AR, Rachakatla RS, Choi M, Merchav S, Luo Y, Rao MS, Velagaleti G, Troyer D (2006) Human umbilical cord matrix stem cells: preliminary characterization and effect of transplantation in a rodent model of Parkinson's disease. Stem Cells 24(3):781–792. doi:10.1634/stemcells.2005-0330
4. Erices A, Conget P, Minguell JJ (2000) Mesenchymal progenitor cells in human umbilical cord blood. Br J Haematol 109(1):235–242
5. Gronthos S, Mankani M, Brahim J, Robey PG, Shi S (2000) Postnatal human dental pulp stem cells (DPSCs) in vitro and in vivo. Proc Natl Acad Sci U S A 97(25):13625–13630. doi:10.1073/pnas.240309797
6. Sonoyama W, Liu Y, Yamaza T, Tuan RS, Wang S, Shi S, Huang GT (2008) Characterization of the apical papilla and its residing stem cells from human immature permanent teeth: a pilot study. J Endod 34(2):166–171. doi:10.1016/j.joen.2007.11.021
7. Seo BM, Miura M, Gronthos S, Bartold PM, Batouli S, Brahim J, Young M, Robey PG, Wang CY, Shi S (2004) Investigation of multipotent postnatal stem cells from human periodontal ligament. Lancet 364(9429):149–155. doi:10.1016/S0140-6736(04)16627-0
8. Morsczeck C, Gotz W, Schierholz J, Zeilhofer F, Kuhn U, Mohl C, Sippel C, Hoffmann KH (2005) Isolation of precursor cells (PCs) from human dental follicle of wisdom teeth. J Int Soc Matrix Biol 24(2):155–165. doi:10.1016/j.matbio.2004.12.004
9. Xiong J, Mrozik K, Gronthos S, Bartold PM (2012) Epithelial cell rests of Malassez contain unique stem cell populations capable of undergoing

epithelial-mesenchymal transition. Stem Cells Dev 21(11):2012–2025. doi:10.1089/scd.2011.0471
10. Miura M, Gronthos S, Zhao M, Lu B, Fisher LW, Robey PG, Shi S (2003) SHED: stem cells from human exfoliated deciduous teeth. Proc Natl Acad Sci U S A 100(10):5807–5812. doi:10.1073/pnas.0937635100
11. Silverio KG, Rodrigues TL, Coletta RD, Benevides L, Da Silva JS, Casati MZ, Sallum EA, Nociti FH Jr (2010) Mesenchymal stem cell properties of periodontal ligament cells from deciduous and permanent teeth. J Periodontol 81(8):1207–1215. doi:10.1902/jop.2010.090729
12. Shi S, Gronthos S (2003) Perivascular niche of postnatal mesenchymal stem cells in human bone marrow and dental pulp. J Bone Miner Res 18(4):696–704. doi:10.1359/jbmr.2003.18.4.696
13. Waddington RJ, Youde SJ, Lee CP, Sloan AJ (2009) Isolation of distinct progenitor stem cell populations from dental pulp. Cells Tissues Organs 189(1–4):268–274. doi:10.1159/000151447
14. Kaukua N, Shahidi MK, Konstantinidou C, Dyachuk V, Kaucka M, Furlan A, An Z, Wang L, Hultman I, Ahrlund-Richter L, Blom H, Brismar H, Lopes NA, Pachnis V, Suter U, Clevers H, Thesleff I, Sharpe P, Ernfors P, Fried K, Adameyko I (2014) Glial origin of mesenchymal stem cells in a tooth model system. Nature 513(7519):551–554. doi:10.1038/nature13536
15. Hilkens P, Meschi N, Lambrechts P, Bronckaers A, Lambrichts I (2015) Dental stem cells in pulp regeneration: near future or long road ahead? Stem Cells Dev 24(14):1610–1622. doi:10.1089/scd.2014.0510
16. Pittenger MF, Mackay AM, Beck SC, Jaiswal RK, Douglas R, Mosca JD, Moorman MA, Simonetti DW, Craig S, Marshak DR (1999) Multilineage potential of adult human mesenchymal stem cells. Science 284(5411):143–147
17. Dominici M, Le Blanc K, Mueller I, Slaper-Cortenbach I, Marini F, Krause D, Deans R, Keating A, Prockop D, Horwitz E (2006) Minimal criteria for defining multipotent mesenchymal stromal cells. The International Society for Cellular Therapy position statement. Cytotherapy 8(4):315–317. doi:10.1080/14653240600855905
18. Akpinar G, Kasap M, Aksoy A, Duruksu G, Gacar G, Karaoz E (2014) Phenotypic and proteomic characteristics of human dental pulp derived mesenchymal stem cells from a natal, an exfoliated deciduous, and an impacted third molar tooth. Stem Cells Int 2014:457059. doi:10.1155/2014/457059
19. Pisciotta A, Carnevale G, Meloni S, Riccio M, De Biasi S, Gibellini L, Ferrari A, Bruzzesi G, De Pol A (2015) Human dental pulp stem cells (hDPSCs): isolation, enrichment and comparative differentiation of two sub-populations. BMC Dev Biol 15:14. doi:10.1186/s12861-015-0065-x
20. Tamaki Y, Nakahara T, Ishikawa H, Sato S (2013) In vitro analysis of mesenchymal stem cells derived from human teeth and bone marrow. Odontol Soc Nippon Dent Univ 101(2):121–132. doi:10.1007/s10266-012-0075-0
21. Ishizaka R, Hayashi Y, Iohara K, Sugiyama M, Murakami M, Yamamoto T, Fukuta O, Nakashima M (2013) Stimulation of angiogenesis, neurogenesis and regeneration by side population cells from dental pulp. Biomaterials 34(8):1888–1897. doi:10.1016/j.biomaterials.2012.10.045
22. Bakopoulou A, Leyhausen G, Volk J, Koidis P, Geurtsen W (2013) Comparative characterization of STRO-1(neg)/CD146(pos) and STRO-1(pos)/CD146(pos) apical papilla stem cells enriched with flow cytometry. Arch Oral Biol 58(10):1556–1568. doi:10.1016/j.archoralbio.2013.06.018
23. Yang X, Walboomers XF, van den Beucken JJ, Bian Z, Fan M, Jansen JA (2009) Hard tissue formation of STRO-1-selected rat dental pulp stem cells in vivo. Tissue Eng A 15(2):367–375. doi:10.1089/ten.tea.2008.0133
24. Zhang W, Walboomers XF, Van Kuppevelt TH, Daamen WF, Van Damme PA, Bian Z, Jansen JA (2008) In vivo evaluation of human dental pulp stem cells differentiated towards multiple lineages. J Tissue Eng Regen Med 2(2–3):117–125. doi:10.1002/term.71
25. Zhang W, Walboomers XF, Shi S, Fan M, Jansen JA (2006) Multilineage differentiation potential of stem cells derived from human dental pulp after cryopreservation. Tissue Eng 12(10):2813–2823. doi:10.1089/ten.2006.12.2813
26. Nakatsuka R, Nozaki T, Uemura Y, Matsuoka Y, Sasaki Y, Shinohara M, Ohura K, Sonoda Y (2010) 5-Aza-2′-deoxycytidine treatment induces skeletal myogenic differentiation of mouse dental pulp stem cells. Arch Oral Biol 55(5):350–357. doi:10.1016/j.archoralbio.2010.03.003
27. Song M, Kim H, Choi Y, Kim K, Chung C (2012) Skeletal myogenic differentiation of human periodontal ligament stromal cells isolated from orthodontically extracted premolars. Korean J Orthod 42(5):249–254. doi:10.4041/kjod.2012.42.5.249
28. Kerkis I, Kerkis A, Dozortsev D, Stukart-Parsons GC, Gomes Massironi SM, Pereira LV, Caplan AI, Cerruti HF (2006) Isolation and characterization of a population of immature dental pulp stem cells expressing OCT-4 and other embryonic stem cell markers. Cells Tissues Organs 184(3–4):105–116. doi:10.1159/000099617
29. Oswald J, Boxberger S, Jorgensen B, Feldmann S, Ehninger G, Bornhauser M, Werner C (2004) Mesenchymal stem cells can be differentiated into endothelial cells in vitro. Stem Cells 22(3):377–384. doi:10.1634/stemcells.22-3-377
30. Takahashi M, Suzuki E, Oba S, Nishimatsu H, Kimura K, Nagano T, Nagai R, Hirata Y (2010) Adipose tissue-derived stem cells inhibit neointimal formation in a paracrine fashion in rat femoral artery.

Am J Phys Heart Circ Phys 298(2):H415–H423. doi:10.1152/ajpheart.00391.2009
31. Bai K, Huang Y, Jia X, Fan Y, Wang W (2010) Endothelium oriented differentiation of bone marrow mesenchymal stem cells under chemical and mechanical stimulations. J Biomech 43(6):1176–1181. doi:10.1016/j.jbiomech.2009.11.030
32. Chen MY, Lie PC, Li ZL, Wei X (2009) Endothelial differentiation of Wharton's jelly-derived mesenchymal stem cells in comparison with bone marrow-derived mesenchymal stem cells. Exp Hematol 37(5):629–640. doi:10.1016/j.exphem.2009.02.003
33. Bronckaers A, Hilkens P, Martens W, Gervois P, Ratajczak J, Struys T, Lambrichts I (2014) Mesenchymal stem/stromal cells as a pharmacological and therapeutic approach to accelerate angiogenesis. Pharmacol Ther 143(2):181–196. doi:10.1016/j.pharmthera.2014.02.013
34. d'Aquino R, Graziano A, Sampaolesi M, Laino G, Pirozzi G, De Rosa A, Papaccio G (2007) Human postnatal dental pulp cells co-differentiate into osteoblasts and endotheliocytes: a pivotal synergy leading to adult bone tissue formation. Cell Death Differ 14(6):1162–1171. doi:10.1038/sj.cdd.4402121
35. Marchionni C, Bonsi L, Alviano F, Lanzoni G, Di Tullio A, Costa R, Montanari M, Tazzari PL, Ricci F, Pasquinelli G, Orrico C, Grossi A, Prati C, Bagnara GP (2009) Angiogenic potential of human dental pulp stromal (stem) cells. Int J Immunopathol Pharmacol 22(3):699–706
36. Iohara K, Zheng L, Wake H, Ito M, Nabekura J, Wakita H, Nakamura H, Into T, Matsushita K, Nakashima M (2008) A novel stem cell source for vasculogenesis in ischemia: subfraction of side population cells from dental pulp. Stem Cells 26(9):2408–2418. doi:10.1634/stemcells.2008-0393
37. Sakai VT, Zhang Z, Dong Z, Neiva KG, Machado MA, Shi S, Santos CF, Nor JE (2010) SHED differentiate into functional odontoblasts and endothelium. J Dent Res 89(8):791–796. doi:10.1177/0022034510368647
38. Amorim BR, Silverio KG, Casati MZ, Sallum EA, Kantovitz KR, Nociti FH, Jr (2016) Neuropilin controls endothelial differentiation by mesenchymal stem cells from the periodontal ligament. J Periodontol:1–14. doi:10.1902/jop.2016.150603
39. Bakopoulou A, Kritis A, Andreadis D, Papachristou E, Leyhausen G, Koidis P, Geurtsen W, Tsiftsoglou A (2015) Angiogenic potential and secretome of human apical papilla mesenchymal stem cells in various stress microenvironments. Stem Cells Dev 24(21):2496–2512. doi:10.1089/scd.2015.0197
40. Zhang Z, Nor F, Oh M, Cucco C, Shi S, Nor JE (2016) Wnt/beta-catenin signaling determines the vasculogenic fate of post-natal mesenchymal stem cells. Stem Cells. doi:10.1002/stem.2334
41. Janebodin K, Zeng Y, Buranaphatthana W, Ieronimakis N, Reyes M (2013) VEGFR2-dependent angiogenic capacity of pericyte-like dental pulp stem cells. J Dent Res 92(6):524–531. doi:10.1177/0022034513485599
42. Barros MA, Martins JF, Maria DA, Wenceslau CV, De Souza DM, Kerkis A, Camara NO, Balieiro JC, Kerkis I (2015) Immature dental pulp stem cells showed renotropic and pericyte-like properties in acute renal failure in rats. Cell Med 7(3):95–108. doi:10.3727/215517914X680038
43. Arthur A, Rychkov G, Shi S, Koblar SA, Gronthos S (2008) Adult human dental pulp stem cells differentiate toward functionally active neurons under appropriate environmental cues. Stem Cells 26(7):1787–1795. doi:10.1634/stemcells.2007-0979
44. Kiraly M, Porcsalmy B, Pataki A, Kadar K, Jelitai M, Molnar B, Hermann P, Gera I, Grimm WD, Ganss B, Zsembery A, Varga G (2009) Simultaneous PKC and cAMP activation induces differentiation of human dental pulp stem cells into functionally active neurons. Neurochem Int 55(5):323–332. doi:10.1016/j.neuint.2009.03.017
45. Gervois P, Struys T, Hilkens P, Bronckaers A, Ratajczak J, Politis C, Brone B, Lambrichts I, Martens W (2014) Neurogenic maturation of human dental pulp stem cells following neurosphere generation induces morphological and electrophysiological characteristics of functional neurons. Stem Cells Dev. doi:10.1089/scd.2014.0117
46. Osathanon T, Sawangmake C, Nowwarote N, Pavasant P (2014) Neurogenic differentiation of human dental pulp stem cells using different induction protocols. Oral Dis 20(4):352–358. doi:10.1111/odi.12119
47. Lee JH, Um S, Song IS, Kim HY, Seo BM (2014) Neurogenic differentiation of human dental stem cells in vitro. J Korean Assoc Oral Maxillofac Surg 40(4):173–180. doi:10.5125/jkaoms.2014.40.4.173
48. Jarmalaviciute A, Tunaitis V, Strainiene E, Aldonyte R, Ramanavicius A, Venalis A, Magnusson KE, Pivoriunas A (2013) A new experimental model for neuronal and glial differentiation using stem cells derived from human exfoliated deciduous teeth. J Mol Neurosci: MN. doi:10.1007/s12031-013-0046-0
49. Nourbakhsh N, Soleimani M, Taghipour Z, Karbalaie K, Mousavi SB, Talebi A, Nadali F, Tanhaei S, Kiyani GA, Nematollahi M, Rabiei F, Mardani M, Bahramiyan H, Torabinejad M, Nasr-Esfahani MH, Baharvand H (2011) Induced in vitro differentiation of neural-like cells from human exfoliated deciduous teeth-derived stem cells. Int J Dev Biol 55(2):189–195. doi:10.1387/ijdb.103090nn
50. Wang J, Wang X, Sun Z, Wang X, Yang H, Shi S, Wang S (2010) Stem cells from human-exfoliated deciduous teeth can differentiate into dopaminergic neuron-like cells. Stem Cells Dev 19(9):1375–1383. doi:10.1089/scd.2009.0258
51. Morsczeck C, Vollner F, Saugspier M, Brandl C, Reichert TE, Driemel O, Schmalz G (2010)

Comparison of human dental follicle cells (DFCs) and stem cells from human exfoliated deciduous teeth (SHED) after neural differentiation in vitro. Clin Oral Investig 14(4):433–440. doi:10.1007/s00784-009-0310-4
52. Fortino VR, Chen RS, Pelaez D, Cheung HS (2014) Neurogenesis of neural crest-derived periodontal ligament stem cells by EGF and bFGF. J Cell Physiol 229(4):479–488. doi:10.1002/jcp.24468
53. Bueno C, Ramirez C, Rodriguez-Lozano FJ, Tabares-Seisdedos R, Rodenas M, Moraleda JM, Jones JR, Martinez S (2013) Human adult periodontal ligament-derived cells integrate and differentiate after implantation into the adult mammalian brain. Cell Transplant 22(11):2017–2028. doi:10.3727/096368912X657305
54. Techawattanawisal W, Nakahama K, Komaki M, Abe M, Takagi Y, Morita I (2007) Isolation of multipotent stem cells from adult rat periodontal ligament by neurosphere-forming culture system. Biochem Biophys Res Commun 357(4):917–923. doi:10.1016/j.bbrc.2007.04.031
55. Abe S, Hamada K, Miura M, Yamaguchi S (2012) Neural crest stem cell property of apical pulp cells derived from human developing tooth. Cell Biol Int 36(10):927–936. doi:10.1042/CBI20110506
56. Martens W, Sanen K, Georgiou M, Struys T, Bronckaers A, Ameloot M, Phillips J, Lambrichts I (2014) Human dental pulp stem cells can differentiate into Schwann cells and promote and guide neurite outgrowth in an aligned tissue-engineered collagen construct in vitro. FASEB J 28(4):1634–1643. doi:10.1096/fj.13-243980
57. Li X, Gong P, Liao D (2010) In vitro neural/glial differentiation potential of periodontal ligament stem cells. AMS 6(5):678–685. doi:10.5114/aoms.2010.17080
58. Hilkens P, Fanton Y, Martens W, Gervois P, Struys T, Politis C, Lambrichts I, Bronckaers A (2014) Pro-angiogenic impact of dental stem cells in vitro and in vivo. Stem Cell Res 12(3):778–790. doi:10.1016/j.scr.2014.03.008
59. Bronckaers A, Hilkens P, Fanton Y, Struys T, Gervois P, Politis C, Martens W, Lambrichts I (2013) Angiogenic properties of human dental pulp stem cells. PLoS One 8(8), e71104. doi:10.1371/journal.pone.0071104
60. Yeasmin S, Ceccarelli J, Vigen M, Carrion B, Putnam AJ, Tarle SA, Kaigler D (2014) Stem cells derived from tooth periodontal ligament enhance functional angiogenesis by endothelial cells. Tissue Eng A 20(7–8):1188–1196. doi:10.1089/ten.TEA.2013.0512
61. Yu S, Zhao Y, Ma Y, Ge L (2016) Profiling the secretome of human stem cells from dental apical papilla. Stem Cells Dev 25(6):499–508. doi:10.1089/scd.2015.0298
62. Yuan C, Wang P, Zhu L, Dissanayaka WL, Green DW, Tong EH, Jin L, Zhang C (2015) Coculture of stem cells from apical papilla and human umbilical vein endothelial cell under hypoxia increases the formation of three-dimensional vessel-like structures in vitro. Tissue Eng A 21(5–6):1163–1172. doi:10.1089/ten.TEA.2014.0058
63. Dissanayaka WL, Hargreaves KM, Jin L, Samaranayake LP, Zhang C (2015) The interplay of dental pulp stem cells and endothelial cells in an injectable peptide hydrogel on angiogenesis and pulp regeneration in vivo. Tissue Eng A 21(3–4):550–563. doi:10.1089/ten.TEA.2014.0154
64. Sugimura-Wakayama Y, Katagiri W, Osugi M, Kawai T, Ogata K, Sakaguchi K, Hibi H (2015) Peripheral nerve regeneration by secretomes of stem cells from human exfoliated deciduous teeth. Stem Cells Dev 24(22):2687–2699. doi:10.1089/scd.2015.0104
65. Aranha AM, Zhang Z, Neiva KG, Costa CA, Hebling J, Nor JE (2010) Hypoxia enhances the angiogenic potential of human dental pulp cells. J Endod 36(10):1633–1637. doi:10.1016/j.joen.2010.05.013
66. Vanacker J, Viswanath A, De Berdt P, Everard A, Cani PD, Bouzin C, Feron O, Diogenes A, Leprince JG, des Rieux A (2014) Hypoxia modulates the differentiation potential of stem cells of the apical papilla. J Endod 40(9):1410–1418. doi:10.1016/j.joen.2014.04.008
67. Muller HD, Cvikl B, Gruber R, Watzek G, Agis H (2012) Prolyl hydroxylase inhibitors increase the production of vascular endothelial growth factor in dental pulp-derived cells. J Endod 38(11):1498–1503. doi:10.1016/j.joen.2012.08.003
68. Agis H, Watzek G, Gruber R (2012) Prolyl hydroxylase inhibitors increase the production of vascular endothelial growth factor by periodontal fibroblasts. J Periodontal Res 47(2):165–173. doi:10.1111/j.1600-0765.2011.01415.x
69. Kim MK, Park HJ, Kim YD, Ryu MH, Takata T, Bae SK, Bae MK (2014) Hinokitiol increases the angiogenic potential of dental pulp cells through ERK and p38MAPK activation and hypoxia-inducible factor-1alpha (HIF-1alpha) upregulation. Arch Oral Biol 59(2):102–110. doi:10.1016/j.archoralbio.2013.10.009
70. Nosrat IV, Widenfalk J, Olson L, Nosrat CA (2001) Dental pulp cells produce neurotrophic factors, interact with trigeminal neurons in vitro, and rescue motoneurons after spinal cord injury. Dev Biol 238(1):120–132. doi:10.1006/dbio.2001.0400
71. Mead B, Logan A, Berry M, Leadbeater W, Scheven BA (2014) Paracrine-mediated neuroprotection and neuritogenesis of axotomised retinal ganglion cells by human dental pulp stem cells: comparison with human bone marrow and adipose-derived mesenchymal stem cells. PLoS One 9(10), e109305. doi:10.1371/journal.pone.0109305
72. Song M, Jue SS, Cho YA, Kim EC (2015) Comparison of the effects of human dental pulp stem cells and human bone marrow-derived mesenchymal

stem cells on ischemic human astrocytes in vitro. J Neurosci Res 93(6):973–983. doi:10.1002/jnr.23569

73. Apel C, Forlenza OV, de Paula VJ, Talib LL, Denecke B, Eduardo CP, Gattaz WF (2009) The neuroprotective effect of dental pulp cells in models of Alzheimer's and Parkinson's disease. J Neural Transm 116(1):71–78. doi:10.1007/s00702-008-0135-3
74. Nosrat IV, Smith CA, Mullally P, Olson L, Nosrat CA (2004) Dental pulp cells provide neurotrophic support for dopaminergic neurons and differentiate into neurons in vitro; implications for tissue engineering and repair in the nervous system. Eur J Neurosci 19(9):2388–2398. doi:10.1111/j.0953-816X.2004.03314.x
75. Nesti C, Pardini C, Barachini S, D'Alessandro D, Siciliano G, Murri L, Petrini M, Vaglini F (2011) Human dental pulp stem cells protect mouse dopaminergic neurons against MPP+ or rotenone. Brain Res 1367:94–102. doi:10.1016/j.brainres.2010.09.042
76. Yamamoto T, Osako Y, Ito M, Murakami M, Hayashi Y, Horibe H, Iohara K, Takeuchi N, Okui N, Hirata H, Nakayama H, Kurita K, Nakashima M (2016) Trophic effects of dental pulp stem cells on schwann cells in peripheral nerve regeneration. Cell Transplant 25(1):183–193. doi:10.3727/0963689 15X688074
77. Uccelli A, Moretta L, Pistoia V (2008) Mesenchymal stem cells in health and disease. Nat Rev Immunol 8(9):726–736. doi:10.1038/nri2395
78. Pierdomenico L, Bonsi L, Calvitti M, Rondelli D, Arpinati M, Chirumbolo G, Becchetti E, Marchionni C, Alviano F, Fossati V, Staffolani N, Franchina M, Grossi A, Bagnara GP (2005) Multipotent mesenchymal stem cells with immunosuppressive activity can be easily isolated from dental pulp. Transplantation 80(6):836–842
79. Demircan PC, Sariboyaci AE, Unal ZS, Gacar G, Subasi C, Karaoz E (2011) Immunoregulatory effects of human dental pulp-derived stem cells on T cells: comparison of transwell co-culture and mixed lymphocyte reaction systems. Cytotherapy 13(10):1205–1220. doi:10.3109/14653249.2011.60 5351
80. Murakami M, Horibe H, Iohara K, Hayashi Y, Osako Y, Takei Y, Nakata K, Motoyama N, Kurita K, Nakashima M (2013) The use of granulocyte-colony stimulating factor induced mobilization for isolation of dental pulp stem cells with high regenerative potential. Biomaterials 34(36):9036–9047. doi:10.1016/j.biomaterials.2013.08.011
81. Tang R, Ding G (2011) Swine dental pulp stem cells inhibit T-cell proliferation. Transplant Proc 43(10):3955–3959. doi:10.1016/j.transproceed.2011.08.102
82. Wada N, Menicanin D, Shi S, Bartold PM, Gronthos S (2009) Immunomodulatory properties of human periodontal ligament stem cells. J Cell Physiol 219(3):667–676. doi:10.1002/jcp.21710
83. Zhao Y, Wang L, Jin Y, Shi S (2012) Fas ligand regulates the immunomodulatory properties of dental pulp stem cells. J Dent Res 91(10):948–954. doi:10.1177/0022034512458690
84. Tomic S, Djokic J, Vasilijic S, Vucevic D, Todorovic V, Supic G, Colic M (2011) Immunomodulatory properties of mesenchymal stem cells derived from dental pulp and dental follicle are susceptible to activation by toll-like receptor agonists. Stem Cells Dev 20(4):695–708. doi:10.1089/scd.2010.0145
85. Ding G, Niu J, Liu Y (2015) Dental pulp stem cells suppress the proliferation of lymphocytes via transforming growth factor-beta1. Hum Cell 28(2):81–90. doi:10.1007/s13577-014-0106-y
86. Bianco P, Robey PG (2001) Stem cells in tissue engineering. Nature 414(6859):118–121. doi:10.1038/35102181
87. Wei X, Yang X, Han ZP, Qu FF, Shao L, Shi YF (2013) Mesenchymal stem cells: a new trend for cell therapy. Acta Pharmacol Sin 34(6):747–754. doi:10.1038/aps.2013.50
88. Nakashima M, Iohara K, Murakami M (2013) Dental pulp stem cells and regeneration. Endod Top 28(1):38–50
89. Buzhor E, Leshansky L, Blumenthal J, Barash H, Warshawsky D, Mazor Y, Shtrichman R (2014) Cell-based therapy approaches: the hope for incurable diseases. Regen Med 9(5):649–672. doi:10.2217/rme.14.35
90. Wayne AS, Baird K, Egeler RM (2010) Hematopoietic stem cell transplantation for leukemia. Pediatr Clin N Am 57(1):1–25. doi:10.1016/j.pcl.2009.11.005
91. Ghieh F, Jurjus R, Ibrahim A, Geagea AG, Daouk H, El Baba B, Chams S, Matar M, Zein W, Jurjus A (2015) The use of stem cells in burn wound healing: a review. Biomed Res Int 2015:684084. doi:10.1155/2015/684084
92. Park YJ, Cha S, Park YS (2016) Regenerative applications using tooth derived stem cells in other than tooth regeneration: a literature review. Stem Cells Int 2016:9305986. doi:10.1155/2016/9305986
93. Daltoe FP, Mendonca PP, Mantesso A, Deboni MC (2014) Can SHED or DPSCs be used to repair/regenerate non-dental tissues? A systematic review of in vivo studies. Braz Oral Res 28: 1–7
94. Estrela C, Alencar AH, Kitten GT, Vencio EF, Gava E (2011) Mesenchymal stem cells in the dental tissues: perspectives for tissue regeneration. Braz Dent J 22(2):91–98
95. Potdar PD, Jethmalani YD (2015) Human dental pulp stem cells: applications in future regenerative medicine. World J Stem Cells 7(5):839–851. doi:10.4252/wjsc.v7.i5.839
96. La Noce M, Paino F, Spina A, Naddeo P, Montella R, Desiderio V, De Rosa A, Papaccio G, Tirino V, Laino L (2014) Dental pulp stem cells: state of the art and

suggestions for a true translation of research into therapy. J Dent 42(7):761–768. doi:10.1016/j.jdent.2014.02.018
97. Su-Min Lee QZ, Le AD (2014) Dental stem cells: sources and potential applications. Curr Oral Health Rep 1(1):34–42
98. Gandia C, Arminan A, Garcia-Verdugo JM, Lledo E, Ruiz A, Minana MD, Sanchez-Torrijos J, Paya R, Mirabet V, Carbonell-Uberos F, Llop M, Montero JA, Sepulveda P (2008) Human dental pulp stem cells improve left ventricular function, induce angiogenesis, and reduce infarct size in rats with acute myocardial infarction. Stem Cells 26(3):638–645. doi:10.1634/stemcells.2007-0484
99. Shen CY, Li L, Feng T, Li JR, Yu MX, Lu Q, Li H (2015) Dental pulp stem cells derived conditioned medium promotes Angiogenesis in Hindlimb Ischemia. Tissue Eng Regen Med 12(1):59–68. doi:10.1007/s13770-014-9053-7
100. Fang CZ, Yang YJ, Wang QH, Yao Y, Zhang XY, He XH (2013) Intraventricular injection of human dental pulp stem cells improves hypoxic-ischemic brain damage in neonatal rats. PLoS One 8(6), e66748. doi:10.1371/journal.pone.0066748
101. Sakai K, Yamamoto A, Matsubara K, Nakamura S, Naruse M, Yamagata M, Sakamoto K, Tauchi R, Wakao N, Imagama S, Hibi H, Kadomatsu K, Ishiguro N, Ueda M (2012) Human dental pulp-derived stem cells promote locomotor recovery after complete transection of the rat spinal cord by multiple neuro-regenerative mechanisms. J Clin Invest 122(1):80–90. doi:10.1172/JCI59251
102. Inoue T, Sugiyama M, Hattori H, Wakita H, Wakabayashi T, Ueda M (2013) Stem cells from human exfoliated deciduous tooth-derived conditioned medium enhance recovery of focal cerebral ischemia in rats. Tissue Eng A 19(1–2):24–29. doi:10.1089/ten.TEA.2011.0385
103. Yamagata M, Yamamoto A, Kako E, Kaneko N, Matsubara K, Sakai K, Sawamoto K, Ueda M (2013) Human dental pulp-derived stem cells protect against hypoxic-ischemic brain injury in neonatal mice. Stroke 44(2):551–554. doi:10.1161/STROKEAHA.112.676759
104. Taghipour Z, Karbalaie K, Kiani A, Niapour A, Bahramian H, Nasr-Esfahani MH, Baharvand H (2012) Transplantation of undifferentiated and induced human exfoliated deciduous teeth-derived stem cells promote functional recovery of rat spinal cord contusion injury model. Stem Cells Dev 21(10):1794–1802. doi:10.1089/scd.2011.0408
105. Yamamoto A, Sakai K, Matsubara K, Kano F, Ueda M (2014) Multifaceted neuro-regenerative activities of human dental pulp stem cells for functional recovery after spinal cord injury. Neurosci Res 78:16–20. doi:10.1016/j.neures.2013.10.010
106. Yamaza T, Kentaro A, Chen C, Liu Y, Shi Y, Gronthos S, Wang S, Shi S (2010) Immunomodulatory properties of stem cells from human exfoliated deciduous teeth. Stem Cell Res Ther 1(1):5. doi:10.1186/scrt5
107. Wada N, Gronthos S, Bartold PM (2013) Immunomodulatory effects of stem cells. Periodontol 2000 63(1):198–216. doi:10.1111/prd.12024
108. Govindasamy V, Ronald VS, Abdullah AN, Nathan KR, Ab Aziz ZA, Abdullah M, Musa S, Kasim NH, Bhonde RR (2011) Differentiation of dental pulp stem cells into islet-like aggregates. J Dent Res 90(5):646–652. doi:10.1177/0022034510396879
109. Kanafi MM, Rajeshwari YB, Gupta S, Dadheech N, Nair PD, Gupta PK, Bhonde RR (2013) Transplantation of islet-like cell clusters derived from human dental pulp stem cells restores normoglycemia in diabetic mice. Cytotherapy 15(10):1228–1236. doi:10.1016/j.jcyt.2013.05.008
110. Lee JS, An SY, Kwon IK, Heo JS (2014) Transdifferentiation of human periodontal ligament stem cells into pancreatic cell lineage. Cell Biochem Funct 32(7):605–611. doi:10.1002/cbf.3057
111. Ishkitiev N, Yaegaki K, Kozhuharova A, Tanaka T, Okada M, Mitev V, Fukuda M, Imai T (2013) Pancreatic differentiation of human dental pulp CD117(+) stem cells. Regen Med 8(5):597–612. doi:10.2217/rme.13.42
112. Carnevale G, Riccio M, Pisciotta A, Beretti F, Maraldi T, Zavatti M, Cavallini GM, La Sala GB, Ferrari A, De Pol A (2013) In vitro differentiation into insulin-producing beta-cells of stem cells isolated from human amniotic fluid and dental pulp. Dig Liver Dis 45(8):669–676. doi:10.1016/j.dld.2013.02.007
113. Kerkis I, Ambrosio CE, Kerkis A, Martins DS, Zucconi E, Fonseca SA, Cabral RM, Maranduba CM, Gaiad TP, Morini AC, Vieira NM, Brolio MP, Sant'Anna OA, Miglino MA, Zatz M (2008) Early transplantation of human immature dental pulp stem cells from baby teeth to golden retriever muscular dystrophy (GRMD) dogs: local or systemic? J Transl Med 6:35. doi:10.1186/1479-5876-6-35
114. Yang R, Chen M, Lee CH, Yoon R, Lal S, Mao JJ (2010) Clones of ectopic stem cells in the regeneration of muscle defects in vivo. PLoS One 5(10), e13547. doi:10.1371/journal.pone.0013547
115. Graziano A, d'Aquino R, Laino G, Papaccio G (2008) Dental pulp stem cells: a promising tool for bone regeneration. Stem Cell Rev 4(1):21–26. doi:10.1007/s12015-008-9013-5
116. Graziano A, d'Aquino R, Cusella-De Angelis MG, De Francesco F, Giordano A, Laino G, Piattelli A, Traini T, De Rosa A, Papaccio G (2008) Scaffold's surface geometry significantly affects human stem cell bone tissue engineering. J Cell Physiol 214(1):166–172. doi:10.1002/jcp.21175
117. De Mendonca Costa A, Bueno DF, Martins MT, Kerkis I, Kerkis A, Fanganiello RD, Cerruti H, Alonso N, Passos-Bueno MR (2008) Reconstruction of large cranial defects in nonimmunosuppressed experimental design with human dental pulp stem

cells. J Craniofac Surg 19(1):204–210. doi:10.1097/scs.0b013e31815c8a54

118. Fecek C, Yao D, Kacorri A, Vasquez A, Iqbal S, Sheikh H, Svinarich DM, Perez-Cruet M, Chaudhry GR (2008) Chondrogenic derivatives of embryonic stem cells seeded into 3D polycaprolactone scaffolds generated cartilage tissue in vivo. Tissue Eng A 14(8):1403–1413. doi:10.1089/tea.2007.0293
119. Iohara K, Zheng L, Ito M, Tomokiyo A, Matsushita K, Nakashima M (2006) Side population cells isolated from porcine dental pulp tissue with self-renewal and multipotency for dentinogenesis, chondrogenesis, adipogenesis, and neurogenesis. Stem Cells 24(11):2493–2503. doi:10.1634/stemcells.2006-0161
120. Reynolds AJ, Jahoda CA (2004) Cultured human and rat tooth papilla cells induce hair follicle regeneration and fiber growth. Differentiation 72(9–10):566–575. doi:10.1111/j.1432-0436.2004.07209010.x
121. Nishino Y, Ebisawa K, Yamada Y, Okabe K, Kamei Y, Ueda M (2011) Human deciduous teeth dental pulp cells with basic fibroblast growth factor enhance wound healing of skin defect. J Craniofac Surg 22(2):438–442. doi:10.1097/SCS.0b013e318207b507
122. Yamamura Y, Yamada H, Sakurai T, Ide F, Inoue H, Muramatsu T, Mishima K, Hamada Y, Saito I (2013) Treatment of salivary gland hypofunction by transplantation with dental pulp cells. Arch Oral Biol 58(8):935–942. doi:10.1016/j.archoralbio.2013.02.015
123. Gomes JA, Geraldes Monteiro B, Melo GB, Smith RL, Cavenaghi Pereira da Silva M, Lizier NF, Kerkis A, Cerruti H, Kerkis I (2010) Corneal reconstruction with tissue-engineered cell sheets composed of human immature dental pulp stem cells. Invest Ophthalmol Vis Sci 51(3):1408–1414. doi:10.1167/iovs.09-4029
124. Syed-Picard FN, Du Y, Lathrop KL, Mann MM, Funderburgh ML, Funderburgh JL (2015) Dental pulp stem cells: a new cellular resource for corneal stromal regeneration. Stem Cells Transl Med 4(3):276–285. doi:10.5966/sctm.2014-0115
125. Huang H, Wang Y, Pan B, Yang X, Wang L, Chen J, Ma D, Yang C (2013) Simple bipolar hosts with high glass transition temperatures based on 1,8-disubstituted carbazole for efficient blue and green electrophosphorescent devices with "ideal" turn-on voltage. Chemistry 19(5):1828–1834. doi:10.1002/chem.201202329
126. Mead B, Logan A, Berry M, Leadbeater W, Scheven BA (2013) Intravitreally transplanted dental pulp stem cells promote neuroprotection and axon regeneration of retinal ganglion cells after optic nerve injury. Invest Ophthalmol Vis Sci 54(12):7544–7556. doi:10.1167/iovs.13-13045
127. Horibe H, Murakami M, Iohara K, Hayashi Y, Takeuchi N, Takei Y, Kurita K, Nakashima M (2014) Isolation of a stable subpopulation of mobilized dental pulp stem cells (MDPSCs) with high proliferation, migration, and regeneration potential is independent of age. PLoS One 9(5), e98553. doi:10.1371/journal.pone.0098553
128. Pejcic A, Kojovic D, Mirkovic D, Minic I (2013) Stem cells for periodontal regeneration. Balkan J Med Genet 16(1):7–12. doi:10.2478/bjmg-2013-0012
129. Bansal R, Jain A, Mittal S (2015) Current overview on challenges in regenerative endodontics. J Conserv Dent 18(1):1–6. doi:10.4103/0972-0707.148861
130. Nakashima M, Iohara K (2011) Regeneration of dental pulp by stem cells. Adv Dent Res 23(3):313–319. doi:10.1177/0022034511405323
131. Demarco FF, Conde MC, Cavalcanti BN, Casagrande L, Sakai VT, Nor JE (2011) Dental pulp tissue engineering. Braz Dent J 22(1):3–13
132. Huang GT, Yamaza T, Shea LD, Djouad F, Kuhn NZ, Tuan RS, Shi S (2010) Stem/progenitor cell-mediated de novo regeneration of dental pulp with newly deposited continuous layer of dentin in an in vivo model. Tissue Eng A 16(2):605–615. doi:10.1089/ten.TEA.2009.0518
133. Casagrande L, Cordeiro MM, Nor SA, Nor JE (2011) Dental pulp stem cells in regenerative dentistry. Odontol Soc Nippon Dent Univ 99(1):1–7. doi:10.1007/s10266-010-0154-z
134. Mantesso A, Sharpe P (2009) Dental stem cells for tooth regeneration and repair. Expert Opin Biol Ther 9(9):1143–1154. doi:10.1517/14712590903103795
135. Volponi AA, Pang Y, Sharpe PT (2010) Stem cell-based biological tooth repair and regeneration. Trends Cell Biol 20(12):715–722. doi:10.1016/j.tcb.2010.09.012
136. Shi S, Bartold PM, Miura M, Seo BM, Robey PG, Gronthos S (2005) The efficacy of mesenchymal stem cells to regenerate and repair dental structures. Orthod Craniofacial Res 8(3):191–199. doi:10.1111/j.1601-6343.2005.00331.x
137. Nakashima M, Akamine A (2005) The application of tissue engineering to regeneration of pulp and dentin in endodontics. J Endod 31(10):711–718
138. Iohara K, Nakashima M, Ito M, Ishikawa M, Nakasima A, Akamine A (2004) Dentin regeneration by dental pulp stem cell therapy with recombinant human bone morphogenetic protein 2. J Dent Res 83(8):590–595
139. Gronthos S, Brahim J, Li W, Fisher LW, Cherman N, Boyde A, DenBesten P, Robey PG, Shi S (2002) Stem cell properties of human dental pulp stem cells. J Dent Res 81(8):531–535
140. He H, Yu J, Liu Y, Lu S, Liu H, Shi J, Jin Y (2008) Effects of FGF2 and TGFbeta1 on the differentiation of human dental pulp stem cells in vitro. Cell Biol Int 32(7):827–834. doi:10.1016/j.cellbi.2008.03.013
141. Almushayt A, Narayanan K, Zaki AE, George A (2006) Dentin matrix protein 1 induces cytodifferentiation of dental pulp stem cells into odontoblasts. Gene Ther 13(7):611–620. doi:10.1038/sj.gt.3302687
142. Prescott RS, Alsanea R, Fayad MI, Johnson BR, Wenckus CS, Hao J, John AS, George A (2008)

In vivo generation of dental pulp-like tissue by using dental pulp stem cells, a collagen scaffold, and dentin matrix protein 1 after subcutaneous transplantation in mice. J Endod 34(4):421–426. doi:10.1016/j.joen.2008.02.005
143. Tran Hle B, Doan VN (2015) Human dental pulp stem cells cultured onto dentin derived scaffold can regenerate dentin-like tissue in vivo. Cell Tissue Bank 16(4):559–568. doi:10.1007/s10561-015-9503-z
144. Cordeiro MM, Dong Z, Kaneko T, Zhang Z, Miyazawa M, Shi S, Smith AJ, Nor JE (2008) Dental pulp tissue engineering with stem cells from exfoliated deciduous teeth. J Endod 34(8):962–969. doi:10.1016/j.joen.2008.04.009
145. Taba M Jr, Jin Q, Sugai JV, Giannobile WV (2005) Current concepts in periodontal bioengineering. Orthod Craniofac Res 8(4):292–302. doi:10.1111/j.1601-6343.2005.00352.x
146. Park JY, Jeon SH, Choung PH (2011) Efficacy of periodontal stem cell transplantation in the treatment of advanced periodontitis. Cell Transplant 20(2):271–285. doi:10.3727/096368910X519292
147. Trubiani O, Orsini G, Zini N, Di Iorio D, Piccirilli M, Piattelli A, Caputi S (2008) Regenerative potential of human periodontal ligament derived stem cells on three-dimensional biomaterials: a morphological report. J Biomed Mater Res A 87(4):986–993. doi:10.1002/jbm.a.31837
148. Liu Y, Zheng Y, Ding G, Fang D, Zhang C, Bartold PM, Gronthos S, Shi S, Wang S (2008) Periodontal ligament stem cell-mediated treatment for periodontitis in miniature swine. Stem Cells 26(4):1065–1073. doi:10.1634/stemcells.2007-0734
149. He H, Yu J, Cao JEL, Wang D, Zhang H, Liu H (2011) Biocompatibility and osteogenic capacity of periodontal ligament stem cells on nHAC/PLA and HA/TCP scaffolds. J Biomater Sci Polym Ed 22(1–3):179–194. doi:10.1163/092050609X12587018007767
150. Lin Y, Gallucci GO, Buser D, Bosshardt D, Belser UC, Yelick PC (2011) Bioengineered periodontal tissue formed on titanium dental implants. J Dent Res90(2):251–256.doi:10.1177/0022034510384872
151. Gault P, Black A, Romette JL, Fuente F, Schroeder K, Thillou F, Brune T, Berdal A, Wurtz T (2010) Tissue-engineered ligament: implant constructs for tooth replacement. J Clin Periodontol 37(8):750–758. doi:10.1111/j.1600-051X.2010.01588.x
152. Feng F, Akiyama K, Liu Y, Yamaza T, Wang TM, Chen JH, Wang BB, Huang GT, Wang S, Shi S (2010) Utility of PDL progenitors for in vivo tissue regeneration: a report of 3 cases. Oral Dis 16(1):20–28
153. Yamada Y, Ito K, Nakamura S, Ueda M, Nagasaka T (2011) Promising cell-based therapy for bone regeneration using stem cells from deciduous teeth, dental pulp, and bone marrow. Cell Transplant 20(7):1003–1013. doi:10.3727/096368910X539128
154. Ji YM, Jeon SH, Park JY, Chung JH, Choung YH, Choung PH (2010) Dental stem cell therapy with calcium hydroxide in dental pulp capping. Tissue Eng A 16(6):1823–1833. doi:10.1089/ten.TEA.2009.0054
155. Sonoyama W, Liu Y, Fang D, Yamaza T, Seo BM, Zhang C, Liu H, Gronthos S, Wang CY, Wang S, Shi S (2006) Mesenchymal stem cell-mediated functional tooth regeneration in swine. PLoS One 1, e79. doi:10.1371/journal.pone.0000079
156. Nor JE (2006) Tooth regeneration in operative dentistry. Oper Dent 31(6):633–642. doi:10.2341/06-000
157. Khorsand A, Eslaminejad MB, Arabsolghar M, Paknejad M, Ghaedi B, Rokn AR, Moslemi N, Nazarian H, Jahangir S (2013) Autologous dental pulp stem cells in regeneration of defect created in canine periodontal tissue. J Oral Implantol 39(4):433–443. doi:10.1563/AAID-JOI-D-12-00027
158. Iohara K, Imabayashi K, Ishizaka R, Watanabe A, Nabekura J, Ito M, Matsushita K, Nakamura H, Nakashima M (2011) Complete pulp regeneration after pulpectomy by transplantation of CD105+ stem cells with stromal cell-derived factor-1. Tissue Eng A 17(15–16):1911–1920. doi:10.1089/ten.TEA.2010.0615
159. Guo W, Gong K, Shi H, Zhu G, He Y, Ding B, Wen L, Jin Y (2012) Dental follicle cells and treated dentin matrix scaffold for tissue engineering the tooth root. Biomaterials 33(5):1291–1302. doi:10.1016/j.biomaterials.2011.09.068
160. Guo W, Chen L, Gong K, Ding B, Duan Y, Jin Y (2012) Heterogeneous dental follicle cells and the regeneration of complex periodontal tissues. Tissue Eng A 18(5–6):459–470. doi:10.1089/ten.TEA.2011.0261
161. Shen YY, Chen K, Xu N (2010) Osteogenic capacity of human deciduous dental pulp stem cells in vitro. Nan Fang Yi Ke Da Xue Xue Bao 30(1):96–99
162. Hwang YC, Hwang IN, Oh WM, Park JC, Lee DS, Son HH (2008) Influence of TGF-beta1 on the expression of BSP, DSP, TGF-beta1 receptor I and Smad proteins during reparative dentinogenesis. J Mol Histol 39(2):153–160. doi:10.1007/s10735-007-9148-8
163. Laino G, d'Aquino R, Graziano A, Lanza V, Carinci F, Naro F, Pirozzi G, Papaccio G (2005) A new population of human adult dental pulp stem cells: a useful source of living autologous fibrous bone tissue (LAB). J Bone Miner Res 20(8):1394–1402. doi:10.1359/JBMR.050325
164. Ueno A, Yamashita K, Miyoshi K, Horiguchi T, Ruspita I, Abe K, Noma T (2006) Soluble matrix from osteoblastic cells induces mineralization by dental pulp cells. J Med Invest 53(3–4):297–302
165. Hosoya A, Nakamura H, Ninomiya T, Hoshi K, Yoshiba K, Yoshiba N, Takahashi M, Okabe T, Sahara N, Yamada H, Kasahara E, Ozawa H (2007)

Hard tissue formation in subcutaneously transplanted rat dental pulp. J Dent Res 86(5):469–474
166. Otaki S, Ueshima S, Shiraishi K, Sugiyama K, Hamada S, Yorimoto M, Matsuo O (2007) Mesenchymal progenitor cells in adult human dental pulp and their ability to form bone when transplanted into immunocompromised mice. Cell Biol Int 31(10):1191–1197. doi:10.1016/j.cellbi.2007.04.001
167. Graziano A, d'Aquino R, Laino G, Proto A, Giuliano MT, Pirozzi G, De Rosa A, Di Napoli D, Papaccio G (2008) Human CD34+ stem cells produce bone nodules in vivo. Cell Prolif 41(1):1–11. doi:10.1111/j.1365-2184.2007.00497.x
168. d'Aquino R, De Rosa A, Lanza V, Tirino V, Laino L, Graziano A, Desiderio V, Laino G, Papaccio G (2009) Human mandible bone defect repair by the grafting of dental pulp stem/progenitor cells and collagen sponge biocomplexes. Eur Cell Mater 18:75–83
169. Akkouch A, Zhang Z, Rouabhia M (2014) Engineering bone tissue using human dental pulp stem cells and an osteogenic collagen-hydroxyapatite-poly (L-lactide-co-epsilon-caprolactone) scaffold. J Biomater Appl 28(6):922–936. doi:10.1177/0885328213486705
170. Annibali S, Bellavia D, Ottolenghi L, Cicconetti A, Cristalli MP, Quaranta R, Pilloni A (2014) Micro-CT and PET analysis of bone regeneration induced by biodegradable scaffolds as carriers for dental pulp stem cells in a rat model of calvarial "critical size" defect: preliminary data. J Biomed Mater Res B Appl Biomater 102(4):815–825. doi:10.1002/jbm.b.33064
171. Mangano C, De Rosa A, Desiderio V, d'Aquino R, Piattelli A, De Francesco F, Tirino V, Mangano F, Papaccio G (2010) The osteoblastic differentiation of dental pulp stem cells and bone formation on different titanium surface textures. Biomaterials 31(13):3543–3551. doi:10.1016/j.biomaterials.2010.01.056
172. Maraldi T, Riccio M, Pisciotta A, Zavatti M, Carnevale G, Beretti F, La Sala GB, Motta A, De Pol A (2013) Human amniotic fluid-derived and dental pulp-derived stem cells seeded into collagen scaffold repair critical-size bone defects promoting vascularization. Stem cell Res Ther 4(3):53. doi:10.1186/scrt203
173. Kolind K, Kraft D, Boggild T, Duch M, Lovmand J, Pedersen FS, Bindslev DA, Bunger CE, Foss M, Besenbacher F (2014) Control of proliferation and osteogenic differentiation of human dental-pulp-derived stem cells by distinct surface structures. Acta Biomater 10(2):641–650. doi:10.1016/j.actbio.2013.11.006
174. Ikeda H, Sumita Y, Ikeda M, Ikeda H, Okumura T, Sakai E, Nishimura M, Asahina I (2011) Engineering bone formation from human dental pulp- and periodontal ligament-derived cells. Ann Biomed Eng 39(1):26–34. doi:10.1007/s10439-010-0115-2
175. Yang X, van der Kraan PM, Bian Z, Fan M, Walboomers XF, Jansen JA (2009) Mineralized tissue formation by BMP2-transfected pulp stem cells. J Dent Res 88(11):1020–1025. doi:10.1177/0022034509346258
176. Liu HC, Ling-Ling E, Wang DS, Su F, Wu X, Shi ZP, Lv Y, Wang JZ (2011) Reconstruction of alveolar bone defects using bone morphogenetic protein 2 mediated rabbit dental pulp stem cells seeded on nano-hydroxyapatite/collagen/poly(L-lactide). Tissue Eng A 17(19–20):2417–2433. doi:10.1089/ten.TEA.2010.0620
177. Ge S, Zhao N, Wang L, Yu M, Liu H, Song A, Huang J, Wang G, Yang P (2012) Bone repair by periodontal ligament stem cellseeded nanohydroxyapatite-chitosan scaffold. Int J Nanomedicine 7:5405–5414. doi:10.2147/IJN.S36714
178. Su F, Liu SS, Ma JL, Wang DS, Ling-Ling E, Liu HC (2015) Enhancement of periodontal tissue regeneration by transplantation of osteoprotegerin-engineered periodontal ligament stem cells. Stem Cell Res Ther 6:22. doi:10.1186/s13287-015-0023-3
179. Yu BH, Zhou Q, Wang ZL (2014) Periodontal ligament versus bone marrow mesenchymal stem cells in combination with Bio-Oss scaffolds for ectopic and in situ bone formation: a comparative study in the rat. J Biomater Appl 29(2):243–253. doi:10.1177/0885328214521846
180. Seo BM, Sonoyama W, Yamaza T, Coppe C, Kikuiri T, Akiyama K, Lee JS, Shi S (2008) SHED repair critical-size calvarial defects in mice. Oral Dis 14(5):428–434
181. Zheng Y, Liu Y, Zhang CM, Zhang HY, Li WH, Shi S, Le AD, Wang SL (2009) Stem cells from deciduous tooth repair mandibular defect in swine. J Dent Res88(3):249–254. doi:10.1177/0022034509333804
182. Abe S, Yamaguchi S, Watanabe A, Hamada K, Amagasa T (2008) Hard tissue regeneration capacity of apical pulp derived cells (APDCs) from human tooth with immature apex. Biochem Biophys Res Commun 371(1):90–93. doi:10.1016/j.bbrc.2008.04.016
183. Martens W, Bronckaers A, Politis C, Jacobs R, Lambrichts I (2013) Dental stem cells and their promising role in neural regeneration: an update. Clin Oral Investig 17(9):1969–1983. doi:10.1007/s00784-013-1030-3
184. Miletich I, Sharpe PT (2004) Neural crest contribution to mammalian tooth formation. Birth Defects Res C Embryo Today 72(2):200–212. doi:10.1002/bdrc.20012
185. Chai Y, Jiang X, Ito Y, Bringas P Jr, Han J, Rowitch DH, Soriano P, McMahon AP, Sucov HM (2000) Fate of the mammalian cranial neural crest during tooth and mandibular morphogenesis. Development 127(8):1671–1679
186. Ibarretxe G, Crende O, Aurrekoetxea M, Garcia-Murga V, Etxaniz J, Unda F (2012) Neural crest stem

cells from dental tissues: a new hope for dental and neural regeneration. Stem Cells Int 2012:103503. doi:10.1155/2012/103503

187. Graham A, Begbie J, McGonnell I (2004) Significance of the cranial neural crest. Dev Dyn 229(1):5–13. doi:10.1002/dvdy.10442
188. Arthur A, Shi S, Zannettino AC, Fujii N, Gronthos S, Koblar SA (2009) Implanted adult human dental pulp stem cells induce endogenous axon guidance. Stem Cells 27(9):2229–2237. doi:10.1002/stem.138
189. Kiraly M, Kadar K, Horvathy DB, Nardai P, Racz GZ, Lacza Z, Varga G, Gerber G (2011) Integration of neuronally predifferentiated human dental pulp stem cells into rat brain in vivo. Neurochem Int 59(3):371–381. doi:10.1016/j.neuint.2011.01.006
190. Leong WK, Henshall TL, Arthur A, Kremer KL, Lewis MD, Helps SC, Field J, Hamilton-Bruce MA, Warming S, Manavis J, Vink R, Gronthos S, Koblar SA (2012) Human adult dental pulp stem cells enhance poststroke functional recovery through non-neural replacement mechanisms. Stem Cells Transl Med 1(3):177–187. doi:10.5966/sctm.2011-0039
191. Yang KL, Chen MF, Liao CH, Pang CY, Lin PY (2009) A simple and efficient method for generating Nurr1-positive neuronal stem cells from human wisdom teeth (tNSC) and the potential of tNSC for stroke therapy. Cytotherapy 11(5):606–617. doi:10.1080/14653240902806994
192. Sugiyama M, Iohara K, Wakita H, Hattori H, Ueda M, Matsushita K, Nakashima M (2011) Dental pulp-derived CD31(-)/CD146(-) side population stem/progenitor cells enhance recovery of focal cerebral ischemia in rats. Tissue Eng A 17(9–10):1303–1311. doi:10.1089/ten.TEA.2010.0306
193. Nagpal A, Kremer KL, Hamilton-Bruce MA, Kaidonis X, Milton AG, Levi C, Shi S, Carey L, Hillier S, Rose M, Zacest A, Takhar P, Koblar SA (2016) TOOTH (The Open study Of dental pulp stem cell Therapy in Humans): study protocol for evaluating safety and feasibility of autologous human adult dental pulp stem cell therapy in patients with chronic disability after stroke. Int J Stroke. doi:10.1177/1747493016641111
194. Tseng LS, Chen SH, Lin MT, Lin YC (2015) Transplantation of human dental pulp-derived stem cells protects against heatstroke in mice. Cell Transplant 24(5):921–937. doi:10.3727/096368914X678580
195. Li X, Yang C, Li L, Xiong J, Xie L, Yang B, Yu M, Feng L, Jiang Z, Guo W, Tian W (2015) A therapeutic strategy for spinal cord defect: human dental follicle cells combined with aligned PCL/PLGA electrospun material. Biomed Res Int 2015:197183. doi:10.1155/2015/197183
196. De Berdt P, Vanacker J, Ucakar B, Elens L, Diogenes A, Leprince JG, Deumens R, des Rieux A (2015) Dental apical papilla as therapy for spinal cord injury. J Dent Res 94(11):1575–1581. doi:10.1177/0022034515604612
197. De Berdt P, Vanacker J, Ucakar B, Elens L, Diogenes A, Leprince JG, Deumens R, des Rieux A (2015) Dental apical papilla as therapy for spinal cord injury. J Dent Res 94(11):1575–1581. doi:10.1177/0022034515604612
198. Huang AH, Snyder BR, Cheng PH, Chan AW (2008) Putative dental pulp-derived stem/stromal cells promote proliferation and differentiation of endogenous neural cells in the hippocampus of mice. Stem Cells 26(10):2654–2663. doi:10.1634/stemcells.2008-0285
199. Sasaki R, Aoki S, Yamato M, Uchiyama H, Wada K, Ogiuchi H, Okano T, Ando T (2011) PLGA artificial nerve conduits with dental pulp cells promote facial nerve regeneration. J Tissue Eng Regen Med 5(10):823–830. doi:10.1002/term.387
200. Sasaki R, Aoki S, Yamato M, Uchiyama H, Wada K, Okano T, Ogiuchi H (2008) Tubulation with dental pulp cells promotes facial nerve regeneration in rats. Tissue Eng A 14(7):1141–1147. doi:10.1089/ten.tea.2007.0157
201. Sasaki R, Matsumine H, Watanabe Y, Takeuchi Y, Yamato M, Okano T, Miyata M, Ando T (2014) Electrophysiologic and functional evaluations of regenerated facial nerve defects with a tube containing dental pulp cells in rats. Plast Reconstr Surg 134(5):970–978. doi:10.1097/PRS.0000000000000602
202. Iohara K, Murakami M, Takeuchi N, Osako Y, Ito M, Ishizaka R, Utunomiya S, Nakamura H, Matsushita K, Nakashima M (2013) A novel combinatorial therapy with pulp stem cells and granulocyte colony-stimulating factor for total pulp regeneration. Stem Cells Transl Med 2(7):521–533. doi:10.5966/sctm.2012-0132
203. Kumar A, Bhattacharyya S, Rattan V (2015) Effect of uncontrolled freezing on biological characteristics of human dental pulp stem cells. Cell Tissue Bank 16(4):513–522. doi:10.1007/s10561-015-9498-5
204. Davies OG, Smith AJ, Cooper PR, Shelton RM, Scheven BA (2014) The effects of cryopreservation on cells isolated from adipose, bone marrow and dental pulp tissues. Cryobiology 69(2):342–347. doi:10.1016/j.cryobiol.2014.08.003
205. Baust JM, Corwin WL, VanBuskirk R, Baust JG (2015) Biobanking: the future of cell preservation strategies. Adv Exp Med Biol 864:37–53. doi:10.1007/978-3-319-20579-3_4
206. Marquez-Curtis LA, Janowska-Wieczorek A, McGann LE, Elliott JA (2015) Mesenchymal stromal cells derived from various tissues: biological, clinical and cryopreservation aspects. Cryobiology 71(2):181–197. doi:10.1016/j.cryobiol.2015.07.003
207. Mazur P, Leibo SP, Chu EH (1972) A two-factor hypothesis of freezing injury. Evidence from Chinese hamster tissue-culture cells. Exp Cell Res 71(2):345–355
208. Zhurova M, Woods EJ, Acker JP (2010) Intracellular ice formation in confluent monolayers of human

dental stem cells and membrane damage. Cryobiology 61(1):133–141. doi:10.1016/j.cryobiol.2010.06.007
209. Fleming KK, Hubel A (2006) Cryopreservation of hematopoietic and non-hematopoietic stem cells. Transfus Apher Sci 34(3):309–315. doi:10.1016/j.transci.2005.11.012
210. Papaccio G, Graziano A, d'Aquino R, Graziano MF, Pirozzi G, Menditti D, De Rosa A, Carinci F, Laino G (2006) Long-term cryopreservation of dental pulp stem cells (SBP-DPSCs) and their differentiated osteoblasts: a cell source for tissue repair. J Cell Physiol 208(2):319–325. doi:10.1002/jcp.20667
211. Perry BC, Zhou D, Wu X, Yang FC, Byers MA, Chu TM, Hockema JJ, Woods EJ, Goebel WS (2008) Collection, cryopreservation, and characterization of human dental pulp-derived mesenchymal stem cells for banking and clinical use. Tissue Eng Part C Methods 14(2):149–156. doi:10.1089/ten.tec.2008.0031
212. Woods EJ, Perry BC, Hockema JJ, Larson L, Zhou D, Goebel WS (2009) Optimized cryopreservation method for human dental pulp-derived stem cells and their tissues of origin for banking and clinical use. Cryobiology 59(2):150–157. doi:10.1016/j.cryobiol.2009.06.005
213. Ding G, Wang W, Liu Y, An Y, Zhang C, Shi S, Wang S (2010) Effect of cryopreservation on biological and immunological properties of stem cells from apical papilla. J Cell Physiol 223(2):415–422. doi:10.1002/jcp.22050
214. Vasconcelos RG, Ribeiro RA, Vasconcelos MG, Lima KC, Barboza CA (2012) In vitro comparative analysis of cryopreservation of undifferentiated mesenchymal cells derived from human periodontal ligament. Cell Tissue Bank 13(3):461–469. doi:10.1007/s10561-011-9271-3
215. Harel A (2013) Cryopreservation and cell banking for autologous mesenchymal stem cell-based therapies. Cell Tissue Transpl Ther 5:1–7. doi:10.4137/CTTT.S11249
216. Hunt CJ (2011) Cryopreservation of human stem cells for clinical application: a review. Transfusion Med Hemother 38(2):107–123. doi:10.1159/000326623
217. Rowley SD, Feng Z, Yadock D, Holmberg L, Macleod B, Heimfeld S (1999) Post-thaw removal of DMSO does not completely abrogate infusional toxicity or the need for pre-infusion histamine blockade. Cytotherapy 1(6):439–446. doi:10.1080/0032472031000141303
218. Demirci S, Dogan A, Sisli B, Sahin F (2014) Boron increases the cell viability of mesenchymal stem cells after long-term cryopreservation. Cryobiology 68(1):139–146. doi:10.1016/j.cryobiol.2014.01.010
219. Umemura E, Yamada Y, Nakamura S, Ito K, Hara K, Ueda M (2011) Viable cryopreserving tissue-engineered cell-biomaterial for cell banking therapy in an effective cryoprotectant. Tissue Eng Part C Methods 17(8):799–807. doi:10.1089/ten.tec.2011.0003
220. Ducret M, Fabre H, Farges JC, Degoul O, Atzeni G, McGuckin C, Forraz N, Mallein-Gerin F, Perrier-Groult E (2015) Production of human dental pulp cells with a medicinal manufacturing approach. J Endod 41(9):1492–1499. doi:10.1016/j.joen.2015.05.017
221. Lee SY, Huang GW, Shiung JN, Huang YH, Jeng JH, Kuo TF, Yang JC, Yang WC (2012) Magnetic cryopreservation for dental pulp stem cells. Cells Tissues Organs 196(1):23–33. doi:10.1159/000331247
222. Kamada H, Kaku M, Kawata T, Koseki H, Abedini S, Kojima S, Sumi A, Motokawa M, Fujita T, Ohtani J, Ohwada N, Tanne K (2011) In-vitro and in-vivo study of periodontal ligament cryopreserved with a magnetic field. Am J Orthod Dentofac Orthop 140(6):799–805. doi:10.1016/j.ajodo.2011.04.024
223. Amos MJ, Day P, Littlewood SJ (2009) Autotransplantation of teeth: an overview. Dent Update 36(2):102–104, 107–110, 113
224. Kaku M, Kamada H, Kawata T, Koseki H, Abedini S, Kojima S, Motokawa M, Fujita T, Ohtani J, Tsuka N, Matsuda Y, Sunagawa H, Hernandes RA, Ohwada N, Tanne K (2010) Cryopreservation of periodontal ligament cells with magnetic field for tooth banking. Cryobiology 61(1):73–78. doi:10.1016/j.cryobiol.2010.05.003
225. Abedini S, Kaku M, Kawata T, Koseki H, Kojima S, Sumi H, Motokawa M, Fujita T, Ohtani J, Ohwada N, Tanne K (2011) Effects of cryopreservation with a newly-developed magnetic field programmed freezer on periodontal ligament cells and pulp tissues. Cryobiology 62(3):181–187. doi:10.1016/j.cryobiol.2011.03.001
226. Oh YH, Che ZM, Hong JC, Lee EJ, Lee SJ, Kim J (2005) Cryopreservation of human teeth for future organization of a tooth bank – a preliminary study. Cryobiology 51(3):322–329. doi:10.1016/j.cryobiol.2005.08.008
227. Min KS, Lee HW, Lee HS, Lee JH, Park SH (2010) Comparison of gene expression in human periodontal ligament cells cultured from teeth immediately after extraction and from teeth cryopreserved for 1 week. Cryobiology 60(3):326–330. doi:10.1016/j.cryobiol.2010.02.008
228. Gioventu S, Andriolo G, Bonino F, Frasca S, Lazzari L, Montelatici E, Santoro F, Rebulla P (2012) A novel method for banking dental pulp stem cells. Transfus Apher Sci 47(2):199–206. doi:10.1016/j.transci.2012.06.005
229. Temmerman L, Beele H, Dermaut LR, Van Maele G, De Pauw GA (2010) Influence of cryopreservation on the pulpal tissue of immature third molars in vitro. Cell Tissue Bank 11(3):281–289. doi:10.1007/s10561-009-9148-x
230. Malekfar A, Valli KS, Kanafi MM, Bhonde RR (2016) Isolation and characterization of human den-

tal pulp stem cells from cryopreserved pulp tissues obtained from teeth with irreversible pulpitis. J Endod 42(1):76–81. doi:10.1016/j.joen.2015.10.001
231. Ji EH, Song JS, Kim SO, Jeon M, Choi BJ, Lee JH (2014) Viability of pulp stromal cells in cryopreserved deciduous teeth. Cell Tissue Bank 15(1):67–74. doi:10.1007/s10561-013-9375-z
232. Lindemann D, Werle SB, Steffens D, Garcia-Godoy F, Pranke P, Casagrande L (2014) Effects of cryopreservation on the characteristics of dental pulp stem cells of intact deciduous teeth. Arch Oral Biol 59(9):970–976. doi:10.1016/j.archoralbio.2014.04.008
233. Bressan WS, Lindemann D, Machado J, Pranke P, Casagrande L (2014) Serum-containing medium effect on isolation rate of dental pulp cells from cryopreserved intact deciduous teeth. J Clin Pediatr Dent 38(4):345–348
234. Park BW, Jang SJ, Byun JH, Kang YH, Choi MJ, Park WU, Lee WJ, Rho GJ (2014) Cryopreservation of human dental follicle tissue for use as a resource of autologous mesenchymal stem cells. J Tissue Eng Regen Med. doi:10.1002/term.1945
235. Kang YH, Lee HJ, Jang SJ, Byun JH, Lee JS, Lee HC, Park WU, Lee JH, Rho GJ, Park BW (2015) Immunomodulatory properties and in vivo osteogenesis of human dental stem cells from fresh and cryopreserved dental follicles. Differentiation 90(1–3):48–58. doi:10.1016/j.diff.2015.10.001
236. Seo BM, Miura M, Sonoyama W, Coppe C, Stanyon R, Shi S (2005) Recovery of stem cells from cryopreserved periodontal ligament. J Dent Res 84(10):907–912
237. Crook JM, Hei D, Stacey G (2010) The International Stem Cell Banking Initiative (ISCBI): raising standards to bank on. In Vitro Cell Dev Biol Animal 46(3–4):169–172. doi:10.1007/s11626-010-9301-7
238. ISSCR (2008) Guidelines for the clinical translation of stem cells
239. Eaker S, Armant M, Brandwein H, Burger S, Campbell A, Carpenito C, Clarke D, Fong T, Karnieli O, Niss K, Van't Hof W, Wagey R (2013) Concise review: guidance in developing commercializable autologous/patient-specific cell therapy manufacturing. Stem Cells Transl Med 2(11):871–883. doi:10.5966/sctm.2013-0050
240. Ahrlund-Richter L, De Luca M, Marshak DR, Munsie M, Veiga A, Rao M (2009) Isolation and production of cells suitable for human therapy: challenges ahead. Cell Stem Cell 4(1):20–26. doi:10.1016/j.stem.2008.11.012
241. Alici E, Blomberg P (2010) GMP facilities for manufacturing of advanced therapy medicinal products for clinical trials: an overview for clinical researchers. Curr Gene Ther 10(6):508–515
242. Sensebe L, Gadelorge M, Fleury-Cappellesso S (2013) Production of mesenchymal stromal/stem cells according to good manufacturing practices: a review. Stem Cell Res Ther 4(3):66. doi:10.1186/scrt217
243. Wuchter P, Bieback K, Schrezenmeier H, Bornhauser M, Muller LP, Bonig H, Wagner W, Meisel R, Pavel P, Tonn T, Lang P, Muller I, Renner M, Malcherek G, Saffrich R, Buss EC, Horn P, Rojewski M, Schmitt A, Ho AD, Sanzenbacher R, Schmitt M (2015) Standardization of good manufacturing practice-compliant production of bone marrow-derived human mesenchymal stromal cells for immunotherapeutic applications. Cytotherapy 17(2):128–139. doi:10.1016/j.jcyt.2014.04.002
244. Ducret M, Fabre H, Degoul O, Atzeni G, McGuckin C, Forraz N, Alliot-Licht B, Mallein-Gerin F, Perrier-Groult E, Farges JC (2015) Manufacturing of dental pulp cell-based products from human third molars: current strategies and future investigations. Front Physiol 6:213. doi:10.3389/fphys.2015.00213
245. Takeda T, Tezuka Y, Horiuchi M, Hosono K, Iida K, Hatakeyama D, Miyaki S, Kunisada T, Shibata T, Tezuka K (2008) Characterization of dental pulp stem cells of human tooth germs. J Dent Res 87(7):676–681
246. Hilkens P, Gervois P, Fanton Y, Vanormelingen J, Martens W, Struys T, Politis C, Lambrichts I, Bronckaers A (2013) Effect of isolation methodology on stem cell properties and multilineage differentiation potential of human dental pulp stem cells. Cell Tissue Res 353(1):65–78. doi:10.1007/s00441-013-1630-x
247. Alt E, Yan Y, Gehmert S, Song YH, Altman A, Gehmert S, Vykoukal D, Bai X (2011) Fibroblasts share mesenchymal phenotypes with stem cells, but lack their differentiation and colony-forming potential. Biol Cell 103(4):197–208. doi:10.1042/BC20100117
248. Lv FJ, Tuan RS, Cheung KM, Leung VY (2014) Concise review: the surface markers and identity of human mesenchymal stem cells. Stem Cells 32(6):1408–1419. doi:10.1002/stem.1681
249. Bansal R, Jain A (2015) Current overview on dental stem cells applications in regenerative dentistry. J Nat Sci Biol Med 6(1):29–34. doi:10.4103/0976-9668.149074
250. Gottipamula S, Muttigi MS, Kolkundkar U, Seetharam RN (2013) Serum-free media for the production of human mesenchymal stromal cells: a review. Cell Prolif 46(6):608–627. doi:10.1111/cpr.12063
251. Halme DG, Kessler DA (2006) FDA regulation of stem-cell-based therapies. N Engl J Med 355(16):1730–1735. doi:10.1056/NEJMhpr063086
252. Hemeda H, Giebel B, Wagner W (2014) Evaluation of human platelet lysate versus fetal bovine serum for culture of mesenchymal stromal cells. Cytotherapy 16(2):170–180. doi:10.1016/j.jcyt.2013.11.004
253. van der Valk J, Mellor D, Brands R, Fischer R, Gruber F, Gstraunthaler G, Hellebrekers L, Hyllner J,

Jonker FH, Prieto P, Thalen M, Baumans V (2004) The humane collection of fetal bovine serum and possibilities for serum-free cell and tissue culture. Toxicol In Vitro: Int J Published Assoc BIBRA 18(1):1–12

254. Mannello F, Tonti GA (2007) Concise review: no breakthroughs for human mesenchymal and embryonic stem cell culture: conditioned medium, feeder layer, or feeder-free; medium with fetal calf serum, human serum, or enriched plasma; serum-free, serum replacement nonconditioned medium, or ad hoc formula? All that glitters is not gold! Stem Cells 25(7):1603–1609. doi:10.1634/stemcells.2007-0127
255. Tarle SA, Shi S, Kaigler D (2011) Development of a serum-free system to expand dental-derived stem cells: PDLSCs and SHEDs. J Cell Physiol 226(1):66–73. doi:10.1002/jcp.22304
256. Hirata TM, Ishkitiev N, Yaegaki K, Calenic B, Ishikawa H, Nakahara T, Mitev V, Tanaka T, Haapasalo M (2010) Expression of multiple stem cell markers in dental pulp cells cultured in serum-free media. J Endod 36(7):1139–1144. doi:10.1016/j.joen.2010.03.002
257. Khanna-Jain RVS, Vuorinen A, Sandor GK, Suuronen R, Mannerstrom B, Miettinen S (2012) Growth and differentiation of human dental pulp stem cells maintained in fetal bovine serum, human serum and serum-free/xeno-free culture media. J Stem Cell Res Ther 2(4)
258. Morsczeck C, Ernst W, Florian C, Reichert TE, Proff P, Bauer R, Muller-Richter U, Driemel O (2008) Gene expression of nestin, collagen type I and type III in human dental follicle cells after cultivation in serum-free medium. Oral Maxillofac Surg 12(2):89–92. doi:10.1007/s10006-008-0111-y
259. Okamoto T, Yatsuzuka N, Tanaka Y, Kan M, Yamanaka T, Sakamoto A, Takata T, Akagawa Y, Sato GH, Sato JD, Takada K (1997) Growth and differentiation of periodontal ligament-derived cells in serum-free defined culture. In Vitro Cell Dev Biol Animal 33(4):302–309
260. Bonnamain V, Thinard R, Sergent-Tanguy S, Huet P, Bienvenu G, Naveilhan P, Farges JC, Alliot-Licht B (2013) Human dental pulp stem cells cultured in serum-free supplemented medium. Front Physiol 4:357. doi:10.3389/fphys.2013.00357
261. Xiao L, Tsutsui T (2013) Characterization of human dental pulp cells-derived spheroids in serum-free medium: stem cells in the core. J Cell Biochem 114(11):2624–2636. doi:10.1002/jcb.24610
262. Karbanova J, Soukup T, Suchanek J, Pytlik R, Corbeil D, Mokry J (2011) Characterization of dental pulp stem cells from impacted third molars cultured in low serum-containing medium. Cells Tissues Organs 193(6):344–365. doi:10.1159/000321160
263. Bernardo ME, Avanzini MA, Perotti C, Cometa AM, Moretta A, Lenta E, Del Fante C, Novara F, de Silvestri A, Amendola G, Zuffardi O, Maccario R, Locatelli F (2007) Optimization of in vitro expansion of human multipotent mesenchymal stromal cells for cell-therapy approaches: further insights in the search for a fetal calf serum substitute. J Cell Physiol 211(1):121–130. doi:10.1002/jcp.20911
264. Bieback K (2013) Platelet lysate as replacement for fetal bovine serum in mesenchymal stromal cell cultures. Transfus Med Hemother 40(5):326–335. doi:10.1159/000354061
265. Burnouf T, Strunk D, Koh MB, Schallmoser K (2016) Human platelet lysate: replacing fetal bovine serum as a gold standard for human cell propagation? Biomaterials 76:371–387. doi:10.1016/j.biomaterials.2015.10.065
266. Govindasamy V, Ronald VS, Abdullah AN, Ganesan Nathan KR, Aziz ZA, Abdullah M, Zain RB, Kasim NH, Musa S, Bhonde RR (2011) Human platelet lysate permits scale-up of dental pulp stromal cells for clinical applications. Cytotherapy 13(10):1221–1233. doi:10.3109/14653249.2011.602337
267. Iudicone P, Fioravanti D, Bonanno G, Miceli M, Lavorino C, Totta P, Frati L, Nuti M, Pierelli L (2014) Pathogen-free, plasma-poor platelet lysate and expansion of human mesenchymal stem cells. J Transl Med 12:28. doi:10.1186/1479-5876-12-28
268. Pisciolaro RL, Duailibi MT, Novo NF, Juliano Y, Pallos D, Yelick PC, Vacanti JP, Ferreira LM, Duailibi SE (2015) Tooth tissue engineering: the importance of blood products as a supplement in tissue culture medium for human pulp dental stem cells. Tissue Eng A 21(21–22):2639–2648. doi:10.1089/ten.TEA.2014.0617
269. Rauch C, Feifel E, Amann EM, Spotl HP, Schennach H, Pfaller W, Gstraunthaler G (2011) Alternatives to the use of fetal bovine serum: human platelet lysates as a serum substitute in cell culture media. Altex 28(4):305–316
270. Shahdadfar A, Fronsdal K, Haug T, Reinholt FP, Brinchmann JE (2005) In vitro expansion of human mesenchymal stem cells: choice of serum is a determinant of cell proliferation, differentiation, gene expression, and transcriptome stability. Stem Cells 23(9):1357–1366. doi:10.1634/stemcells.2005-0094
271. Shih DT, Burnouf T (2015) Preparation, quality criteria, and properties of human blood platelet lysate supplements for ex vivo stem cell expansion. New Biotechnol 32(1):199–211. doi:10.1016/j.nbt.2014.06.001
272. Cardaropoli DG, L. (2007) The influence of orthodontic movement on periodontal tissues level. Semin Orthod 13(4):234–245
273. Chen Y, Mohammed A, Oubaidin M, Evans CA, Zhou X, Luan X, Diekwisch TG, Atsawasuwan P (2015) Cyclic stretch and compression forces alter microRNA-29 expression of human periodontal ligament cells. Gene 566(1):13–17. doi:10.1016/j.gene.2015.03.055

274. Krishnan V, Davidovitch Z (2006) Cellular, molecular, and tissue-level reactions to orthodontic force. Am J Orthod Dentofac Orthop 129(4):469. doi:10.1016/j.ajodo.2005.10.007, e461–432
275. Meeran NA (2013) Cellular response within the periodontal ligament on application of orthodontic forces. J Indian Soc Periodontol 17(1):16–20. doi:10.4103/0972-124X.107468
276. Iohara K, Murakami M, Nakata K, Nakashima M (2014) Age-dependent decline in dental pulp regeneration after pulpectomy in dogs. Exp Gerontol 52:39–45. doi:10.1016/j.exger.2014.01.020
277. Feng X, Xing J, Feng G, Huang D, Lu X, Liu S, Tan W, Li L, Gu Z (2014) p16(INK4A) mediates age-related changes in mesenchymal stem cells derived from human dental pulp through the DNA damage and stress response. Mech Ageing Dev 141–142:46–55. doi:10.1016/j.mad.2014.09.004
278. Lossdorfer S, Kraus D, Jager A (2010) Aging affects the phenotypic characteristics of human periodontal ligament cells and the cellular response to hormonal stimulation in vitro. J Periodontal Res 45(6):764–771. doi:10.1111/j.1600-0765.2010.01297.x
279. Ma D, Ma Z, Zhang X, Wang W, Yang Z, Zhang M, Wu G, Lu W, Deng Z, Jin Y (2009) Effect of age and extrinsic microenvironment on the proliferation and osteogenic differentiation of rat dental pulp stem cells in vitro. J Endod 35(11):1546–1553. doi:10.1016/j.joen.2009.07.016
280. Zhang J, An Y, Gao LN, Zhang YJ, Jin Y, Chen FM (2012) The effect of aging on the pluripotential capacity and regenerative potential of human periodontal ligament stem cells. Biomaterials 33(29):6974–6986. doi:10.1016/j.biomaterials.2012.06.032
281. Atari M, Gil-Recio C, Fabregat M, Garcia-Fernandez D, Barajas M, Carrasco MA, Jung HS, Alfaro FH, Casals N, Prosper F, Ferres-Padro E, Giner L (2012) Dental pulp of the third molar: a new source of pluripotent-like stem cells. J Cell Sci 125(Pt 14):3343–3356. doi:10.1242/jcs.096537
282. Kellner M, Steindorff MM, Strempel JF, Winkel A, Kuhnel MP, Stiesch M (2014) Differences of isolated dental stem cells dependent on donor age and consequences for autologous tooth replacement. Arch Oral Biol 59(6):559–567. doi:10.1016/j.archoralbio.2014.02.014
283. James JA, Sayers NM, Drucker DB, Hull PS (1999) Effects of tobacco products on the attachment and growth of periodontal ligament fibroblasts. J Periodontol 70(5):518–525. doi:10.1902/jop.1999.70.5.518
284. Kim SY, Kang KL, Lee JC, Heo JS (2012) Nicotinic acetylcholine receptor alpha7 and beta4 subunits contribute nicotine-induced apoptosis in periodontal ligament stem cells. Mol Cells 33(4):343–350. doi:10.1007/s10059-012-2172-x
285. Ng TK, Carballosa CM, Pelaez D, Wong HK, Choy KW, Pang CP, Cheung HS (2013) Nicotine alters MicroRNA expression and hinders human adult stem cell regenerative potential. Stem Cells Dev 22(5):781–790. doi:10.1089/scd.2012.0434
286. Yanagita M, Kashiwagi Y, Kobayashi R, Tomoeda M, Shimabukuro Y, Murakami S (2008) Nicotine inhibits mineralization of human dental pulp cells. J Endod 34(9):1061–1065. doi:10.1016/j.joen.2008.06.005
287. Cooper PR, Takahashi Y, Graham LW, Simon S, Imazato S, Smith AJ (2010) Inflammation-regeneration interplay in the dentine-pulp complex. J Dent 38(9):687–697. doi:10.1016/j.jdent.2010.05.016
288. Alongi DJ, Yamaza T, Song Y, Fouad AF, Romberg EE, Shi S, Tuan RS, Huang GT (2010) Stem/progenitor cells from inflamed human dental pulp retain tissue regeneration potential. Regen Med 5(4):617–631. doi:10.2217/rme.10.30
289. Dokic J, Tomic S, Cerovic S, Todorovic V, Rudolf R, Colic M (2012) Characterization and immunosuppressive properties of mesenchymal stem cells from periapical lesions. J Clin Periodontol 39(9):807–816. doi:10.1111/j.1600-051X.2012.01917.x
290. Liu Y, Gao Y, Zhan X, Cui L, Xu S, Ma D, Yue J, Wu B, Gao J (2014) TLR4 activation by lipopolysaccharide and Streptococcus mutans induces differential regulation of proliferation and migration in human dental pulp stem cells. J Endod 40(9):1375–1381. doi:10.1016/j.joen.2014.03.015
291. Ma D, Cui L, Gao J, Yan W, Liu Y, Xu S, Wu B (2014) Proteomic analysis of mesenchymal stem cells from normal and deep carious dental pulp. PLoS One 9(5), e97026. doi:10.1371/journal.pone.0097026
292. Ma D, Gao J, Yue J, Yan W, Fang F, Wu B (2012) Changes in proliferation and osteogenic differentiation of stem cells from deep caries in vitro. J Endod 38(6):796–802. doi:10.1016/j.joen.2012.02.014
293. Yazid FB, Gnanasegaran N, Kunasekaran W, Govindasamy V, Musa S (2014) Comparison of immunodulatory properties of dental pulp stem cells derived from healthy and inflamed teeth. Clin Oral Investig 18(9):2103–2112. doi:10.1007/s00784-014-1207-4
294. Pereira LO, Rubini MR, Silva JR, Oliveira DM, Silva IC, Pocas-Fonseca MJ, Azevedo RB (2012) Comparison of stem cell properties of cells isolated from normal and inflamed dental pulps. Int Endod J 45(12):1080–1090. doi:10.1111/j.1365-2591.2012.02068.x
295. Sun HH, Chen B, Zhu QL, Kong H, Li QH, Gao LN, Xiao M, Chen FM, Yu Q (2014) Investigation of dental pulp stem cells isolated from discarded human teeth extracted due to aggressive periodontitis. Biomaterials 35(35):9459–9472. doi:10.1016/j.biomaterials.2014.08.003
296. Werle SB, Lindemann D, Steffens D, Demarco FF, de Araujo FB, Pranke P, Casagrande L (2016) Carious deciduous teeth are a potential source for dental pulp stem cells. Clin Oral Investig 20(1):75–81. doi:10.1007/s00784-015-1477-5

297. Havelek R, Soukup T, Cmielova J, Seifrtova M, Suchanek J, Vavrova J, Mokry J, Muthna D, Rezacova M (2013) Ionizing radiation induces senescence and differentiation of human dental pulp stem cells. Folia Biol 59(5):188–197
298. Muthna D, Soukup T, Vavrova J, Mokry J, Cmielova J, Visek B, Jiroutova A, Havelek R, Suchanek J, Filip S, English D, Rezacova M (2010) Irradiation of adult human dental pulp stem cells provokes activation of p53, cell cycle arrest, and senescence but not apoptosis. Stem Cells Dev 19(12):1855–1862. doi:10.1089/scd.2009.0449
299. Abe S, Hamada K, Yamaguchi S, Amagasa T, Miura M (2011) Characterization of the radioresponse of human apical papilla-derived cells. Stem Cell Res Ther 2(1):2. doi:10.1186/scrt43
300. Murphy SV, Atala A (2014) 3D bioprinting of tissues and organs. Nat Biotechnol 32(8):773–785. doi:10.1038/nbt.2958
301. Jung JW, Lee JS, Cho DW (2016) Computer-aided multiple-head 3D printing system for printing of heterogeneous organ/tissue constructs. Sci Rep 6:21685. doi:10.1038/srep21685
302. Kang HW, Lee SJ, Ko IK, Kengla C, Yoo JJ, Atala A (2016) A 3D bioprinting system to produce human-scale tissue constructs with structural integrity. Nat Biotechnol 34(3):312–319. doi:10.1038/nbt.3413

Index

F. Karimi-Busheri, M. Weinfeld (eds.), *Biobanking and Cryopreservation of Stem Cells*, Advances in Experimental Medicine and Biology 951, DOI 10.1007/978-3-319-45457-3

The manufacturer's authorised representative in the EU is Springer Nature Customer Service Centre GmbH, Europaplatz 3, 69115 Heidelberg, Germany. If you have any concerns regarding our products, please contact ProductSafety@springernature.com

Printed and bound by CPI Group (UK) Ltd, Croydon, CR0 4YY
22/09/2026
02221059-0020